DOE/EIA-0383(2007)

AF696523

Annual Energy Outlook 2007

With Projections to 2030

February 2007

Energy Information Administration
Office of Integrated Analysis and Forecasting
U.S. Department of Energy
Washington, DC 20585

This publication is on the WEB at:
www.eia.doe.gov/oiaf/aeo/

This report was prepared by the Energy Information Administration, the independent statistical and analytical agency within the U.S. Department of Energy. The information contained herein should be attributed to the Energy Information Administration and should not be construed as advocating or reflecting any policy position of the Department of Energy or any other organization.

Preface

The *Annual Energy Outlook 2007* (*AEO2007*), prepared by the Energy Information Administration (EIA), presents long-term projections of energy supply, demand, and prices through 2030. The projections are based on results from EIA's National Energy Modeling System (NEMS).

The report begins with an "Overview" summarizing the *AEO2007* reference case. The next section, "Legislation and Regulations," discusses evolving legislation and regulatory issues, including recently enacted legislation and regulation, such as the new Corporate Average Fuel Economy (CAFE) standards for light-duty trucks finalized by the National Highway Traffic Safety Administration (NHTSA) in March 2006. It also provides an update on the handling of key provisions in the Energy Policy Act of 2005 (EPACT2005) that could not be incorporated in the *Annual Energy Outlook 2006* (*AEO2006*) because of the absence of implementing regulations or funding appropriations. Finally, it provides a summary of how sunset provisions in selected Federal fuel taxes and tax credits are handled in *AEO2007*.

The "Issues in Focus" section includes discussions of the potential for biofuels in U.S. transportation markets, the relationship between oil and natural gas prices, and the impact of rising construction costs on energy markets. It also discusses possible construction of an Alaska natural gas pipeline; renewed interest in nuclear generating capacity; and the demand response to higher energy prices in end-use sectors.

The "Market Trends" section summarizes the *AEO2007* projections for energy markets. The projections for 2006 and 2007 incorporate the short-term projections from EIA's September 2006 *Short-Term Energy Outlook*, where the data are comparable. The analysis in *AEO2007* focuses primarily on a reference case, lower and higher economic growth cases, and lower and higher energy price cases. Results from a number of other alternative cases are also presented, illustrating uncertainties associated with the reference case projections for energy demand, supply, and prices. Readers are encouraged to review the full range of cases, which address many of the uncertainties inherent in long-term projections. Complete tables for the five primary cases are provided in Appendixes A through C. Major results from many of the alternative cases are provided in Appendix D. Appendix E briefly describes NEMS and the alternative cases.

AEO2007 projections generally are based on Federal, State, and local laws and regulations in effect on or before October 31, 2006. The potential impacts of pending or proposed legislation, regulations, and standards (and sections of existing legislation that require implementing regulations or funds that have not been appropriated) are not reflected in the projections.

In general, historical data used in the *AEO2006* projections are based on EIA's *Annual Energy Review 2005*, published in August 2006; however, only partial or preliminary 2005 data were available in some cases. Other historical data, taken from multiple sources, are presented in this report for comparative purposes; documents referenced in the source notes should be consulted for official data values.

AEO2007 is published in accordance with Section 205c of the Department of Energy Organization Act of 1977 (Public Law 95-91), which requires the EIA Administrator to prepare annual reports on trends and projections for energy use and supply.

The projections in the *Annual Energy Outlook 2007* are not statements of what will happen but of what might happen, given the assumptions and methodologies used. The projections are business-as-usual trend estimates, given known technology and technological and demographic trends. *AEO2007* generally assumes that current laws and regulations are maintained throughout the projections. Thus, the projections provide a policy-neutral reference case that can be used to analyze policy initiatives. EIA does not propose, advocate, or speculate on future legislative and regulatory changes. Most laws are assumed to remain as currently enacted; however, the impacts of emerging regulatory changes, when defined, are reflected.

Because energy markets are complex, models are simplified representations of energy production and consumption, regulations, and producer and consumer behavior. Projections are highly dependent on the data, methodologies, model structures, and assumptions used in their development. Behavioral characteristics are indicative of real-world tendencies rather than representations of specific outcomes.

Energy market projections are subject to much uncertainty. Many of the events that shape energy markets are random and cannot be anticipated, including severe weather, political disruptions, strikes, and technological breakthroughs. In addition, future developments in technologies, demographics, and resources cannot be foreseen with certainty. Many key uncertainties in the *AEO2007* projections are addressed through alternative cases.

EIA has endeavored to make these projections as objective, reliable, and useful as possible; however, they should serve as an adjunct to, not a substitute for, a complete and focused analysis of public policy initiatives.

DISCARDED
PROPERTY OF USAF
BASE LIBRARY, FL 4801
HOLLOMAN AFB NM 88330

For sale by the Superintendent of Documents, U.S. Government Printing Office
Internet: bookstore.gpo.gov Phone: toll free (866) 512-1800; DC area (202) 512-1800
Fax: (202) 512-2104 Mail: Stop IDCC, Washington, DC 20402-0001

ISBN 978-0-16-078095-0

Contents

Contents

Contents

Figures (Continued) **Page**

Overview

Overview

Energy Trends to 2030

EIA, in preparing projections for the *AEO2007*, evaluated a wide range of trends and issues that could have major implications for U.S. energy markets between today and 2030. This overview focuses on one case, the reference case, which is presented and compared with the *AEO2006* reference case (see Table 1). Readers are encouraged to review the full range of alternative cases included in other sections of *AEO2007*. As in previous editions of the *Annual Energy Outlook* (*AEO*), the reference case assumes that current policies affecting the energy sector remain unchanged throughout the projection period. Some possible policy changes—notably, the adoption of policies to limit or reduce greenhouse gas emissions—could change the reference case projections significantly.

Trends in energy supply and demand are affected by many factors that are difficult to predict, such as energy prices, U.S. economic growth, advances in technologies, changes in weather patterns, and future public policy decisions. It is clear, however, that energy markets are changing gradually in response to such readily observable factors as the higher energy prices that have been experienced since 2000, the greater influence of developing countries on worldwide energy requirements, recently enacted legislation and regulations in the United States, and changing public perceptions of issues related to the use of alternative fuels, emissions of air pollutants and greenhouse gases, and the acceptability of various energy technologies, among others. Such changes are reflected in the *AEO2007* reference case, which projects increased consumption of biofuels (both ethanol and biodiesel), growth in coal-to-liquids (CTL) capacity and production, growing demand for unconventional transportation technologies (such as flex-fuel, hybrid, and diesel vehicles), growth in nuclear power capacity and generation, and accelerated improvements in energy efficiency throughout the economy.

Despite the rapid growth projected for biofuels and other nonhydroelectric renewable energy sources and the expectation that orders will be placed for new nuclear power plants for the first time in more than 25 years, oil, coal, and natural gas still are projected to provide roughly the same 86-percent share of the total U.S. primary energy supply in 2030 that they did in 2005 (assuming no changes in existing laws and regulations). The expected rapid growth in the use of biofuels and other nonhydropower renewable energy sources begins from a very low current share of total energy use; hydroelectric power production, which accounts for the bulk of current renewable electricity supply, is nearly stagnant; and the share of total electricity supplied from nuclear power falls despite the projected new plant builds, which more than offset retirements, because the overall market for electricity continues to expand rapidly in the projection.

World oil prices since 2000 have been substantially higher than those of the 1990s, as have the prices of natural gas and coal (although coal prices began to rise somewhat later than oil and natural gas prices). The sustained increase in world oil prices caused EIA to reevaluate earlier oil price expectations in producing *AEO2006*. The long-term path of world oil prices in the *AEO2007* reference case is similar to that in the *AEO2006* reference case, although near-term prices in *AEO2007* are somewhat higher than those in *AEO2006*.

In the *AEO2007* reference case, real world crude oil prices, expressed in terms of the average price of imported light, low-sulfur crude oil to U.S. refiners, are projected to decline gradually from their 2006 average level through 2015, as expanded investment in exploration and development brings new supplies to the world market. After 2015, real prices begin to rise as demand continues to grow and higher cost supplies are brought to market. In 2030, the average real price of crude oil is projected to be above \$59 per barrel in 2005 dollars, or about \$95 per barrel in nominal dollars.

The energy price projections for natural gas and coal in the *AEO2007* reference case also are similar to those in *AEO2006*. The real wellhead price of natural

World Oil Price Concept Used in AEO2007

The world oil price in *AEO2007* is defined as the average price of low-sulfur, light crude oil imported into the United States—the same definition used in *AEO2006*. This price is approximately equal to the price of the light, sweet crude oil contract traded on the NYMEX exchange and the price of West Texas Intermediate (WTI) crude oil delivered to Cushing, Oklahoma. Prior to *AEO2006*, the world crude oil price was defined on the basis of the U.S. average imported refiners' acquisition cost of crude oil (IRAC), which represented the weighted average of all imported crude oil. On average, the IRAC price is \$5 to \$8 per barrel less than the price of imported low-sulfur, light crude oil.

gas is projected to decline from current levels through 2015, when new supplies enter the market, but it does not return to the levels of the 1990s. After 2015, the natural gas price rises to nearly $6.00 per thousand cubic feet in 2030 in 2005 dollars (about $9.60 per thousand cubic feet in nominal dollars). For coal, the average minemouth price ranges between $1.08 and $1.18 (2005 dollars) per million British thermal units (Btu) over the projection period; in 2030, the price of coal is projected to be roughly the same as it was in 2005, at $1.15 per million Btu ($1.85 per million Btu in nominal dollars). The 2030 price projection is higher than the *AEO2006* reference case projection of $1.11 per million Btu and much higher than projected in earlier *AEOs*—typically, below $0.90 per million Btu. Greater price increases are avoided, because lower cost production from surface mines in the West is projected to capture a growing share of the U.S. market.

The use of alternative fuels, such as ethanol, biodiesel, and CTL, is projected to increase substantially in the reference case as a result of the higher prices projected for traditional fuels and the support for alternative fuels provided in recently enacted Federal legislation. Ethanol use grows in the *AEO2007* reference case from 4 billion gallons in 2005 to 14.6 billion gallons in 2030 (about 8 percent of total gasoline consumption by volume). Ethanol use for gasoline blending grows to 14.4 billion gallons and E85 consumption to 0.2 billion gallons in 2030. The ethanol supply is expected to be produced from both corn and cellulose feedstocks, both of which are supported by ethanol tax credits included in EPACT2005 [*1*], but domestically grown corn is expected to be the primary source, accounting for 13.6 billion gallons of ethanol production in 2030.

Alternative sources of distillate fuel oil are projected to be key contributors to total supply (particularly, low-sulfur diesel fuels) in 2030. Consumption of biodiesel, also supported by tax credits in EPACT-2005, reaches 0.4 billion gallons in 2030, and distillate fuel oil produced from CTL reaches 5.7 billion gallons in 2030. In total, these two alternative sources of distillate fuel oil account for more than 7 percent of the total distillate pool in 2030.

The *AEO2007* reference case also reflects growing market penetration by unconventional vehicle technologies, such as flex-fuel, hybrid, and diesel vehicles. Sales of flex-fuel vehicles (FFVs), which are capable of using gasoline and E85, reach 2 million per year in 2030, or 10 percent of total sales of new light-duty vehicles. Sales of hybrids, including both full and mild hybrids [*2*], are projected to reach 2 million per year by 2030, accounting for another 10 percent of total light-duty vehicles sales. Diesel vehicles sales reach 1.2 million per year in 2030, or 6 percent of new light-duty vehicle sales. Including other alternative vehicle technologies (such as gaseous, electric, and fuel cell), all the projected sales of alternative vehicle technologies account for nearly 28 percent of projected new light-duty vehicle sales in 2030, up from just over 8 percent in 2005.

In the electric power sector, the last new nuclear generating unit brought on line in the United States began operation in 1996. Since then, changes in U.S. nuclear capacity have resulted only from uprating of existing units and retirements. The *AEO2007* reference case projects total operable nuclear generating capacity of 112.6 gigawatts in 2030, including 3 gigawatts of additional capacity uprates, 9 gigawatts of new capacity built primarily in response to EPACT2005 tax credits, 3.5 gigawatts added in later years in response to higher fossil fuel prices, and 2.6 gigawatts of older plant retirements. As a result of the growth in available capacity, total nuclear generation is projected to grow from 780 billion kilowatthours in 2005 to 896 billion kilowatthours in 2030. Even with the projected increase in nuclear capacity and generation, however, the nuclear share of total electricity generation is expected to fall from 19 percent in 2005 to 15 percent in 2030.

Natural gas consumption is projected to grow to 26.1 trillion cubic feet in 2030, down from the projection of 26.9 trillion cubic feet in 2030 in the *AEO2006* reference case and well below the projections of 30 trillion cubic feet or more included in *AEO* reference cases only a few years ago. The generally higher natural gas prices projected in the *AEO2007* reference case result in lower projected growth of natural gas use for electricity generation over the last decade of the projection period. Total natural gas consumption is almost flat from 2020 through 2030, when growth in residential, commercial, and industrial consumption is offset by a decline in natural gas use for electricity generation as a result of greater coal use.

As in *AEO2006*, coal is projected to play a major role in the *AEO2007* reference case, particularly for electricity generation. Coal consumption is projected to increase from 22.9 quadrillion Btu (1,128 million short tons) in 2005 to more than 34 quadrillion Btu (1,772 million short tons) in 2030, with significant additions of new coal-fired generation capacity over the last decade of the projection period, when rising natural gas prices are projected. The reference case projections for coal consumption are particularly sensitive to the underlying assumption that current energy and environmental policies remain unchanged throughout the projection period. Recent EIA service reports have shown that steps to reduce greenhouse gas emissions through the use of an economy-wide emissions tax or cap-and-trade system could have a significant impact on coal use [*3*].

Economic Growth

U.S. gross domestic product (GDP) is projected to grow at an average annual rate of 2.9 percent from 2005 to 2030 in the *AEO2007* reference case—0.1 percentage point lower than projected for the same period in the *AEO2006* reference case. The main factors influencing the change in long-term GDP are growth in the labor force and labor productivity. The slightly lower rate of growth in the *AEO2007* reference case reflects a slowing of the economy as a result of higher energy prices in the near term.

The projections for key interest rates (the Federal funds rate, the nominal yield on the 10-year Treasury note, and the AA utility bond rate) in the *AEO2007* reference case are slightly lower than those in the *AEO2006* reference case during most of the projection period, based on an expected lower rate of inflation over the long term. The projected value of industrial shipments is also lower in *AEO2007*, reflecting higher energy prices in the early years of the period.

Energy Prices

In the reference case—one of several cases included in *AEO2007*—the average world crude oil price declines

Figure 1. Energy prices, 1980-2030 (2005 dollars per million Btu)

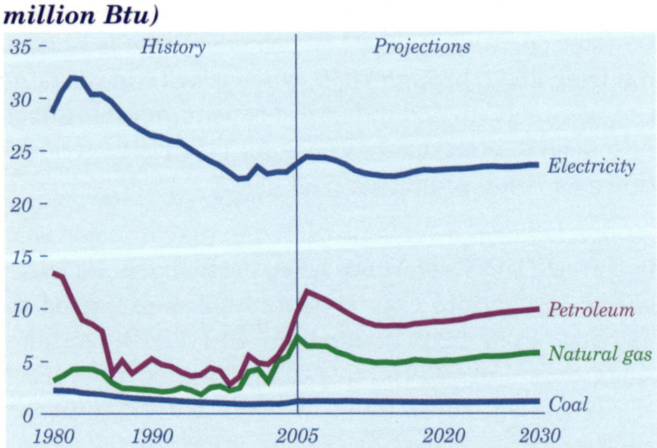

slowly in real terms (2005 dollars), from a 2006 average of more than $69 per barrel ($11.56 per million Btu) to just under $50 per barrel ($8.30 per million Btu) in 2014 as new supplies enter the market, then rises slowly to about $59 per barrel ($9.89 per million Btu) in 2030 (Figure 1). The 2030 world oil price in the *AEO2007* reference case is slightly above the 2030 price in the *AEO2006* reference case. Alternative *AEO2007* cases address higher and lower world oil prices and U.S. natural gas prices.

Oil prices are currently above EIA's estimate of long-run equilibrium prices, a situation that could persist for several more years. Temporary shortages of experienced personnel, equipment, and construction materials in the oil industry; political instability in some major producing regions; and recent strong economic growth in major consuming nations have combined to push oil prices well above equilibrium levels. Although some analysts believe that current high oil prices signal an unanticipated scarcity of petroleum resources, EIA's expectations regarding the ultimate size and cost of both conventional and unconventional liquid resources have not changed since last year's *AEO*.

This year's reference case anticipates substantial increases in conventional oil production in several Organization of the Petroleum Exporting Countries (OPEC) and non-OPEC countries over the next 10 years, as well as substantial development of unconventional production over the next 25 years. The prices in the *AEO2007* reference case are high enough to trigger entry into the market of some alternative energy supplies that are expected to become economically viable in the range of $25 to $50 per barrel. They include oil sands, ultra-heavy oils, gas-to-liquids (GTL), and CTL.

The *AEO2007* reference case represents EIA's current judgment about the expected behavior of OPEC in the mid-term. In the projection, OPEC increases production at a rate that keeps average prices in the range of $50 to $60 per barrel (2005 dollars) through 2030. This would not preclude the possibility that prices could move outside the $50 to $60 range for short periods of time over the next 25 years. OPEC is expected to recognize that allowing oil prices to remain above that level for an extended period could lower the long-run profits of OPEC producers by encouraging more investment in non-OPEC conventional and unconventional supplies and discouraging consumption of liquids worldwide.

The reference case also projects significant long-term supply potential from non-OPEC producers. In several resource-rich regions, with wars ending, new pipelines being built, new exploration and drilling technologies becoming available, and world oil prices rising, access to resources has increased and production has risen. For example, oil production in Angola has nearly doubled since the end of a 27-year civil war in 2002. In Azerbaijan and Kazakhstan, new investment has been stimulated by the 2006 opening of the Baku-Tbilisi-Ceyhan (BTC) pipeline connecting the Caspian and Mediterranean seas, and production in both countries is expected to increase by more than 1 million barrels per day from 2006 to 2010. Brazil's pioneering development of offshore deepwater drilling, coupled with clear government policies, has attracted foreign investment and steadily increased production. In Canada, where the economic viability of the country's oil sands has been enhanced by higher world oil prices and advances in production technology, production from those resources is expected to reach 3.7 million barrels per day in 2030.

In the *AEO2007* reference case, world liquids demand is projected to increase from about 84 million barrels per day in 2005 to 117 million barrels per day in 2030. OPEC liquids production is projected to total 48 million barrels per day in 2030, 40 percent higher than the 34 million barrels per day produced in 2005 and almost 2 million barrels per day above the *AEO2006* reference case projection of 46 million barrels per day in 2030. The Middle East OPEC producers and Venezuela have the resources to boost their output substantially over the period. Non-OPEC liquids production is projected to increase from 50 million barrels per day in 2005 to 70 million in 2030, as compared with the *AEO2006* reference case projection of 72 million barrels per day.

The average U.S. wellhead price for natural gas in the *AEO2007* reference case declines gradually from the current level, as increased drilling brings on new supplies and new import sources become available. The average price falls to just under $5 per thousand cubic feet in 2015 (2005 dollars), then rises gradually to about $6 per thousand cubic feet in 2030 (equivalent to $9.63 per thousand cubic feet in nominal dollars). Imports of liquefied natural gas (LNG), new natural gas production in Alaska, and production from unconventional sources in the lower 48 States are not expected to increase sufficiently to offset the impacts of resource decline and increased demand. The trend in projected wellhead natural gas prices in the *AEO2007* reference case is similar to that in the *AEO2006* reference case.

Minemouth coal prices in the *AEO2007* reference case are higher in most regions of the country than was projected in the *AEO2006* reference case, because of higher mining costs. The largest price increase relative to the *AEO2006* reference case is expected in Appalachia, an area that has been extensively mined, and where mining costs appear to be rising. At the national level, higher Appalachian coal prices are offset over the 25-year projection period by the increasing share of total coal production expected to come from relatively low-cost western mines, such as those in the Powder River Basin in Wyoming.

Average real minemouth coal prices (in 2005 dollars) are expected to fall from $1.15 per million Btu ($23.34 per short ton) in 2005 to $1.08 per million Btu ($21.51 per short ton) in 2019 in the reference case, as prices moderate following a rapid run-up over the past few years. After 2019, new coal-fired power plants are expected to increase total coal demand, and prices are projected to rise to $1.15 per million Btu ($22.60 per short ton) in 2030. The projected 2020 and 2030 prices are 4.2 percent and 1.4 percent higher, respectively, than those in the *AEO2006* reference case. Without adjustment for inflation, the average minemouth price of coal in the *AEO2007* reference case rises to $1.85 per million Btu ($36.38 per ton) in 2030.

The projected price of coal delivered to power plants is also higher in the *AEO2007* reference case than in the *AEO2006* reference case, reflecting higher minemouth prices and higher transportation costs. Increases in diesel fuel prices in recent years have led railroads to implement fuel adjustment charges, which are incorporated in the *AEO2007* reference case. The average delivered price of coal to power plants is projected to increase from $1.53 per million Btu ($30.83 per short ton) in 2005 to $1.69 per million Btu ($33.52 per short ton) in 2030 in 2005 dollars, 7.0 percent higher than in the *AEO2006* reference case. In nominal dollars, the average delivered price of coal to power plants is projected to reach $2.72 per million Btu ($53.98 per short ton) in 2030.

Electricity prices follow the prices of fuels to power plants in the reference case, falling initially as fuel prices retreat after the rapid increases of recent years and then rising slowly. From a peak of 8.3 cents per kilowatthour (2005 dollars) in 2006, average delivered electricity prices decline to a low of 7.7 cents per kilowatthour in 2015 and then increase to 8.1 cents per kilowatthour in 2030. In the *AEO2006* reference case, with lower expectations for delivered fuel prices and the added costs of maintaining reliability, electricity prices increased to 7.7 cents per kilowatthour (2005 dollars) in 2030. Without adjustment for inflation, average delivered electricity prices in the *AEO-2007* reference case are projected to reach 13 cents per kilowatthour in 2030.

Energy Consumption

Total primary energy consumption in the *AEO2007* reference case is projected to increase at an average rate of 1.1 percent per year, from 100.2 quadrillion Btu in 2005 to 131.2 quadrillion Btu in 2030—3.4 quadrillion Btu less than in the *AEO2006* reference case. In 2030, the projected consumption levels for liquid fuels, natural gas, and coal all are lower in the *AEO2007* reference case than in the *AEO2006* reference case. Among the most important factors accounting for the differences are higher energy prices, particularly for coal, but also for natural gas and petroleum in the earlier part of the projection, slightly lower economic growth and greater use of more efficient appliances that reduces energy consumption in the residential and commercial sectors and slows the growth of electricity demand.

As a result of demographic trends and housing preferences, residential delivered energy consumption in the *AEO2007* reference case is projected to grow from 11.6 quadrillion Btu in 2005 to 13.8 quadrillion Btu in 2030, or by 0.7 percent per year (Figure 2). In comparison, the corresponding *AEO2006* projection was 14.0 quadrillion Btu in 2030. Higher projected electricity prices in the *AEO2007* reference case and increases in end-use efficiency for most services contribute to the slightly lower level of residential energy use.

Consistent with projected growth in commercial floorspace in the *AEO2007* reference case, delivered commercial energy consumption is projected to grow from 8.5 quadrillion Btu in 2005 to 12.4 quadrillion Btu in 2030, about the same as the *AEO2006* reference case projection. Higher projected electricity prices, along with revisions to provide better accounting of miscellaneous uses of electricity, lead to lower growth in commercial electricity consumption in the *AEO2007* reference case than was projected in the *AEO2006* reference case. That reduction is offset, however, by a higher projected level of natural gas use in the commercial sector (as compared with the *AEO2006* reference case), because higher electricity prices are expected to prompt more use of combined heat and power (CHP) to satisfy electricity and space conditioning requirements.

After falling to relatively low levels in the early 1980s, industrial energy consumption recovered and peaked in 1997. In the 2000 to 2003 period, industrial sector activity was reduced by an economic recession; in some industrial subsectors, the hurricanes of 2005 also resulted in reduced activity. In the *AEO2007* reference case, the industrial sector is projected to

Reorganization of Fuel Categories in AEO2007

AEO2007 includes, for the first time, a reorganized breakdown of fuel categories that reflects the increasing importance, both now and in the future, of conversion technologies that can produce liquid fuels from natural gas, coal, and biomass. In the past, petroleum production, net imports of petroleum, and refinery gain could be balanced against the supply of liquid fuels and other petroleum products. Now, with other primary energy sources being used to produce significant amounts of liquid fuels, those inputs must be added in order to balance production and supply. Conversely, the use of coal, biomass, and natural gas for liquid fuels production must be accounted for in order to balance net supply against net consumption for each primary fuel. In *AEO2007*, the conversion of nonpetroleum primary fuels to liquid fuels is explicitly modeled, along with petroleum refining, as part of a broadly defined refining activity that is included in the industrial sector. Unlike earlier *AEO*s, *AEO2007* specifically accounts for conversion losses and co-product outputs in the broadly defined refining activity.

Figure 2. Delivered energy consumption by sector, 1980-2030 (quadrillion Btu)

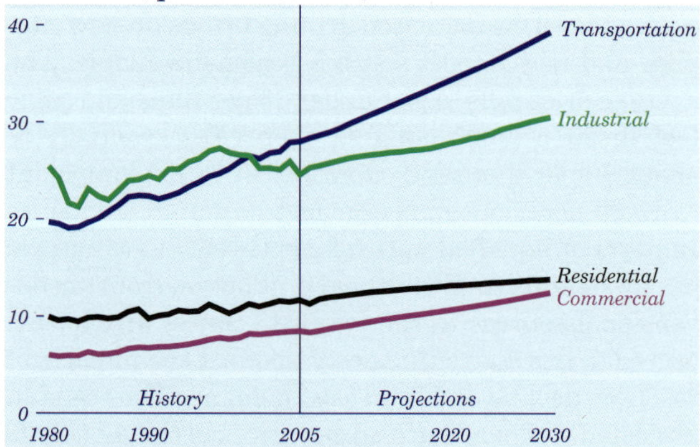

return to more typical output growth rates, and industrial energy consumption is expected to reflect the trend. The industrial value of shipments in the reference case is projected to grow by 2.0 percent per year from 2005 to 2030—more slowly than in the *AEO2006* reference case (2.1 percent per year) due to a slight slowdown in projected investment spending, higher energy prices, and increased competition from imports. Delivered industrial energy consumption in the *AEO2007* reference case is projected to reach 30.5 quadrillion Btu in 2030, significantly lower than the *AEO2006* reference case projection of 32.9 quadrillion Btu.

Total industrial energy consumption is boosted in *AEO2007* by strong growth in the production of nontraditional fuels, such as CTL and biofuels. Approximately 0.9 quadrillion Btu of coal is projected to be used to produce liquids in 2030, up from virtually no CTL production in 2005. Biofuels consumption in the industrial sector is projected to grow from 0.2 quadrillion Btu in 2005 to 0.9 quadrillion Btu in 2030. Much of the nontraditional fuel consumption is accounted for in the refining sector. Excluding energy consumption by refiners from the industrial total reveals that delivered energy consumption in 2030 for nonrefining industrial uses is projected to be only about 3 quadrillion Btu above 2005 levels (24.2 quadrillion Btu in 2030 compared with 21.1 quadrillion Btu in 2005).

Delivered energy consumption in the transportation sector is projected to total 39.3 quadrillion Btu in 2030 in the *AEO2007* reference case, 0.4 quadrillion Btu lower than the *AEO2006* projection. The slightly lower level of consumption predominantly reflects the influence of slower economic growth. Travel demand for light-duty vehicles is a significant determinant of total transportation energy demand, and over the past 20 years it has grown by about 3 percent annually. In the *AEO2007* reference case it is projected to grow at an average rate of 1.9 percent per year through 2030, reflecting demographic factors (for example, a leveling off of the increase in the labor force participation rate for women) and higher energy prices. The projected average fuel economy of new light-duty vehicles in 2030 is 29.2 miles per gallon, or 4 miles per gallon higher than the current average. Projected increases in new vehicle fuel economy are due not only to new Federal CAFE standards for light trucks but also to market-driven increases in the sale of unconventional vehicle technologies, such as flex-fuel, hybrid, and diesel vehicles, and a slowdown in the growth of new light truck sales.

Total electricity consumption, including both purchases from electric power producers and on-site generation, is projected to grow from 3,821 billion kilowatthours in 2005 to 5,478 billion kilowatthours in 2030, increasing at an average annual rate of 1.5 percent in the *AEO2007* reference case. In comparison, total electricity consumption of 5,619 billion kilowatthours in 2030 was projected in the *AEO2006* reference case. A larger portion of the projected growth in electricity use for computers, office equipment, and a variety of electrical appliances is offset in the *AEO2007* reference case by improved efficiency in those and other, more traditional electrical applications.

Total consumption of natural gas in the *AEO2007* reference case is projected to increase from 22.0 trillion cubic feet in 2005 to 26.1 trillion cubic feet in 2030 (Figure 3), with virtually no growth over the last decade of the projection. Compared with *AEO2006*, industrial natural gas use is lower (8.6 trillion cubic feet in 2030 in the *AEO2007* reference case, versus 8.8 trillion cubic feet in the *AEO2006* reference case) as a result of better efforts to account for natural gas demand in the metal durables and balance of manufacturing sectors than in previous *AEO*s. In comparison with *AEO2006*, lower projected natural gas consumption in the residential, industrial, and electric power sectors more than offsets higher projected consumption in the commercial sector in the *AEO2007* reference case (4.2 trillion cubic feet in 2030 in *AEO2007* compared with 4.0 trillion cubic feet in *AEO2006*). The increase results from lower delivered natural gas prices projected for the commercial sector in the *AEO2007* reference case.

Figure 3. Energy consumption by fuel, 1980-2030 (quadrillion Btu)

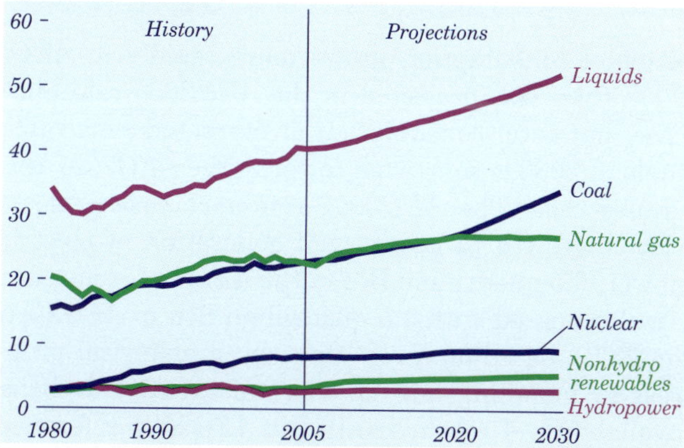

Total coal consumption is projected to increase from 22.9 quadrillion Btu in 2005 to 34.1 quadrillion Btu in 2030 in the *AEO2007* reference case, or from 1,128 million short tons in 2005 to 1,772 million short tons in 2030. As in the *AEO2006* reference case, coal consumption is projected to grow at a faster rate toward the end of the projection period in the *AEO2007* reference case, particularly after 2020, as coal use for new coal-fired generating capacity and for CTL production grows rapidly. In the *AEO2007* reference case, coal consumption in the electric power sector is projected to increase from 25.1 quadrillion Btu in 2020 to 31.1 quadrillion Btu in 2030, and coal use at CTL plants is projected to increase from 0.4 quadrillion Btu in 2020 to 1.8 quadrillion Btu in 2030.

Total consumption of liquid fuels and other petroleum products is projected to grow from 20.7 million barrels per day in 2005 to 26.9 million barrels per day in 2030 in the *AEO2007* reference case (Figure 3), less than the *AEO2006* reference case projection of 27.6 million barrels per day in 2030. In 2030, liquid fuels consumption in the residential sector is slightly higher in the *AEO2007* reference case, due to a lower projection for distillate fuel oil prices; lower in the industrial sector, due to higher liquefied petroleum gas prices and slower growth in industrial production; and lower in the transportation sector, due to slower economic growth.

Total consumption of marketed renewable fuels in the *AEO2007* reference case (including ethanol for gasoline blending, of which 1.2 quadrillion Btu in 2030 is included with liquid fuels consumption) is projected to grow from 6.5 quadrillion Btu in 2005 to 10.2 quadrillion Btu in 2030 (Figure 3). The robust growth is a result of State renewable portfolio standard (RPS) programs, mandates, and goals for renewable electricity generation; technological advances; high petroleum and natural gas prices; and Federal tax credits, including those in EPACT2005.

Ethanol consumption grows more rapidly in *AEO2007* than was projected in the *AEO2006* reference case, but total consumption of marketed renewable fuels in 2030 is somewhat lower in the *AEO2007* reference case. The *AEO2007* reference case projects slower growth in geothermal generation of electric power (0.5 quadrillion Btu in the *AEO2007* reference case compared with 1.5 quadrillion Btu in *AEO2006* in 2030), based on a reevaluation of historical progress in installing new geothermal capacity and the availability of resources. In the *AEO2007* reference case, more than 50 percent of the projected demand for renewables is for grid-connected electricity generation, including CHP, and the rest is for dispersed heating and cooling, industrial uses, and fuel blending.

The *AEO2007* reference case projects 21 percent more ethanol consumption in 2030 than was projected in the *AEO2006* reference case—14.6 billion gallons, compared with 12.1 billion gallons. As corn and biofeedstock supplies increase, and with price advantages over other motor gasoline blending components, ethanol consumption grows from 4.0 billion gallons in 2005 to 11.2 billion gallons in 2012 in the *AEO2007* reference case. This far exceeds the required 7.5 billion gallons in the Renewable Fuel Standard (RFS) that was enacted as part of EPACT2005. Ethanol supply in *AEO2007* is dominated by corn-based production, as a result of its cost advantages and eligibility for tax credits. Production of cellulosic ethanol is projected to total only 0.3 billion gallons in 2030, and ethanol imports are projected to total 0.8 billion gallons—a level consistent with the *AEO2006* reference case projection.

Energy Intensity

Energy intensity, measured as energy use per dollar of GDP (in 2000 dollars), is projected to decline at an average annual rate of 1.8 percent from 2005 to 2030 in the *AEO2007* reference case (Figure 4), about the same rate as in the *AEO2006* reference case (1.7 percent). Although energy use generally increases as the economy grows, continuing improvement in the energy efficiency of the U.S. economy and a shift to less energy-intensive activities are projected to keep the rate of energy consumption growth lower than the GDP growth rate.

Figure 4. Energy use per capita and per dollar of gross domestic product, 1980-2030 (index, 1980 = 1)

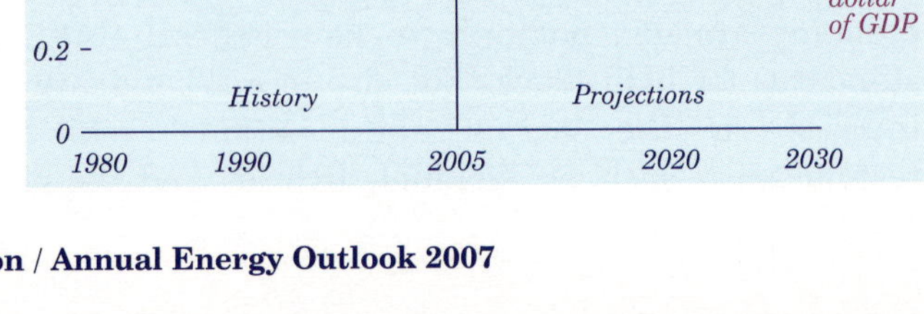

Since 1992, the energy intensity of the U.S. economy has declined on average by 1.9 percent per year, in part because the share of industrial shipments accounted for by the energy-intensive industries has fallen from 30 percent in 1992 to 26 percent in 2005. In the *AEO2007* reference case, the energy-intensive industries' share of total industrial shipments is projected to continue declining, although at a slower rate, to 24 percent in 2030.

Population is a key determinant of energy consumption, influencing demand for travel, housing, consumer goods, and services. Since 1990, both population and energy consumption in the United States have increased by about 18 percent, with annual variations in energy use per capita resulting from variations in weather and economic factors. The age, income, and geographic distribution of the population also affects energy consumption growth. The aging of the population, a gradual shift from the North to the South, and rising per-capita income will influence future trends. Overall, population in the reference case is projected to increase by 23 percent from 2005 to 2030. Over the same period, energy consumption is projected to increase by 31 percent. The result is a projected increase in energy consumption per capita, at an annual rate of 0.3 percent per year from 2005 to 2030—about the same rate as projected in the *AEO2006* reference case.

Recently, as energy prices have risen, the potential for more energy conservation has received increased attention. Although some additional energy conservation is induced by higher energy prices in the *AEO-2007* reference case, no policy-induced conservation measures are assumed beyond those in existing legislation and regulation, nor does the reference case assume behavioral changes beyond those observed in the past.

Electricity Generation

U.S. electricity consumption—including both purchases from electric power producers and on-site generation—is projected to increase steadily in the *AEO2007* reference case, at an average rate of 1.5 percent per year. In comparison, electricity consumption grew by annual rates of 4.2 percent, 2.6 percent, and 2.3 percent in the 1970s, 1980s, and 1990s, respectively. The growth rate in the *AEO2007* projection is lower than was projected in the *AEO2006* reference case, and it leads to lower projections for new plant additions and electricity generation.

In the *AEO2007* reference case, electricity generation from natural-gas-fired power plants is projected to increase from 2005 to 2020, as recently built plants are used more intensively to meet growing demand. Coal-fired generation is projected to increase less rapidly than was projected in the *AEO2006* reference case. After 2020, however, generation from new coal and nuclear plants is expected to displace some natural-gas-fired generation (Figure 5). In the *AEO2007* reference case, 937 billion kilowatthours of electricity is projected to be generated from natural gas in 2030, 6 percent less than the *AEO2006* reference case projection of 993 billion kilowatthours in 2030.

In the *AEO2007* reference case, the natural gas share of electricity generation (including generation in the end-use sectors) is projected to increase from 19 percent in 2005 to 22 percent around 2016, before falling to 16 percent in 2030. The coal share is projected to decline slightly, from 50 percent in 2005 to 49 percent in 2020, before increasing to 57 percent in 2030. Additions to coal-fired generating capacity in the *AEO2007* reference case are projected to total 156 gigawatts from 2005 to 2030 (as compared with 174 gigawatts in the *AEO2006* reference case), including 11 gigawatts at CTL plants and 67 gigawatts at integrated gasification combined-cycle (IGCC) plants. Given the assumed continuation of current energy and environmental policies in the reference case, carbon capture and sequestration (CCS) technology is not projected to come into use during the projection period.

Nuclear generating capacity in the *AEO2007* reference case is projected to increase from 100 gigawatts in 2005 to 112.6 gigawatts in 2030. The increase includes 12.5 gigawatts of capacity at newly built

Figure 5. Electricity generation by fuel, 1980-2030 (billion kilowatthours)

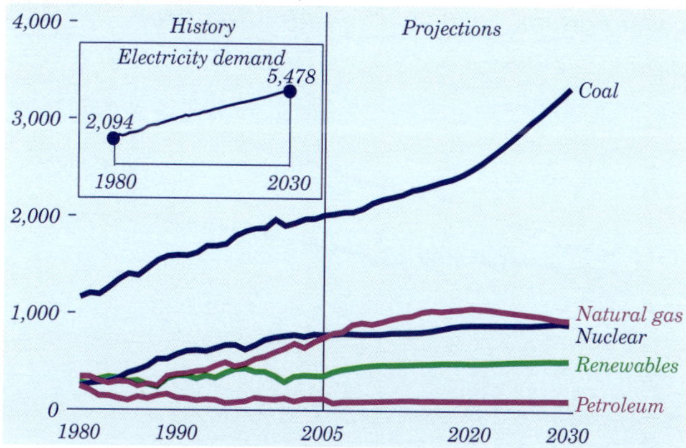

nuclear power plants (more than double the 6 gigawatts of new additions projected in the *AEO2006* reference case) and 3 gigawatts expected from uprates of existing plants, offset by 2.6 gigawatts of retirements.

Rules issued by the Internal Revenue Service (IRS) in 2006 for the EPACT2005 production tax credit (PTC) for new nuclear plants allow the credits to be shared out on a prorated basis to more than the 6 gigawatts of new capacity assumed in the *AEO2006* reference case. In the *AEO2007* reference case it is assumed that the credits will be shared out to 9 gigawatts of new nuclear capacity, and that 3.5 additional gigawatts of capacity will be built without credits. *AEO2007* also reflects the change in the PTC for new nuclear power plants that was included in the Gulf Opportunity Zone Act of 2005 (P.L. 109-135), eliminating the indexing provision in the value of the credit that had been provided in EPACT2005.

Total electricity generation from nuclear power plants is projected to grow from 780 billion kilowatthours in 2005 to 896 billion kilowatthours in 2030 in the *AEO2007* reference case, accounting for about 15 percent of total generation in 2030. Additional nuclear capacity is projected in some of the alternative *AEO2007* cases, particularly those that project higher demand for electricity or even higher fossil fuel prices.

The use of renewable technologies for electricity generation is projected to grow, stimulated by improved technology, higher fossil fuel prices, and extended tax credits in EPACT2005. Like the *AEO2006* reference case, the *AEO2007* reference case also includes the extension and expansion of the Federal PTC for renewable generation through December 31, 2007, as enacted in EPACT2005. Total renewable generation in the *AEO2007* reference case, including CHP and end-use generation, is projected to grow by 1.5 percent per year, from 357 billion kilowatthours in 2005 to 519 billion kilowatthours in 2030. The projection for renewable generation in the *AEO2007* reference case is lower than the comparable *AEO2006* projection, because new, less positive cost and performance characteristics are assumed for several renewable technologies.

In the *AEO2007* reference case, projected emissions of sulfur dioxide (SO_2) from electric power plants in 2030 are 64 percent lower, emissions of nitrogen oxides (NO_x) are 37 percent lower, and emissions of mercury are 70 percent lower than their 2005 levels. The reductions are about the same as those projected in the *AEO2006* reference case.

Energy Production and Imports

Net imports of energy on a Btu basis are projected to meet a growing share of total U.S. energy demand (Figure 6). In the *AEO2007* reference case, net imports are expected to constitute 32 percent of total U.S. energy consumption in 2030 (about the same as in the *AEO2006* reference case), up from 30 percent in 2005. Rising fuel prices over the projection period are expected to spur increases in domestic energy production (Figure 7) and to moderate the growth in demand, thus tempering the projected growth in imports.

The projections for U.S. crude oil production in the *AEO2007* reference case are significantly different from those in the *AEO2006* reference case. U.S. crude oil production in the *AEO2007* reference case is projected to increase from 5.2 million barrels per day in 2005 to a peak of 5.9 million barrels per day in 2017 as a result of increased production offshore, predominantly from the deep waters of the Gulf of

Figure 6. Total energy production and consumption, 1980-2030 (quadrillion Btu)

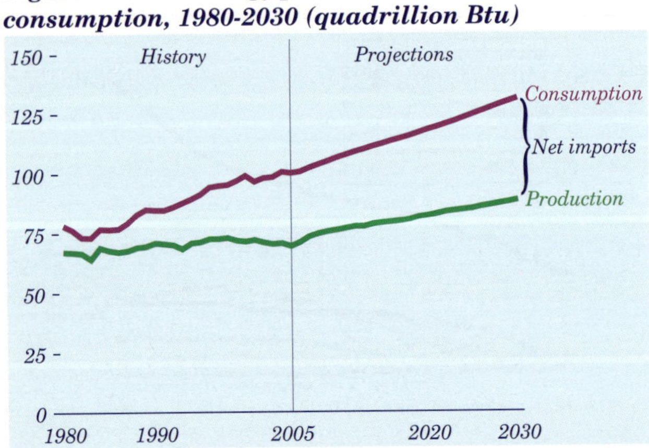

Figure 7. Energy production by fuel, 1980-2030 (quadrillion Btu)

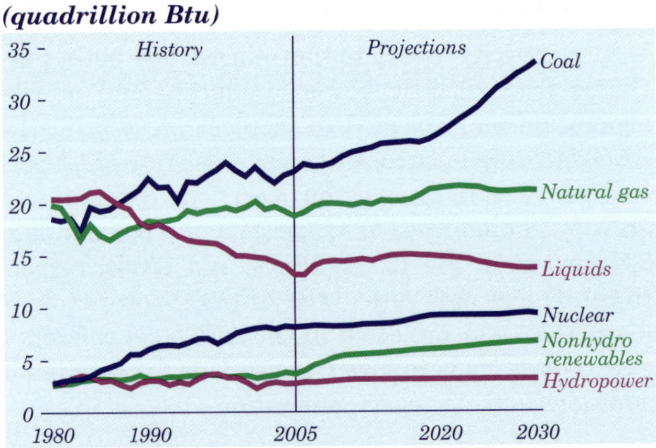

Mexico. Production is subsequently projected to fall to 5.4 million barrels per day in 2030. The *AEO2006* reference case projected a much steeper decline in production from 2017 to 2030, with crude oil production falling from a slightly lower level of 5.8 million barrels per day in 2017 to 4.6 million barrels per day in 2030. The difference is attributable primarily to a slower decline in lower 48 onshore oil production in the *AEO2007* reference case, mostly as a result of increased production from enhanced oil recovery technology and, to a lesser extent, significantly higher resource assumptions for the Bakken Shale formation in the Williston Basin.

Total domestic liquids production, including crude oil, natural gas plant liquids, refinery processing gains, and other refinery inputs, is projected to increase steadily throughout the projection in the *AEO2007* reference case, as growth in refinery processing gains and other refinery inputs offsets the projected decline in crude oil production after 2017. Total supply is projected to grow from 8.3 million barrels per day in 2005 to 10.5 million barrels per day in 2030. In the *AEO2006* reference case, total domestic liquids supply in 2030 was slightly lower, at 10.4 million barrels per day. The higher crude oil production in the *AEO2007* reference case, when compared with the *AEO2006* reference case, is partially offset by lower production of natural gas liquids and lower refinery processing gains.

In the *AEO2007* reference case, the net liquids import share of total supply, including both crude oil and refined products, drops from 60 percent of total liquids supply in 2005 to 54 percent in 2009, before increasing to 61 percent in 2030. In the *AEO2006* reference case, net liquids imports accounted for 62 percent of product supplied in 2030. Net crude oil imports in 2030 are 0.4 million barrels per day lower, and net product imports are 0.5 million barrels per day lower, in the *AEO2007* reference case than projected in the *AEO2006* reference case.

The primary reason for the difference between the *AEO2006* and *AEO2007* projections for net imports of liquid fuels is a lower level of total liquids consumption, by 0.6 million barrels per day in 2030 in the *AEO2007* reference case, and a greater increase in refinery distillation capacity, which increases from 17.1 million barrels per day in 2005 to 20.0 million barrels per day in 2030 in the *AEO2007* reference case, as compared with 19.3 million barrels per day in 2030 in the *AEO2006* reference case. In addition, the *AEO-2007* reference case includes greater investment in heavy oil processing as a result of changes in expected crude slates and pricing differentials. Imports of refined petroleum products account for 20 percent of total net imports in 2030 (about the same as in 2005) in the *AEO2007* reference case, as compared with 22 percent in the *AEO2006* reference case.

Total domestic natural gas production, including supplemental natural gas supplies, increases from 18.3 trillion cubic feet in 2005 to 21.1 trillion cubic feet in 2022, before declining to 20.6 trillion cubic feet in 2030 in the *AEO2007* reference case. In comparison, domestic natural gas production was projected to peak at 21.6 trillion cubic feet in 2019 in the *AEO2006* reference case. Through 2012, natural gas production in the *AEO2007* reference case is generally higher than in the *AEO2006* reference case. After 2012, production in the *AEO2007* reference case is consistently below that in the *AEO2006* reference case. Lower natural gas consumption in the last 18 years of the projection results in lower domestic natural gas production—primarily, offshore and onshore nonassociated conventional production—in the *AEO2007* reference case.

In the *AEO2007* reference case, lower 48 offshore production of natural gas grows from 3.4 trillion cubic feet in 2005 to a peak of 4.6 trillion cubic feet in 2015 as new resources come online in the Gulf of Mexico. After 2015, lower 48 offshore production declines to 3.3 trillion cubic feet in 2030, as investment is inadequate to maintain production levels. In the *AEO2006* reference case, offshore natural gas production was projected to peak at 5.1 trillion cubic feet in 2015 before falling to 4.0 trillion cubic feet in 2030. Onshore nonassociated conventional production of natural gas in the *AEO2007* reference case is higher than was projected in the *AEO2006* reference case through 2012, after which it falls below the projection in the *AEO2006* reference case.

Lower 48 production of unconventional natural gas is expected to be a major contributor to growth in U.S. natural gas supplies. In the *AEO2007* reference case, unconventional natural gas production is projected to account for 50 percent of domestic U.S. natural gas production in 2030 (compared with a 45-percent share in the *AEO2006* reference case). Throughout the projection period, the level of unconventional natural gas production in the *AEO2007* reference case is higher, reaching 10.2 trillion cubic feet in 2030, than in the *AEO2006* reference case (9.5 trillion cubic feet in 2030), due to the addition of the Fayetteville and Woodford shale resources and generally higher natural gas prices.

Construction planning for the Alaska natural gas pipeline is expected to start 3 years later than projected in the *AEO2006* reference case due to startup delays and a longer than anticipated construction time period, with pipeline completion in 2018. After the pipeline goes into operation, Alaska's total natural gas production is projected to increase from 0.5 trillion cubic feet in 2005 to 2.2 trillion cubic feet in 2021 in the *AEO2007* reference case. Although the timing differs, this is the same level that was projected in the *AEO2006* reference case.

With the exception of the last few years of the projection, net pipeline imports of natural gas from Canada and Mexico, predominantly from Canada, in the *AEO2007* reference case are higher than in the *AEO-2006* reference case. Net pipeline imports vary between 2.6 and 3 trillion cubic feet from 2005 to 2013 in the *AEO2007* reference case, then decline to 0.9 trillion cubic feet in 2030—0.3 trillion cubic feet lower than projected in the *AEO2006* reference case. The decline reflects resource depletion in Alberta, growing domestic demand in Canada, and a downward reassessment of the potential for unconventional natural gas production from coal seams and tight formations in Canada.

The *AEO2007* reference case projects that LNG imports will meet much of the increased U.S. demand for natural gas, as was the case in the *AEO2006* reference case. In addition to new terminals, including four that are currently under construction, expansions of three of the four existing onshore U.S. LNG terminals (Cove Point, Maryland; Elba Island, Georgia; and Lake Charles, Louisiana) are included in the *AEO2007* reference case. Because of liquefaction project delays, supply constraints at a number of liquefaction facilities, and rapid growth in global LNG demand, the U.S. LNG market is expected to be tight until 2012. Total net imports of LNG to the United States in the *AEO2007* reference case are projected to increase from 0.6 trillion cubic feet in 2005 to 4.5 trillion cubic feet in 2030 (0.2 trillion cubic feet higher than in the *AEO2006* reference case).

As domestic coal demand grows in the *AEO2007* reference case, U.S. coal production increases at an average rate of 1.6 percent per year, from 1,131 million short tons in 2005 to 1,691 million short tons in 2030, slightly less than in the *AEO2006* reference case. Production from mines west of the Mississippi River is expected to provide the largest share of the incremental coal production. In 2030, almost 68 percent of domestic coal production is projected to originate from States west of the Mississippi.

Typically, trends in U.S. coal production are linked to its use for electricity generation, which currently accounts for more than 90 percent of total coal consumption. Projected coal consumption in the electric power sector in the *AEO2007* reference case is slightly higher than projected in the *AEO2006* reference case (1,570 million short tons versus 1,502 million short tons in 2030), because coal captures a larger share of total electricity generation in the *AEO2007* reference case. Another fast-growing market for coal is CTL. Coal use in CTL plants is projected to grow from 26 million short tons in 2020 to 112 million short tons in 2030. By 2025, coal use for CTL production becomes the second largest use of coal in the *AEO2007* reference case, after electric power generation.

Energy-Related Carbon Dioxide Emissions

Absent the application of CCS technology, which is not expected to come into use without changes in current policies that are not included in the reference case, carbon dioxide (CO_2) emissions from the combustion of fossil fuels are proportional to fuel consumption and carbon content, with coal having the highest carbon content, natural gas the lowest, and petroleum in between. In the *AEO2007* reference case, the coal share of total energy use increases from 23 percent in 2005 to 26 percent in 2030, while the share of natural gas falls from 23 percent to 20 percent, and the liquids share remains at about 40 percent. The combined share of carbon-neutral renewable and nuclear energy is stable from 2005 to 2030 at about 14 percent.

Taken together, projected growth in the absolute level of primary energy consumption and a shift toward a fuel mix with slightly higher average carbon content cause projected energy-related emissions of

Figure 8. U.S. carbon dioxide emissions by sector and fuel, 1990-2030 (million metric tons)

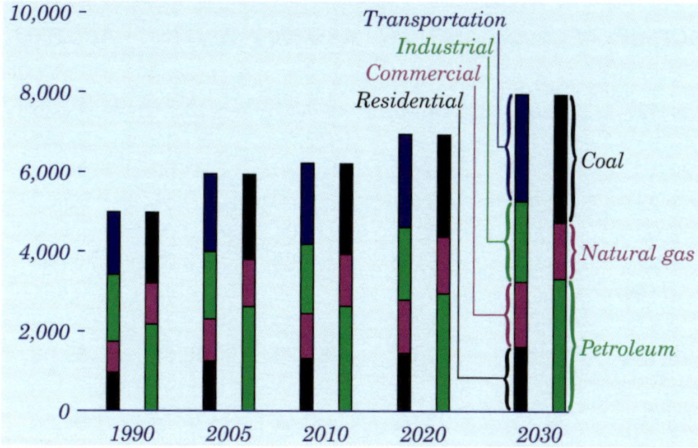

CO_2 to grow by an average of 1.2 percent per year from 2005 to 2030 (Figure 8)—slightly higher than the average annual increase in total energy use. At the same time, the economy becomes less carbon intensive: the percentage increase in CO_2 emissions is about one-third of the projected increase in GDP, and emissions per capita increase by only 9 percent over the 25-year projection period. Projections of energy-related CO_2 emissions in the *AEO2007* reference case are slightly lower than those in the *AEO2006* reference case, consistent with the comparable difference in projections for overall energy use.

Table 1. Total energy supply and disposition in the *AEO2007* and *AEO2006* reference cases, 2005-2030

Energy and economic factors	2005	2010 *AEO2007*	2010 *AEO2006*	2020 *AEO2007*	2020 *AEO2006*	2030 *AEO2007*	2030 *AEO2006*
Primary energy production (quadrillion Btu)							
Petroleum	13.30	14.42	14.83	14.85	14.41	13.71	12.25
Dry natural gas	18.77	19.93	19.13	21.41	22.09	21.15	21.45
Coal	23.20	24.47	25.78	26.61	27.30	33.52	34.10
Nuclear power	8.13	8.23	8.44	9.23	9.09	9.33	9.09
Hydropower	2.71	3.02	3.03	3.08	3.04	3.09	3.04
Biomass	2.71	4.22	3.90	4.69	4.66	5.26	5.07
Other renewable energy	0.76	1.18	1.27	1.33	1.92	1.44	2.61
Other	0.22	0.67	0.97	0.89	1.22	1.12	1.39
Total	**69.80**	**76.13**	**77.36**	**82.09**	**83.73**	**88.63**	**89.00**
Net imports (quadrillion Btu)							
Petroleum	26.94	25.19	26.25	28.92	30.46	34.74	36.56
Natural gas	3.67	4.67	4.45	5.48	5.15	5.59	5.72
Coal/other (- indicates export)	-0.42	-0.19	-0.58	0.93	0.90	1.57	2.02
Total	**30.19**	**29.66**	**30.12**	**35.33**	**36.50**	**41.90**	**44.30**
Consumption (quadrillion Btu)							
Liquid fuels	40.61	41.76	43.14	46.52	48.15	52.17	53.59
Natural gas	22.63	24.73	24.04	27.04	27.70	26.89	27.65
Coal	22.87	24.24	25.09	27.29	27.65	34.14	34.49
Nuclear power	8.13	8.23	8.44	9.23	9.09	9.33	9.09
Hydropower	2.71	3.02	3.03	3.08	3.04	3.09	3.04
Biomass	2.38	3.30	3.25	3.64	3.73	4.06	4.09
Other renewable energy	0.76	1.18	1.27	1.33	1.92	1.44	2.61
Net electricity imports	0.08	0.04	0.07	0.04	0.05	0.04	0.05
Total	**100.19**	**106.50**	**108.34**	**118.16**	**121.32**	**131.16**	**134.60**
Liquid fuels (million barrels per day)							
Domestic crude oil production	5.18	5.67	5.88	5.89	5.55	5.39	4.57
Other domestic production	3.04	4.03	3.98	4.49	4.87	5.08	5.82
Net imports	12.57	11.79	12.36	13.56	14.47	16.37	17.29
Consumption	20.75	21.59	22.18	24.03	24.82	26.95	27.57
Natural gas (trillion cubic feet)							
Production	18.30	19.42	18.65	20.86	21.52	20.61	20.90
Net imports	3.57	4.55	4.35	5.35	5.02	5.45	5.57
Consumption	21.98	24.02	23.35	26.26	26.92	26.12	26.86
Coal (million short tons)							
Production	1,131	1,189	1,261	1,323	1,355	1,691	1,703
Net imports	-21	-7	-26	41	36	68	83
Consumption	1,128	1,195	1,233	1,377	1,390	1,772	1,784
Prices (2005 dollars)							
Imported low-sulfur, light crude oil (dollars per barrel)	56.76	57.47	48.50	52.04	52.00	59.12	58.42
Imported crude oil (dollars per barrel)	49.19	51.20	45.12	46.47	46.14	51.63	51.27
Domestic natural gas at wellhead (dollars per thousand cubic feet)	7.51	5.76	5.15	5.22	5.02	5.98	6.07
Domestic coal at minemouth (dollars per short ton)	23.34	24.20	22.80	21.58	20.72	22.60	22.29
Average electricity price (cents per kilowatthour)	8.1	8.1	7.5	7.9	7.4	8.1	7.7
Economic indicators							
Real gross domestic product (billion 2000 dollars)	11,049	12,790	13,043	17,077	17,541	22,494	23,112
GDP chain-type price index (index, 2000=1.000)	1.127	1.253	1.235	1.495	1.597	1.815	2.048
Real disposable personal income (billion 2000 dollars)	8,105	9,568	9,622	13,000	13,057	17,535	17,562
Value of manufacturing shipments (billion 2000 dollars)	5,763	6,298	6,355	7,779	7,778	9,502	9,578
Primary energy intensity (thousand Btu per 2000 dollar of GDP)	**9.07**	**8.33**	**8.31**	**6.92**	**6.92**	**5.83**	**5.82**
Carbon dioxide emissions (million metric tons)	**5,945**	**6,214**	**6,365**	**6,944**	**7,119**	**7,950**	**8,114**

Notes: Quantities are derived from historical volumes and assumed thermal conversion factors. Other production includes liquid hydrogen, methanol, and some inputs to refineries. Net imports of petroleum include crude oil, petroleum products, unfinished oils, alcohols, ethers, and blending components. Other net imports include coal coke and electricity.

Sources: AEO2007 National Energy Modeling System, run AEO2007.D112106A; and AEO2006 National Energy Modeling System, run AEO2006.D111905A.

Legislation and Regulations

Legislation and Regulations

Introduction

Because analyses by EIA are required to be policy-neutral, the projections in *AEO2007* generally are based on Federal and State laws and regulations in effect on or before October 31, 2006 (although there are exceptions to this rule, as discussed below). **The potential impacts of pending or proposed legislation, regulations, and standards—or of sections of legislation that have been enacted but that require implementing regulations or appropriation of funds that are not provided or specified in the legislation itself—are not reflected in the projections.**

Selected examples of Federal and State legislation incorporated in the projections include the following:

- The new CAFE standards finalized in March 2006, which establish higher minimum fuel economy performance requirements by vehicle footprint for light-duty trucks
- EPACT2005, which includes mandatory energy conservation standards; creates numerous tax credits for businesses and individuals; creates an RFS and eliminates the oxygen content requirement; extends royalty relief for offshore oil and natural gas producers; and extends and expands the PTC for electricity generated from renewable fuels
- The Military Construction Appropriations Act of 2005, which contains provisions to support construction of the Alaska natural gas pipeline, including Federal loan guarantees during construction
- The Working Families Tax Relief Act of 2004, which includes tax deductions for qualified clean-fuel and electric vehicles; and changes in the rules governing oil and natural gas well depletion
- The American Jobs Creation Act of 2004, which includes incentives and tax credits for biodiesel fuels and a modified depreciation schedule for the Alaska natural gas pipeline
- The Maritime Security Act of 2002, which amended the Deepwater Port Act of 1974 to include offshore natural gas facilities
- State RPS programs, including the California RPS passed on September 12, 2002
- The Clean Air Act Amendments of 1990 (CAAA-90), which included new standards for motor gasoline and diesel fuel and for heavy-duty vehicle emissions
- The National Appliance Energy Conservation Act of 1987
- State programs for restructuring of the electricity industry.

AEO2007 assumes that State taxes on gasoline, diesel, jet fuel, and E85 [*4*] will increase with inflation and that Federal taxes on those fuels will remain at the nominal rate established in 2003 (the last time the Federal taxes were changed). *AEO2007* also assumes that the ethanol tax credit, as modified under the American Jobs Creation Act of 2004, will be extended when it expires in 2010 and will remain in force indefinitely. Although the ethanol tax credit includes a "sunset" clause that limits its duration, historically it has been extended regularly, and *AEO-2007* assumes its continuation throughout the projection [*5*]. *AEO2007* also includes the biodiesel tax credits that were created under the American Jobs Creation Act and extended through 2008 under EPACT2005; however, they are not assumed to be extended further, because they have minimal history of legislative extension.

The Safe, Accountable, Flexible, Efficient Transportation Equity Act: A Legacy for Users (SAFETEA-LU) increased the Federal tax on compressed natural gas used in vehicles to 18.3 cents per equivalent gallon of gasoline and provided a credit of 50 cents per gallon through September 2009. *AEO2007* assumes that State and Federal taxes on compressed natural gas for vehicles will continue at 2006 levels in nominal terms and that the tax credit will not be extended.

The PTC for wind, geothermal, landfill gas (LFG), and some types of hydroelectric and biomass-fueled plants, established initially by the Energy Policy Act of 1992 [*6*] also is represented in *AEO2007*. Only new plants that come on line before January 1, 2008, are eligible to receive the credit. *AEO2007* does not assume extension of the PTC, which has been allowed to expire in the past, even though it has typically been renewed retroactively. In most of the extensions, the credit has been modified significantly: additional resources have been included, resources previously eligible have been excluded, and the structure and treatment of the credit itself have been changed.

Selected examples of Federal and State regulations incorporated in *AEO2007* include the following:

- New stationary diesel regulations issued by the U.S. Environmental Protection Agency (EPA) on July 11, 2006, which limit emissions of NO_x, particulate matter, SO_2, carbon monoxide, and hydrocarbons to the same levels required by the EPA's nonroad diesel engine regulations
- The Clean Air Interstate Rule (CAIR) and Clean Air Mercury Rule (CAMR), promulgated by the EPA in March 2005 and published in the *Federal Register* as final rules in May 2005, which will limit emissions of SO_2, NO_x, and mercury from power plants in the United States
- New boiler limits established by the EPA on February 26, 2004, which limit emissions of hazardous air pollutants from industrial, commercial, and institutional boilers and process heaters by requiring that they comply with a Maximum Achievable Control Technology floor.

AEO2007 does not include consideration of California Assembly Bill (A.B.) 32 (discussed below), which mandates a 25-percent reduction in California's greenhouse gas emissions by 2020. Implementing regulations have not been drafted and are not due to be finalized until January 2012.

In addition, California's A.B. 1493, which establishes greenhouse gas emissions standards for light-duty vehicles, is not considered in the *AEO2007* reference case. A.B. 1493 was signed into law in July 2002, and regulations were released by the California Air Resources Board in August 2004 and approved by California's Office of Administrative Law in September 2005; however, the automotive industry has filed suit to block their implementation, and the Board has not yet obtained a Clean Air Act waiver from the EPA, which is required before the regulations can be implemented.

In August 2006, seven northeastern States released a model rule for implementation of the Regional Greenhouse Gas Initiative (RGGI) [*7*], as discussed below, clarifying what had been laid out in December 2005 when the States entered into the agreement [*8*]. The RGGI, which would cap greenhouse gas emissions from power producers, requires each State to enact legislation for accomplishing the emissions reductions. Although the State RGGI caps and timelines are known, many aspects of their implementation remain uncertain, because the participating States have not yet enacted the necessary legislation. Therefore, the RGGI provisions are not modeled in the *AEO2007* reference case.

AEO2007 does include the CAAA90 requirement of a phased-in reduction in vehicle emissions of regulated pollutants. It also reflects "Tier 2" Motor Vehicle Emissions Standards and Gasoline Sulfur Control Requirements finalized by the EPA in February 2000 under CAAA90. The Tier 2 standards for reformulated gasoline (RFG) were required by 2004, but because they included allowances for small refineries, they will not be fully realized for conventional gasoline until 2008. *AEO2007* also incorporates the ultra-low-sulfur diesel fuel (ULSD) regulation finalized by the EPA in December 2000, which requires the production of at least 80 percent ULSD (less than or equal to 15 parts sulfur per million) highway diesel between June 2006 and June 2010 and 100 percent ULSD thereafter. It also includes the rules for nonroad diesel issued by the EPA on May 11, 2004, regulating nonroad diesel engine emissions and sulfur content in fuel.

More detailed information on recent and proposed legislative and regulatory developments is provided below.

EPACT2005: Status of Provisions

EPACT2005 was signed into law by President Bush on August 8, 2005, and became Public Law 109-058 [*9*]. A number of provisions from EPACT2005 were included in the *AEO2006* projections [*10*]. Many others were not considered in *AEO2006*—particularly, those that require funding appropriations or further specification by Federal agencies or Congress before implementation.

A number of the EPACT2005 provisions not included in *AEO2006* could affect the projections. In the preparation of *AEO2007* their status was reviewed, and where possible, additional provisions were included in the projections; however, *AEO2007* still excludes those EPACT2005 provisions whose impacts are highly uncertain or that address a level of detail beyond that modeled in NEMS. Furthermore, EIA does not try to anticipate policy responses to the many studies required by EPACT2005 nor predict the impacts of research and development (R&D) funding authorizations included in the bill.

The following summary examines the status of EPACT2005 provisions that initially could not be included in *AEO2006* but potentially could be modeled in NEMS. It focuses on provisions that are newly included in *AEO2007*, as well as those that might be added in future *AEO*s. The discussion below does not provide a complete summary of all the sections of EPACT2005. More extensive summaries are available from other sources [*11*].

End-Use Demand Provisions

This section summarizes provisions of EPACT2005 that affect the end-use demand sectors.

Buildings

EPACT2005 includes provisions with the potential to affect energy demand in the residential and commercial buildings sector. Many are included in Title I, "Energy Efficiency." Others can be found in the renewable energy, R&D, and tax titles. Most of the provisions that have been funded or for which implementing regulations have been put in place since the publication of *AEO2006*, cannot be modeled in NEMS. The status of those provisions that could potentially be included in NEMS is summarized below.

Section 122 of Title I, "Weatherization Assistance," authorizes $600 million to weatherize low-income households. The weatherization program, in existence since 1976, uses Federal funds to increase the energy efficiency of low-income houses. In fiscal year (FY) 2006, funding for this program was $242 million. FY 2007 funding proposed by the U.S. House of Representatives is set at $250 million. The increase in funding could allow up to 3,200 more homes to be weatherized in FY 2007 than in FY 2006. The *AEO2007* reference case includes increases in energy efficiency in existing building envelopes to account for programs such as weatherization. At current funding levels, roughly 100,000 homes are weatherized each year. The impact of this section is considered in *AEO2007*.

Section 204 of Title II, "Use of Photovoltaic Energy in Public Buildings," authorizes funds for the establishment of a photovoltaic (PV) energy commercialization program to procure, install, and evaluate PV solar electric systems in public buildings. No funding has been appropriated to date for this measure. It is not included in *AEO2007*.

Section 206 of Title II, "Renewable Energy Security," authorizes funds for the establishment of rebates for the purchase of renewable energy systems, including PV, ground-source heat pumps, and solar water heaters. This program was to be in place starting in calendar year 2006 and last through 2010; however, no funding has been appropriated for the measure to date, and it is not included in *AEO2007*.

Section 783 of Title VII, "Federal Procurement of Stationary, Portable, and Micro Fuel Cells," authorizes funds for Federal procurement of stationary, portable, and micro fuel cells. No funding has been appropriated for the measure to date, and it is not considered in *AEO2007*.

Industrial

EPACT2005 includes few provisions that would specifically affect industrial sector energy demand. Provisions in the R&D titles that may affect industrial energy consumption over the long term are not included in *AEO2007*.

Section 108 requires that federally funded projects involving cement or concrete increase the amount of recovered mineral component (e.g., fly ash or blast furnace slag) used in the cement. Such use of mineral components is a standard industry practice, and increasing the amount could reduce both the quantity of energy used for cement clinker production and the level of process-related CO_2 emissions. The proportion of mineral component is not specified in the legislation but is to be determined by Federal procurement rules; however, as of mid-September 2006 the rules had not been promulgated. Section 108 also requires that the energy-saving impact of the rules be assessed by the EPA, in cooperation with the U.S. Department of Energy (DOE) and Department of Transportation (DOT), within 30 months of enactment. Because regulations have not been promulgated, this section is not considered in *AEO2007*. When the regulations are promulgated, their estimated impacts could be modeled in NEMS.

Section 1321 provides for the extension of tax credits for producers of coke or coke gas, effective for tax years beginning after December 31, 2005. Otherwise, the status of Section 1321 is unchanged. Because the bulk of the credits will go to plants already operating or under construction, there is likely to be little impact on coke plant capacity. Consequently, the provision is expected to have no impact on the *AEO2007* projections.

Coal Gasification Provisions

This section provides updates to the funding and implementation status of key tax incentive provisions in Title XIII of EPACT2005 related to coal gasification that were not addressed in *AEO2006*.

Section 1307 creates an investment tax credit program for qualifying advanced clean coal projects, funded at $1.3 billion. The section also includes an additional $350 million for qualifying gasification projects. The gasification credit for any taxable year is equal to 20 percent of the basis of any equipment to be used in the gasification process that is placed in service during the year as part of a gasification project that has been certified by DOE as eligible for the credit. The amount eligible for credit is limited to $650 million per project. Only domestic projects that employ domestic gasification applications are eligible. Applicants must, among other criteria, satisfy certain financial requirements, prove that a market exists for the project's products, and demonstrate competency in the development and operation of the project. Credits are not allowed for gasification projects receiving credits under the program for advanced coal projects. A certificate of eligibility is valid for 10 fiscal years, beginning on October 1, 2005.

In February 2006, the IRS issued guidance for the Section 1307 program. Certifications are to be issued and credits allocated to projects in annual allocation rounds. The first round of submissions began on February 21, 2005, and closed on October 2, 2006. Overall, the period for submission of applications is to run for 3 years, starting on February 21, 2006. As of August 2006, 49 applications had been received, 27 of which fell under the gasification technology program and were for CTL plants in 17 States. The 27 projects are valued at $30 billion and request tax credits of $2.7 billion. Selection of projects to receive the credits is scheduled for the end of November 2006.

Credits will be allocated first to projects that have CO_2 capture capability, use renewable fuel, or have project teams with experience that demonstrates successful operation of the gasification technology. If the requested allocations exceed $350 million, the credits will be allocated to the projects that provide the highest ratio of synthetic gas supplied to the requested allocation of credits. Any remaining credits will be applied to non-priority projects that provide the highest amount of nameplate capacity. If funds remain in the program, additional rounds will be conducted in 2007 and 2008. The $1.3 billion in tax credits for the advanced clean coal program was accounted for in *AEO2006* in the NEMS Electricity Market Module. CTL projects are eligible for the gasification credits, because gasification is the first step in the CTL process; however, because the level of interest in coal gasification projects was not known at the time, the gasification program credits were not included in *AEO2006*. Given the extent of interest in the program to date, they are included in the Electricity Market Module for *AEO2007*.

Oil and Natural Gas Provisions

This section provides updates to the funding and implementation status of key oil and natural gas provisions of EPACT2005 that were not addressed in *AEO2006*. Most of the oil and natural gas provisions in EPACT2005 are included in Title III, "Oil and Gas." Others, covering R&D, are included in Title IX.

The Federal Energy Regulatory Commission (FERC) was authorized by Section 312 to allow natural gas storage facilities to charge market-based rates if it was believed that they would not exert market power. On June 15, 2006, FERC finalized rules implementing the provisions that would allow an applicant for interstate natural gas storage facilities to request authority to charge market-based rates even if a lack of market power had not been demonstrated. The rules are intended to mitigate natural gas price volatility by encouraging the development of new natural gas storage capacity. They apply in circumstances where market-based rates are in the public interest and necessary to encourage the construction of storage capacity and to ensure that customers are adequately protected, even in circumstances where market power may not have been demonstrated. In previous *AEO*s, storage rates were allowed to vary from regulation-based rates, depending on market conditions.

In compliance with Section 354, DOE established a competitive program to provide grants for cost-shared projects to enhance oil and natural gas recovery through CO_2 injection, while at the same time sequestering CO_2 produced from the combustion of fossil fuels in power plants and large industrial processes. Reports issued by DOE indicate that an additional 89 billion barrels of oil could be recovered in the United States through CO_2 injection. Under the program, grants of up to $3 million will be provided to each project selected. On September 6, 2006, DOE announced the selection of the first project to receive one of the grants, a project sponsored by the University of Alabama-Birmingham to implement a demonstration project in the Citronelle oilfield in Mobile County, AL. The total project cost is estimated at $6 million, with DOE's maximum share at just under $3 million. Estimates indicate that an additional 64 million barrels of oil could be recovered from the Citronelle field by this technique.

The implementation of Section 354 was not included in previous *AEO*s, because NEMS does not represent project-level activities and because of the considerable uncertainty surrounding the eventual scope of the program. For *AEO2007*, however, additional oil resources have been added to account for increased use of CO_2-enhanced oil recovery technology.

Section 311 clarified the role of FERC as the final decisionmaking body on any issues concerning onshore facilities that export, import, or process LNG. On October 7, 2005, FERC established mandatory procedures requiring prospective applicants for LNG terminals, related jurisdictional pipelines, and other related natural gas facilities to begin the Commission's pre-filing review process at least 6 months before filing an application to site and/or construct such a facility. The procedures, which also apply to applications for modifications of existing or authorized LNG terminals, are designed to encourage applicants to cooperate with State and local officials.

In March 2005 and June 2006, FERC and DOE, in cooperation with DOT and the U.S. Department of Homeland Security, conducted three public forums on LNG designed to promote public education and encourage cooperation between State and Federal officials in areas where LNG terminals are being considered for construction. They were held in Boston, MA; Astoria, OR; and Los Angeles, CA. An additional forum is planned for Houston, TX, in the 4th quarter of 2006, fully satisfying the Section 317 requirement that a minimum of three such forums be held. Although this provision is not explicitly represented in the *AEO2007* NEMS, the model includes an assumption that there are no major regulatory impediments to the siting of new LNG facilities.

Section 301 authorized DOE to increase the capacity of the Strategic Petroleum Reserve (SPR) to 1 billion barrels from its current capacity of 727 million barrels. DOE has announced plans to add additional storage capacity to its SPR storage sites in Big Hill, TX; Bayou Choctaw, LA; West Hackberry, LA; and one new site in Richton, MS. DOE filed a draft site selection Environmental Impact Statement with the EPA on May 19, 2006, for the selection of a new site, and comments have been received. In order for the additional storage capacity to be authorized, constructed, and ultimately filled, further actions by Congress and the Executive Branch will be required; therefore, it is not considered in *AEO2007*.

Section 369 requires DOE to initiate a process for the leasing of Federal lands for research on oil shale, tar sands, and other unconventional fuels. Several industry research proposals were evaluated, and on January 17, 2006, the U.S. Department of the Interior's Bureau of Land Management announced the selection of six applicants for oil shale leases to receive further consideration. Because the lease applications are still under consideration, this provision is not accounted for in *AEO2007*.

Coal Provisions

This section provides updates to the funding and implementation status of provisions in EPACT2005 that will affect coal supply and prices but were not addressed in *AEO2006*. Many of the provisions can be found in Titles IV and XIII of EPACT2005.

A number of coal-related provisions that were authorized by EPACT2005 but not included in *AEO2006* continue to be excluded from *AEO2007*. They include four loan guarantee or cost-sharing programs. Section 411 authorized a loan guarantee for a coal project in the Upper Great Plains, which must employ both renewable and advanced IGCC technologies. A loan guarantee for the Clean Coal Project in Healy, AK, authorized by Section 412, also is excluded from *AEO2007*. In Section 413, EPACT2005 authorized a cost-sharing program in support of a high-altitude (at least 4,000 feet) Western IGCC Demonstration Project. Finally, a loan guarantee for an IGCC plant located in a deregulated region was authorized by Section 414.

These provisions have spurred some activity and interest. For instance, Xcel Energy, which has proposed building a facility in Colorado with 300 to 350 megawatts of generating capacity, is a potential applicant for the Western IGCC Demonstration Project. On August 7, 2006, DOE released its plans to form a program office with functions that include the drafting of application guidelines for the various loan programs. It will also be charged with the task of awarding the loan guarantees. Although NEMS has the capability to represent these coal provisions, Congress had not appropriated funds for the provisions as of September 1, 2006, and they are not considered in *AEO2007*.

Nuclear Energy Provisions

EPACT2005 includes numerous provisions that address nuclear power generation. This section provides updates to the funding and implementation status of nuclear power generation provisions in EPACT2005 that were not addressed in *AEO2006*.

Section 1306 of Title 13 extends the PTC of 1.8 cents per kilowatthour (not adjusted for inflation) to any nuclear power plant with a "new" design that has a construction start date before January 1, 2014, and enters commercial operation by January 1, 2021. Under this program, the owner of the eligible plant can reduce its tax liability by up to 1.8 cents for each kilowatthour of plant output. For the purposes of this law, construction begins when a utility "that has applied for or been granted a combined operating license . . . initiates the pouring of safety-related concrete for the reactor building." The IRS published an initial set of guidelines for the program in May 2006 and eventually will publish a set of formal rules that will become part of the Tax Code. In EPACT-2005, the per-kilowatthour tax credit was indexed to the rate of inflation; however, the indexing provision was eliminated in the Gulf Opportunity Zone Act of 2005 (P.L. 109-135). Consequently, the credit would be constant in nominal dollars over time. Because the earliest date at which the first new nuclear unit eligible for the tax credit could become operational is about 2015, the "de-indexing" of the credit has the effect of reducing its real value by about 25 to 30 percent.

There are at least three limitations on the amount of tax credits a utility can receive. First, tax credits in any given year are limited to a maximum of $125 per kilowatt ($125 million for a 1,000-megawatt unit). Second, the tax credit can be applied only in the first 8 years of a plant's operation. Third, the credit is limited to a maximum of 6 gigawatts of new nuclear capacity nationally. If the total capacity qualifying for the tax credit exceeds 6 gigawatts, the amount of the credit per kilowatthour will be reduced proportionally. *AEO2007* assumes that up to 9 gigawatts of new capacity will receive the Title 13 PTC at 1.2 cents per kilowatthour. (*AEO2006* assumed that 6 gigawatts would receive the full 1.8 cents per kilowatthour.) *AEO2007* also assumes that participating utilities will be able to take all the tax credits in each of the first 8 years of their qualifying units' operation.

Title 17 of EPACT2005 allows the Government to guarantee loans used to construct new energy technologies "that reduce or avoid greenhouse gases," including new nuclear power plants. The Secretary of Energy can guarantee a loan of up to 80 percent of the project's cost; however, DOE will not guarantee more than 80 percent of the total debt. Thus, if a utility decided to fund a project with 80 percent debt and 20 percent equity, DOE would only guarantee up to 64 percent of the project's total cost. Such loan guarantees would affect the economics of nuclear power, because they would reduce the effective interest rates on the debt and allow utilities to use much more debt financing.

The Secretary of Energy will choose the projects that will receive the loan guarantees. The factors to be considered in the selection of projects include:

- A relatively low probability of failure
- The extent to which the project avoids, reduces, or sequesters air pollutants or emissions of greenhouse gases
- The extent to which the project will advance the goals of the President's Advanced Energy Initiative
- The extent to which the technology is ready to be employed commercially in the United States and can yield a commercially viable product.

Because of the lack of appropriating legislation, this program is not included in *AEO2007*.

Fuel Economy Standards for New Light Trucks

In March 2006, NHTSA finalized CAFE standards requiring higher fuel economy performance for light-duty trucks in MY 2008 through 2011 [*12*]. Unlike the proposed CAFE standards discussed in *AEO2006* [*13*], which would have established minimum fuel economy requirements by six footprint size classes, the final reformed CAFE standards specify a continuous mathematical function that determines minimum fuel economy requirements by vehicle footprint, defined as the wheelbase (the distance from the front axle to the center of the rear axle) times the average track width (the distance between the center lines of the tires) of the vehicle in square feet.

As shown in Figure 9, the new fuel economy standards vary by model year (MY) and by vehicle footprint. By eliminating the categories laid out in the proposed rule, the final rule removes the opportunity for manufacturers to reduce fuel economy requirements by altering vehicle sizes just enough to reach lower target levels. Instead, under a continuous function approach, each footprint value has an assigned fuel economy target, and small changes in vehicle footprint are not rewarded with large decreases in target values.

In addition to reforming the structure of the light truck CAFE program, NHTSA has also increased the gross vehicle weight rating (GVWR) of light trucks covered under CAFE. NHTSA defines light-duty trucks as trucks with a GVWR of 10,000 pounds or less, including pickups, vans, truck-based station wagons, and sport utility vehicles (SUVs). Current CAFE standards apply to light-duty trucks that have a GVWR of 8,500 pounds or less.

Starting in MY 2011, light truck CAFE standards will also apply to medium-duty passenger vehicles (MDPVs), which are defined as complete heavy-duty vehicles less than 10,000 pounds GVWR that are designed primarily for transportation of passengers. The definition of an MDPV does not include vehicles sold as incomplete trucks (i.e., a truck cab on chassis); vehicles that have a seating capacity of more than 12 persons; vehicles designed for more than 9 persons in seating rearward of the driver's seat; or vehicles equipped with an open cargo area (e.g., a pickup truck box or bed) of 6 feet or more in interior length. Hence, the definition of an MDPV essentially includes SUVs, short-bed pickup trucks, and passenger vans that are within the specified weight and weight-rated ranges. This implies that, starting in MY 2011, all SUVs greater than 8,500 GVWR that are currently excluded from CAFE consideration and all passenger vans less than 10,000 pounds GVWR will be included in determining a manufacturer's light truck CAFE compliance.

To provide manufacturers adequate time to adjust their product plans to the new provision, NHTSA is

Figure 9. Reformed CAFE standards (miles per gallon) for light trucks, by model year and vehicle footprint (square feet)

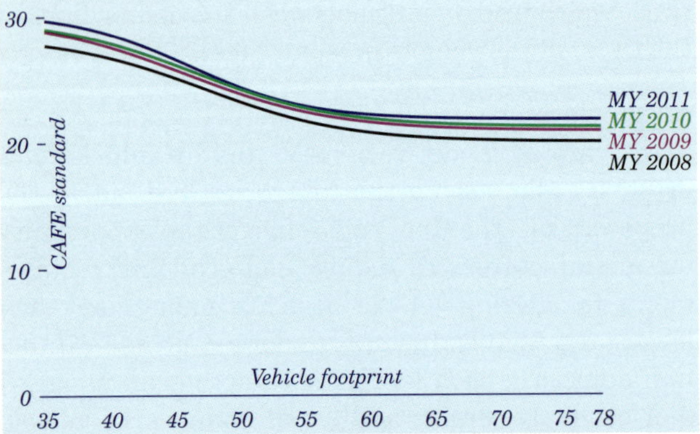

making the new definition effective beginning in MY 2011. As a result, the change will not have an immediate impact on MY 2008-2010 vehicles. In addition, NHTSA is permitting manufacturers to rely on either the old or the revised definition of light trucks until MY 2011.

NHTSA has also amended the "flat floor provision" to include only vehicles that have at least three rows of seats, of which the second and third rows can be detached or folded to create a flat cargo surface. Manufacturers currently offering minivans with folding seats will be able to take advantage of the new definition immediately. The new CAFE standards continue to exclude most medium- and heavy-duty pickups and most medium- and heavy-duty cargo vans that are used primarily for agricultural and commercial purposes. The change in the definition of a light truck can also have an impact on the product mix that a manufacturer will offer, because some light trucks under the current definition could be categorized as cars under the new definition, with a higher CAFE requirement.

The reformed CAFE standards impose a unique fuel economy standard on each manufacturer, based on the product mix sold in a given MY. For MY 2008 through 2010, manufacturers have the option of complying with either the new reformed CAFE standard or an unreformed CAFE standard. The unreformed CAFE standard requires manufacturers to meet an average light truck fleet standard of 22.5 miles per gallon in MY 2008, 23.1 miles per gallon in MY 2009, and 23.5 miles per gallon in MY 2010. All light truck manufacturers must adhere to the new reformed standards for MY 2011 and subsequent years.

Each manufacturer is subject to an identical fuel economy target for light truck models with the same footprint. Moreover, the same formula is applied to determine each manufacturer's required CAFE level, using the fuel economy targets for different footprints, the targets specific for each model, and the production levels of each model. Individual manufacturers face different required CAFE levels only to the extent that they produce different volumes of vehicles by footprint.

To determine compliance with the reformed CAFE standard, each manufacturer's production-weighted average fuel economy will be calculated and compared to the calculated reformed CAFE. If the weighted average fuel economy of all the manufacturer's models is at least equal to the manufacturer's calculated reformed CAFE, then the manufacturer will be in compliance with the reformed CAFE standard. If its actual fleet-wide average fuel economy is greater than its required CAFE level, the manufacturer will earn credits equal to the difference, which can be applied to any of the three preceding or subsequent model years. With this allowance, manufacturers will not be penalized for occasionally failing to meet the targets (due to market conditions, for example) but only for persistent failure to meet them. If the average fuel economy of a manufacturer's annual car or truck production falls below the defined standard, the manufacturer will be required to pay a penalty proportional to its total production for the U.S. domestic market.

The new CAFE standards are captured in the *AEO-2007* projections. For MY 2008 through 2011, manufacturers are assumed to adhere to the increases in unreformed light truck standards. For MY 2011, the *AEO2007* applies a fleet-wide standard of 24 miles per gallon, based loosely on the change between 2010 and 2011 in the proposed footprint-based standards. Because no further changes in fuel economy standards beyond 2011 are assumed, the projected increase in light truck fuel economy after 2011 reflects projected technology adoption resulting from other market forces.

Regulation of Emissions from Stationary Diesel Engines

On July 11, 2006, the EPA issued regulations covering emissions from stationary diesel engines [*14*]—New Source Performance Standards that limit emissions of NO_x, particulate matter, SO_2, carbon monoxide, and hydrocarbons to the same levels required for nonroad diesel engines [*15*]. The regulation affects new, modified, and reconstructed diesel engines. Beginning with MY 2007 [*16*], engine manufacturers must specify that new engines less than 3,000 horsepower meet the same emissions standard as nonroad diesel engines. For engines greater than 3,000 horsepower, the standard will be fully effective in 2011 [*17*]. Stationary diesel engine fuel will also be subject to the same standard as nonroad diesel engine fuel, which reduces the sulfur content of the fuel to 500 parts per million by mid-2007 and 15 parts per million by mid-2010.

Stationary diesel engines are used to generate electricity, to power pumps and compressors, and in irrigation systems. It has been estimated that there were 663,780 such engines larger than 50 horsepower in use in 1998 [*18*]. The EPA estimates that 81,500 engines will be subject to the controls by 2015 and that total pollutant reductions will be more than 68,000 tons per year.

The new standards for stationary diesel engines are included in *AEO2007*, but they are unlikely to affect the projections materially. The nonroad diesel standards were incorporated in the *AEO* projections previously, beginning with *AEO2005*.

Federal and State Ethanol and Biodiesel Requirements

EPACT2005 requires that the use of renewable motor fuels be increased from the 2004 level of just over 4 billion gallons to a minimum of 7.5 billion gallons in 2012, after which the requirement grows at a rate equal to the growth of the gasoline pool [*19*]. The law does not require that every gallon of gasoline or diesel fuel be blended with renewable fuels. Refiners are free to use renewable fuels, such as ethanol and biodiesel, in geographic regions and fuel formulations that make the most sense, as long as they meet the overall standard. Conventional gasoline and diesel can be blended with renewables without any change to the petroleum components, although fuels used in areas with air quality problems are likely to require adjustment to the base gasoline or diesel fuel if they are to be blended with renewables.

Before EPACT2005, a major portion of the RFG pool was blended with methyl tertiary butyl ether (MTBE) to meet required oxygen levels, increase volume, improve octane, and maintain compatibility with existing petroleum product pipelines without a large increase in gasoline volatility. The oxygen content was required under CAAA90 [*20*]. Ethanol is the only other economically feasible oxygenate, but it is incompatible with existing pipelines because of its affinity for water and causes substantial increases in gasoline volatility. Because MTBE was easier to blend and ship, refiners preferred to meet oxygen requirements with MTBE. Over the past several years, however, various State and local governments have banned the use of MTBE, and some have even brought lawsuits against MTBE producers over concerns that spilled MTBE and gasoline containing MTBE were polluting groundwater.

In EPACT2005, Congress repealed the oxygen requirement for Federal RFG but declined to prohibit defective product claims against producers and blenders of MTBE. Refiners believed that the lack of an oxygenate requirement would increase their liability in future groundwater contamination cases and voluntarily eliminated MTBE from the gasoline pool in the summer of 2006.

Several of the largest MTBE-consuming States had already banned the use of MTBE and switched to ethanol-blended gasoline by the time EPACT2005 was passed. California, New York, and Connecticut implemented MTBE bans in 2004 [*21*]. Ethanol distillers, petroleum refiners, and petroleum product terminal operators invested in process changes and additional tanks to accommodate the ethanol. Despite the flexibility allowed by the EPACT2005 RFS and its repeal of the oxygen content requirement, refiners began using ethanol in all RFG in summer 2006.

Overall levels of ethanol and biodiesel use are projected to exceed the EPACT2005 requirement in all *AEO2007* cases, given the projected prices for corn and crude oil, the lack of viable substitutes for MTBE, and extension of the tax credit for ethanol blending [*22*]. EPACT2005 requires the use of 250 million gallons per year of ethanol produced from cellulose after 2013. Production of cellulosic ethanol rises only to the minimum requirement in the *AEO2007* reference case, because the projected capital costs of cellulosic ethanol plants are significantly higher than those of corn ethanol plants.

An older Federal energy law has been used specifically to promote biodiesel. The Energy Policy Act of 1992 required certain vehicle fleets to purchase alternative-fueled light vehicles, but the vehicles were not actually required to run on alternative fuels. The Energy Conservation Reauthorization Act of 1998 allowed the purchase of 450 gallons of pure biodiesel to offset the requirement to purchase one alternative-fueled light vehicle [*23*]. In *AEO2007*, biodiesel demand for Federal fleet purchase offsets is projected to be 7.4 million gallons per year in 2012 and 8.8 million gallons per year in 2030.

Several States have their own requirements for ethanol and biodiesel in their motor fuel supplies, which are reflected in *AEO2007*. Minnesota, a major producer of ethanol, has required all gasoline to contain at least 7.7 percent ethanol since 1997 [*24*]. Hawaii requires 85 percent of its gasoline to contain 10 percent ethanol, effective on April 2, 2006 [*25*]. The intention of the law is to spur local production of ethanol from sugar, but the ethanol could also come from the U.S. mainland or from Brazil.

Minnesota was also the first State to require biodiesel blending into diesel fuel, at 2 percent by volume [*26*]. The requirement became effective in mid-2005, when two new biodiesel plants, each with 30 million gallons per year capacity, began operation in the State. The law was waived several times because of quality problems with the biodiesel, but it is again in effect. Washington requires 2 percent ethanol in gasoline and 2 percent biodiesel in diesel fuel no later than November 30, 2008. The requirement will increase to 5 percent once the State can produce biodiesel equal to 3 percent of its diesel demand [*27*]. Louisiana enacted a requirement for 2 percent ethanol in gasoline and 2 percent biodiesel in diesel fuel, once sufficient capacity is built in-State [*28*, *29*]. Assuming that Louisiana's 2-percent and Washington's 5-percent requirements are triggered, Louisiana, Minnesota, and Washington will require 102 million gallons of biodiesel in 2012 and 146 million gallons in 2030.

The Federal and State policies on renewable fuels have various effects on gasoline supply and price. The substitution of ethanol for MTBE in RFG reduces the yield of gasoline and gasoline components from a given refinery configuration. In the long run, refiners are expected to make additional investments to get back some of the gasoline capacity they lost.

Because ethanol currently is economically competitive as a gasoline blending component in Minnesota, its use in that State is not dependent on the ethanol content requirement, which is estimated to have no adverse impact on gasoline prices. Hawaii, on the other hand, must either produce ethanol from costly sugar or ship ethanol from the U.S. mainland or Brazil. Because both options are expected to be expensive, it is likely that Hawaii's program will raise gasoline prices. The biodiesel requirements in Minnesota, Louisiana, and Washington may increase the availability of diesel fuel in the short run and are likely to increase diesel prices after the Federal motor fuels excise tax credits for blending biodiesel expire. In the longer run, renewable fuels requirements do not affect the availability of gasoline and diesel fuel, because refiners are expected to adjust refinery expansion plans in light of these mandates.

Federal Fuels Taxes and Tax Credits

The *AEO2007* reference case and alternative cases generally assume compliance with current laws and regulations affecting the energy sector. Some provisions of the U.S. Tax Code are scheduled to expire, or may be subject to adjustment, before the end of the projection period. In general, scheduled expirations and adjustments provided in legislation or regulations are assumed to occur, unless there is significant historical evidence to support an alternative assumption. This section examines the *AEO2007* treatment of three provisions that could have significant impacts on U.S. energy markets: the gasoline excise tax, biofuel (ethanol and biodiesel) tax credits, and the PTC for electricity generation from certain renewable resources.

Excise Taxes on Highway Fuels

Excise taxes on highway fuels have been a dedicated source of funding for the Federal Highway Trust Fund since its creation in 1956. The Federal Government levies a tax of 18.4 cents per gallon on domestic gasoline sales and 24.4 cents per gallon on diesel fuel. The tax levels were last adjusted in 2003. Since 1932, when the first Federal excise tax on gasoline was imposed, it has been adjusted by Congress almost 20 times.

Because the statutes do not specify that the Federal excise taxes on highway fuels will be adjusted for inflation, and because they have not been adjusted at regular intervals in the past, they are assumed to remain at current levels in nominal terms through 2030. This assumption can, however, result in seemingly inconsistent results. For example, both the Federal Highway Administration and the Congressional Budget Office (CBO) project that the Highway Account in the Highway Trust Fund will have a negative balance by 2009, based on their respective receipts and outlays [*30, 31*]. Because EIA does not track expenditures on specific transportation infrastructure requirements, the *AEO2007* projections for vehicle miles traveled are not affected by the loss of funding for upkeep of the Nation's transit system, including maintenance of highways and bridges, which would be necessary to support the projected levels of vehicle use.

In addition to the Federal excise tax on highway fuels, the States and some local governments also levy excise or sales taxes on highway fuels. State and local fuel taxes are kept constant in real terms in *AEO2007*, based on analysis of aggregate historical adjustments to State and local fuel taxes, and reflecting the calculation of State sales taxes as a percentage of the sales price of the fuel [*32*].

Biofuels Tax Credits

The ethanol tax credit provides a credit against Federal gasoline taxes that is worth 51 cents for every gallon of ethanol blended into the gasoline pool. For a typical gasoline blend with 10 percent ethanol, the credit reduces the Federal excise tax (18.4 cents per gallon) by 5.1 cents, resulting in an effective tax rate of 13.3 cents per gallon for the blender. Currently, the ethanol tax credit is scheduled to expire in 2010; however, it has been in effect since 1978, and while it has been adjusted both up and down, it has consistently been extended [*33*]. *AEO2007* assumes that reauthorizations will continue throughout the projections.

Biodiesel also receives a tax credit, at $1.00 per gallon for biodiesel produced from virgin oils and 50 cents per gallon for biodiesel produced from recycled oils. The credit is scheduled to expire in 2008, and *AEO2007* assumes that it will not be reauthorized. The biodiesel tax credit was established by the American Jobs Creation Act of 2004, with a 2006 expiration date. It was extended to 2008 in EPACT2005, after the industry had sought an extension to 2010 [*34*]. If the credit is reauthorized after 2008, it will have a significant impact on biodiesel production.

Production Tax Credit for Renewable Electricity Generation

A PTC of 0.95 to 1.9 cents per kilowatthour [*35*] is provided for sales of electricity generated from certain renewable resources at qualifying facilities for the first 10 years of their operation. The PTC is adjusted by the IRS each year, based on the annual inflation rate. First established in 1992, the PTC has been allowed to expire three times, followed by after-the-fact reauthorizations [*36*]. It has been modified significantly with each extension, including changes in the qualifying resources (adding some, removing others), the value and duration of the credit for certain resources, and the interaction with other aspects of the Tax Code (such as the alternative minimum tax). While the *AEO2007* reference case assumes that the PTC will expire at the end of 2007, both *AEO2007* and previous *AEO*s include alternative cases that consider the impacts of a PTC extension.

Electricity Prices in Transition

The push by some States to restructure electricity markets progressed rapidly throughout the late 1990s. Although the energy crisis in California during

2000 and 2001 slowed the momentum, 19 States and the District of Columbia currently have some form of restructuring in place. In addition, Washington State, which has not restructured its electricity market, allows its largest industrial customers to choose their suppliers.

Many States put in place special regulations to protect customers during the transition. For most, this meant a specified period of guaranteed price stability in the form of rate cuts or rate freezes, after which the market was expected to be sufficiently competitive to reduce the need for price regulation. Low transitional rates in most cases were mandated by State utility commissions and offered by regulated utilities to customers who could not or did not choose a competitive supplier—a service often referred to as Standard Offer Service (SOS). Some States required utilities to offer a separate service, often called Provider of Last Resort (POLR) service, for customers who left, or were dropped by, their competitive suppliers. POLR service sometimes offered less price protection than SOS.

The late 1990s saw a promising start to competition. The fuel prices paid by generators were low enough for competitive electricity suppliers to offer rates slightly lower than SOS prices. From 2000 on, however, rapidly increasing fuel prices caused many competitive suppliers to go out of business, because the price of wholesale electricity rose above the price at which they had contracted to sell it.

Since 2004 many State-mandated transition periods with fixed prices have been coming to a close, with competitive retail markets still not developed for large groups of customers. Most residential and small commercial customers have no offers from competitive suppliers, leading many State utility commissions to consider the possibility of extending regulated, cost-of-service rates for SOS customers. Most of those States are now trying to jump-start competitive markets by having electricity suppliers bid for the right to sell energy to SOS customers. Table 2 summarizes the changes that have been made to SOS pricing in key regions and States since the start of restructuring. It also shows the percentages of retail load currently being sold directly to consumers by competitive retailers.

Most States initially required distribution utilities to offer SOS at a discount from regulated rates throughout the transition period, while a few States experimented with options that encouraged some competition. Texas and Massachusetts required utilities to offer both SOS and POLR service. The SOS provided rate stability and price reductions; the price of POLR service was determined by competitive bid. New York offered rate cuts for only 1 year and required most of its large SOS energy users to pay hourly market prices. In Maine, winners of competitive bids supplied SOS load—a method that was soon adopted by Pennsylvania for its largest utility. Both States still had mandated rate caps, however, so that in years when fuel prices were too high for load to be served at prices below capped rates, too few suppliers bid to provide SOS at competitive prices. Maine responded by raising rate caps, which has allowed the auction program for SOS to attract multiple bidders and competitive suppliers to attract more retail customers.

In 2002, New Jersey held the first auction to supply Basic Generation Service (its name for SOS) for the last year of its designated transition period. The auction attracted sufficient bidders, and New Jersey has continued to hold an annual descending clock auction to supply SOS. In a descending clock auction the bidding starts high, and prices "tick down" when supply is greater than demand. The auction ends with the price at which the amount of supply equals demand. Other States have considered the descending clock auction as a means of providing SOS competitively to customers who do not have access or have not chosen retail competitive suppliers. Illinois, which adopted the method, recently held an auction for its 2007 SOS load.

Other States have decided to jump-start competition as transition periods end, rather than extend rate caps. In the East, Maryland (starting in 2004), the District of Columbia and Massachusetts (since 2005), and Delaware and New Hampshire (since 2006) have required utilities to submit requests for proposals to serve load for SOS customers and have chosen the lowest bidding supplier. Pennsylvania has been negotiating with more utilities to offer SOS for competitive bid. Currently, the State has a proposed rulemaking out for comment that seeks to require each utility at the termination of its transition period to pass through the cost of competitively bid SOS.

In Ohio, FirstEnergy has tried to hold an auction for the supply of its SOS obligation but has not attracted many bidders. In Texas, where SOS customers were automatically transferred to retail affiliates at the start of competition, utilities whose districts have at least 40 percent of their load supplied competitively can now offer SOS if it is bid out competitively. In addition to bidding out SOS, New York, Maryland, and New Jersey require large commercial and industrial customers to pay hourly market prices if they have not chosen a competitive supplier; subsequently, most large customers in the three States have chosen competitive suppliers that offer price hedges to decrease possible price volatility or, in the case of New York, have bought hedging products separate from energy supply.

Each State has a slightly different requirement for the provision of SOS, but usually the competitive proposals are to supply load for periods of several months to 3 years, depending on the customer group or the amount of load in each customer group. The supply decrement or "tranche" is chosen on the basis of the lowest bid. Providing load in this manner is thought to allow prices to be determined competitively, but with much less volatility than would occur if energy were bought hourly on the open market. SOS loads for residential and small commercial customers

Table 2. Changes in Standard Offer Supply price determinations by supply region and State

Electricity supply region	*State*	*Competitive (non-SOS) portion of retail load*	*SOS price determination, transition period*	*SOS price determination, post-transition period*
ECAR	*MI*	*10%*	*Rate reductions (6/00-1/06).*	*Rate case.*
	OH	*17%*	*Rate reductions (1/01-12/05).*	*Rate case, new rate caps, some competitive bid.*
	Some PA, MD, and VA load		*See State rules under MAAC and SERC.*	
ERCOT	*TX*	*42%*	*SOS: rate reductions, competitive bid by utility if 40% retail load purchased competitively. POLR: competitive bid (1/01-1/07).*	*POLR for any requesting customer. Energy charges calculated at 130% of average ERCOT spot market prices: hourly for small customers, 15-minute intervals for large customers (1/07-12/08).*
MAAC	*DE* *DC* *MD*	*8%* *59%* *28%*	*Rate reductions and caps (10/99-12/08, depending on State and utility).*	*Competitive bid. Large MD SOS customers pay hourly market rates.*
	NJ	*12%*	*Rate case/price caps (8/1/99-7/31/02).*	*Competitive auction: 8/1/02. Large customers pay hourly market rates.*
	PA	*7%*	*Rate reductions and caps, shopping credits (1/99-12/10 depending on utility).*	*Some competitive bid for PECO and some other utilities (since 1/01).*
MAIN	*IL*	*19%*	*Rate reductions and caps (10/1/99-12/31/06).*	*Competitive auction (since 1/07).*
NPCC-NY	*NY*	*38%*	*Rate reductions (5/99-7/01). Large commercial and industrial customers in two major utilities put on hourly market rates.*	*Rate case for small customers. All large customers pay hourly market rates (since 9/05).*
NPCC-New England	*CT* *RI*	*2%* *11%*	*Rate reductions (7/97-12/03).*	*Generation charges passed through with an administrative charge (11/04-11/09).*
	ME	*38%*	*Competitive bid (3/00-5/05).*	*Competitive bid (since 5/05).*
	MA	*28%*	*SOS: rate reductions. POLR: competitive bid (3/98-3/05).*	*Competitive bid: SOS customers moved to POLR (since 3/05).*
	NH	*1%*	*Rate case (8/98-4/06).*	*Competitive bid (since 5/06).*
SERC	*VA*	*0.02%*	*Rate caps (1/02-12/10).*	*Not decided.*
WECC-NWP	*MT* *OR* *WA*	*21%* *3%* *2%*	*No SOS: regulated supply for small customers, supplier contract for large customers.*	—
WECC-Rocky Mountain, AZ/NM/SNV	*AZ*	*0%*	*Rate reductions (10/99-12/02).*	*Rate case with competitive bid for 50% of load (since 1/03).*
	NV	*0%*	*Rate case.*	*Not decided.*
WECC-CA	*CA*	*11%*	*Rate reductions (3/98-3/01). Suspension of competition (9/01).*	—

usually are fixed for longer periods than are loads for customers who use larger amounts of electricity.

In *AEO2007*, electricity prices are projected for 13 electricity supply regions. The weighted average of the prices constitutes the national electricity price projection. For competitive regions, price projections are based on marginal price calculations to simulate the pricing methods of hourly spot markets. It is assumed that a region will take 10 years after the implementation of competitive markets to become fully competitive, and so the amount of competitive load increases by 10 percent each year until 100 percent of electricity load is priced by marginal energy calculations. Until then, part of the load (as well as any other load from regulated States) is priced using cost-of-service calculations. Reliability costs and taxes are added to the weighted average of hourly marginal energy costs and are passed directly to the consumer. Transition price cuts and freezes have been factored into the *AEO2007* cases, although most have been phased out as initial transition periods have come to an end.

In regulated areas, unless a utility has an automatic fuel adjustment clause, customers do not immediately experience increases or decreases in generating costs, since utilities must wait until the next rate case in order to change rates. As a result, time lags between changes in electricity costs and changes in final prices to consumers are factored into the projections of regulated prices.

In past *AEO*s it was assumed that prices in fully competitive regions would reflect spot market prices and would be passed on to consumers immediately. The end of price reductions and caps in many States, along with the increase in competitively bid SOS load, is expected to push competitive regions closer to that representation of competition; however, most customers in fully competitive regions will not experience price changes immediately in response to changes in market generation costs.

In the interest of balancing the growth of competitive markets with price stability for customers, regulators in some States have mandated that SOS contracts be based on spot market prices but fixed for some period of time. Also, competitive supply often is offered at fixed prices for the contract period. Consequently, for *AEO2007*, lags have been built into the calculation of competitive energy prices to simulate the delay from the time suppliers experience cost changes to the time consumers experience price changes as a result of the length of fixed-price contracts for SOS and competitive retail service. Markets in deregulated regions are expected to become increasingly competitive over the long term, and it is assumed that the lag between the time when energy suppliers pay for energy on the spot market and the time when customer charges reflect those costs will be 6 months. For the short term, the lag is assumed to average 1 year in some regions.

State Renewable Energy Requirements and Goals: Update Through 2006

AEO2006 provided a review of renewable energy programs that were in effect in 23 States at the end of 2005 [*37*]. Since then (as of September 1, 2006), no new State programs have been adopted; however, several States with renewable energy programs in place have made changes as they have gained experience and identified areas for improvement. Revisions made over the past year range from clarification or modification of program definitions, such as which resources qualify, to substantial increases in targets for renewable electricity generation or capacity. The following paragraphs provide an overview of substantive changes in the design or implementation of State renewable energy programs.

The Arizona Corporation Commission currently is engaged in a rulemaking process for the State's energy portfolio standard (EPS), scheduled to run through the end of 2006 [*38*], which could lead to substantial changes in the Arizona program [*39*]. The most significant change proposed is an increase in the State's renewable electricity generation target. Pending final approval by the Commission and the Arizona Attorney General, the EPS target would increase from 1.25 percent of affected electricity sales to 15 percent. The new requirement would also allow trading of renewable energy credits among utilities to facilitate compliance. In addition, several new resources would be qualified to meet program requirements, including new small hydroelectric facilities (less than 10 megawatts) and geothermal power.

The original legislative authority for California's RPS, Senate Bill (S.B.) 1078, established a target of 20 percent renewable electricity generation by 2017. Subsequently, the California Energy Commission and California Public Utility Commission set an administrative goal of 20 percent by 2010 and 33 percent by 2020 [*40*]; however, key funding mechanisms were still tied to the legislative 2017 target [*41*]. On

September 26, 2006, Governor Schwarzenegger approved S.B. 107, which codifies the target of 20 percent by 2010 and calls for a formal study of the 2020 target [*42*]. S.B. 107 also modifies requirements for electricity generation from other States to qualify for the California RPS. Out-of-State generators are now limited to 10 percent of associated supplemental energy payments (SEPs) but have fewer restrictions on physical deliveries of power into the California market.

Connecticut has received new statutory authority to expand the area in which qualifying credits can be generated for the State's RPS program and to use renewable energy credits in lieu of physical energy delivery for program compliance [*43*]. In addition to the New England Independent System Operator territory, credits generated in New York, Pennsylvania, New Jersey, Delaware, and Maryland may also be used to satisfy program requirements, upon a finding that each State has a comparable RPS program.

With one of the oldest RPS programs, Maine has passed an additional requirement that 10 percent of all electricity generation *growth* must come from renewable resources [*44*]. Maine's existing target, 30 percent of total generation, had already been exceeded when the original RPS-enabling statute was enacted. The new law presumably will require the addition of new generating resources to meet the incremental requirement.

Changes in the Massachusetts RPS program, although more incremental than structural, have received significant notice among the affected parties. The changes refine the rules governing the types of biomass electricity generation facility that can qualify for the RPS program [*45*]. Previous regulations did not allow generation from "retooled" biomass plants —those in service before 1998 but subsequently upgraded to meet current environmental specifications—to qualify for the RPS, except by waiver. The changes allow that portion of the output from retooled biomass plants that is in excess of historical generation levels to qualify. This clarification is particularly significant given the importance of biomass electricity generation in meeting the Massachusetts target. In 2004, the latest year for which data are available, 35 percent of the compliance target came from biomass generation [*46*].

Nevada has issued a number of new rules within the context of the current statutory authority for the State's EPS [*47*]. Perhaps most significant is the establishment of a credit trading system to facilitate compliance by individual utilities. Credit trading is a common feature of State RPS policy, which allows utilities to purchase compliance credits from other utilities that have excess renewable electricity generation, in lieu of actually generating renewable electric power. Energy efficiency programs can now also be used to offset a portion of Nevada's renewable energy target.

The New Jersey Board of Public Utilities adopted regulations in 2006 that increase the State's renewable electricity generation target from 6.5 percent of sales by 2008 to 22.5 percent by 2021 [*48*]. The new requirement includes 17.88 percent of sales from "Class I" renewable resources, 2.5 percent of sales from "Class II" resources, and the remainder (2.12 percent of sales) from solar resources. Solar generation in excess of the target may be used to meet Class I or II requirements, and excess Class I generation may be used to meet Class II requirements. Class I facilities can use a broad range of renewable resources, including wind, ocean, geothermal, LFG, and approved biomass resources. Class II facilities include hydropower facilities less than 30 megawatts and approved "resource recovery" facilities (trash incinerators).

Wisconsin has passed new legislation increasing the State's RPS target from 2.2 percent of electricity sales by 2012 to 10 percent by 2015 [*49*]. Under the new legislation, the Wisconsin Public Service Commission is required to provide a report by 2016 indicating whether the goal of 10 percent has been achieved and, if not, what steps are required to achieve it.

The *AEO2007* reference case includes new renewable electric power projects that have been identified. It does not include additional renewable projects that might be required for full compliance with some State programs, because it is not clear whether those requirements will be enforced, in light of provisions for granting of compliance waivers, alternative compliance mechanisms, and other discretionary enforcement options. A case where compliance with nondiscretionary enforcement is assumed projects that most State renewable energy targets should be achievable, with varying impacts on regional electricity markets.

Some regions with State targets could see substantially more renewable electricity generation with nondiscretionary compliance than is projected in the *AEO2007* reference case. State standards in the Mid-Atlantic and New England regions could result in approximately 350 percent and 20 percent more renewable generation by 2030, respectively, than projected in the reference case. Biomass is expected to predominate as the fuel of choice in those regions, which lack exploitable geothermal resources and have only limited low-cost wind resources. While the total increase in renewable generation in New York is just over 10 percent by 2030, generation from nonhydropower renewable resources is nearly double the reference case projection.

In other regions, the impact of the standards is projected to be less pronounced. For example, Texas, the Southwest, and the Northwest have either largely met their renewable electricity requirements with existing and planned capacity or are projected to build sufficient renewable capacity based on economic merits within the reference case. Aggregated nationally, State renewable energy standards would result in approximately 30 percent more electricity generation from nonhydropower renewables in 2030 than is projected in the *AEO2007* reference case.

Although this analysis projects that most States would meet their RPS targets without triggering compliance "safety valves" (such as alternative compliance payments), it also suggests that limitations on the funding of California's RPS program could cause that State not to reach its legislated targets [*50*]. Under current law, California utilities may apply for SEPs from the State to cover above-market costs of acquiring renewable energy resources. The SEPs are funded through a dedicated surcharge on consumer utility bills. As of September 2006, the California Energy Commission, which is responsible for administering the SEP program, had not awarded any SEPs and had developed a current account of around $300 million. Funding authorizations through 2011 should provide an additional $77 million per year in new funds. The surcharge authority must be renewed by 2012.

With the expiration of the Federal PTC at the end of 2007, as assumed in this case, and limits on supplemental funding (without which compliance is waived), California is projected to achieve a nonhydropower renewable electricity generation share of 12 percent by 2012. Thereafter, the State's qualifying renewable generation is projected to grow only to the extent that such power is economically competitive without the SEP. This projection may underestimate overall compliance with the California RPS program, however, to the extent that recently passed program modifications facilitate increased use of resources from other States.

State Regulations on Airborne Emissions: Update Through 2006

Implementation of the Clean Air Interstate Rule and Clean Air Mercury Rule

In May 2005, the EPA published two final rules aimed at reducing emissions from coal-fired power plants. CAIR [*51*] requires 28 States and the District of Columbia to reduce emissions of SO_2 and/or NO_x. CAMR [*52*] requires the States to reduce emissions of mercury from new and existing coal-fired plants [*53*].

The two rules cap emissions at the regional and national levels; however, each State can decide how to meet its own cap, as long as the minimum program milestones are met. For CAIR, the States have until March 2007 to submit implementation plans to the EPA, which then will have until September 2007 to review the plans and identify modifications, if necessary. For CAMR, the States must present their plans by November 2007, and the EPA then will have 6 months to accept the plans or require modifications.

Both CAIR and CAMR provide States the flexibility to participate in a regional cap and trade program. Several States, including most of those in the Northeast, have said as of September 2006 that they will not participate in the cap and trade program for mercury emissions under CAMR [*54*], because they plan to adopt more stringent standards. In addition, some States plan to place mandatory restrictions on individual coal-fired plants in order to reduce the possibility that localized areas will continue to have high levels of mercury emissions. Those restrictions differ from the Federal plan of enforcing only statewide caps.

Final decisions regarding the structure of State programs and participation in the regional trading program will not be made until after November 2007. Currently, both CAIR and CAMR are represented as regional cap and trade programs in *AEO2007*. This approach will be reevaluated when the final State programs have been submitted and reviewed by the EPA.

Regional Greenhouse Gas Initiative

The governors of the seven States participating in the RGGI—Connecticut, Delaware, Maine, New Hampshire, New Jersey, New York, and Vermont—have committed to enact legislation individually for achieving the desired emission reductions under the agreement. The group originally consisted of nine States, but Massachusetts and Rhode Island have withdrawn. In Maryland, recently adopted legislation requires the State to join the RGGI by June 2007 [*55*]. Pennsylvania, the District of Columbia, and several Canadian provinces are observers to the program.

When the original RGGI agreement was signed in December 2005, each participating State agreed to cap its greenhouse gas emissions from power production beginning in January 2009. The States were provided CO_2 allocations based on their average emissions for the 3-year period from 2000 to 2002, with exceptions. States that had built or were anticipating new plants between 2002 and 2009 were allowed additional allowances to reflect the level of emissions expected in January 2009. The governors of the seven States currently participating have already agreed to their respective allowance allotments.

For the seven northeastern States, the annual cap is approximately 121 million short tons, representing a 6.1-percent increase over their combined CO_2 emissions in 2000. After January 2009, the RGGI requires each participating State to hold its emissions at or below its CO_2 allotment. The caps remain unchanged until the end of 2014, after which they are reduced by 2.5 percent annually. Thus, by the end of 2018, CO_2 emissions in the participating States will be 10 percent below the levels at which the allocations were issued.

The August 2006 model rule clarifies several provisions on how States can achieve their emission reductions. It also provides compliance flexibility if prices rise beyond what is anticipated, although threshold levels have not been determined. One-quarter of potential revenue from the auction or sale of emission credits must go to consumer benefits or strategic energy purposes. This broad category includes energy price discounts, renewable and low-carbon energy investments, and energy efficiency programs. Also, CO_2 emission reductions by power producers before the January 2009 start date will be credited for use during the cap period.

Other States and provinces may participate in the RGGI through carbon offset programs. If the price of credits remains below $7 (2005 dollars) per short ton of CO_2, power producers may account for 3.3 percent of their emissions through offset programs in any State or province, including capture of landfill methane and sulfur hexafluoride, afforestation, end-use efficiency programs, and agricultural emission reductions. For each ton of CO_2 avoided or sequestered in the projects, the power producer will be provided one emission credit for use or sale. In order for an offset program to be eligible, it cannot be part of any other State mandate and must be attributable only to the RGGI. If the price of CO_2 credits is sustained above $7 for more than 12 months, power producers will be able to offset up to 5 percent of their CO_2 emissions. If credit prices surpass $10 for a sustained 12-month period, then producers will be able to offset 10 percent of their emissions and may participate in international credit markets.

The individual States still must enact their own legislation to achieve the RGGI milestones. State legislation will determine compliance issues, such as credit allocations, enforcement methods, and options for exiting the agreement. Each State will be responsible for issuing its own allowances. Some States may choose to sell them at a certain price; others may hold auctions. They may also be given away, or the States may use a combination of methods.

Although the State RGGI caps and timelines are known, many aspects of their implementation remain uncertain, because the participating States have not yet enacted the necessary legislation. Therefore, the RGGI provisions are not modeled in *AEO2007*.

California Greenhouse Gas Legislation

A.B. 32, "California Global Warming Solutions Act of 2006," which was signed into law by Governor Arnold Schwarzenegger on September 27, 2006 [*56*], calls for a 25-percent reduction in CO_2 emissions by 2020. The first major controls, for the industrial sector, are scheduled to take effect in 2012. The plan grants the California Air Resources Board lead authority for establishing how much industry groups contribute to global warming pollution, assigning emission targets, and setting noncompliance penalties. It sets a 2009 date for establishing how the system will work and then allows 3 years for the State's industries to prepare for the 2012 startup of mandatory emissions reductions [*57*].

It is not yet known what sources of greenhouse gas emissions will be subject to the restrictions, although the bill states that all major sources of CO_2 will be included. The bill does not mention the transportation sector, which is covered in separate legislation. A.B. 32 also specifies that all emissions from the generation of power consumed within the State are expected to be subject to the new laws. Because California imports power from neighboring States, emissions in those States may also be affected. In addition, California collaborates on its greenhouse gas policy with the States of Washington and Oregon through the West Coast Governors' Global Warming Initiative [*58*].

A.B. 32 delegates most of the responsibility for implementation and enforcement to the California Air Resources Board. Although the bill indicates that the reduction program will rely on market-based compliance mechanisms, it does not indicate the course of action that will be taken to reduce emissions. Reliance on a market-based compliance mechanism suggests the possible use of a credit trading program. If this is the case, issues such as credit distribution, offset allowances, price caps, and other restrictions will be decided by January 2009.

The Air Resources Board will also coordinate enforcement issues with the State's Public Utilities Commission and Energy Resources Conservation and Development Commission. Regulations on the monitoring of greenhouse gas emissions must be in place by 2008, when accurate reports on emissions from all major sources will be mandatory. Final regulations for the emissions reduction program will be presented in January 2011 and will become operative in January 2012. Because the program specifics have not been developed, A.B. 32 is not modeled in *AEO2007*.

Issues in Focus

Introduction

This section of the *AEO* provides in-depth discussions on topics of special interest that may affect the projections, including significant changes in assumptions and recent developments in technologies for energy production, energy consumption, and energy supply. In view of recent increases in energy prices, this year's topics include discussions of the underlying cost factors in key industries and how consumers respond to higher energy prices. The potential impacts of developing oil and natural gas resources in the Outer Continental Shelf (OCS), developments related to an Alaska natural gas pipeline, and key issues for the development of new nuclear and biomass-to-liquids technologies are also discussed.

World Oil Prices in *AEO2007*

Over the long term, the *AEO2007* projection for world oil prices—defined as the average price of imported low-sulfur, light crude oil to U.S. refiners—is similar to the *AEO2006* projection. In the near term, however, *AEO2007* projects prices that are $8 to $10 higher than those in *AEO2006* [*59*].

The *AEO2007* reference case remains optimistic about the long-term supply potential of non-OPEC producers. In the reference case, increased non-OPEC and OPEC supplies are expected to cause a price decline from 2006 levels to under $50 per barrel (2005 dollars) in 2014. After that, a gradual rise in oil prices, averaging 1.1 percent per year in constant dollar terms or about 3.0 percent in nominal terms, is expected through 2030. The *AEO2007* reference case world oil price in 2030 is $59 per barrel in 2005 dollars, or about $95 per barrel in nominal terms.

Any long-term projection of world oil prices is highly uncertain. Above-ground factors that contribute to price uncertainty include the extent of access to oil resources, investment constraints, the economic and other objectives of countries where major reserves and resources are located, the cost and availability of substitutes, and economic and policy developments that affect the demand for oil. Below-ground factors contributing to oil price uncertainty include the extent of reserves and resources and the physical and engineering challenges of producing oil.

The three world oil price paths in *AEO2007* are shown in Figure 10. Compared with the reference case, the world oil price in 2030 is 69 percent (about $41 per barrel) higher in the high price case and 40 percent (about $23 per barrel) lower in the low price case. As a result, world oil consumption in 2030 is 14 percent lower in the high price case and 9 percent higher in the low price case than in the reference case. Prices in the low price case decline from 2006 levels to $34 per barrel in 2016 and remain relatively stable in real dollar terms thereafter, rising only slightly to $36 per barrel in 2030. In the high price case, the world oil price dips somewhat in 2007 from 2006 levels, then increases steadily to $101 per barrel (2005 dollars) in 2030. The *AEO2007* high and low oil price cases illustrate alternative oil market futures, but they do not bound the set of all possible outcomes.

The high and low oil price cases in *AEO2007* are based on different assumptions about world oil supply. The *AEO2007* reference case uses the mean estimates of oil and natural gas resources published by the U.S. Geological Survey (USGS) [*60*]. The high price case assumes that the worldwide crude oil resource is 15 percent smaller and is more costly to produce than assumed in the reference case. The low price case assumes that the worldwide resource is 15 percent larger and is cheaper to produce than assumed in the reference case.

The *AEO2007* reference case represents EIA's current best judgment regarding the expected behavior of key members of OPEC. In the reference case, OPEC members increase production at a rate that keeps world oil prices in the range of $50 to $60 per barrel (2005 dollars) over the projection period, reflecting a view that allowing oil prices to remain above that level for an extended period could lower their long-run profits by encouraging more investment in non-OPEC conventional and unconventional supplies and discouraging consumption of liquids worldwide.

Figure 10. World oil prices in three AEO2007 cases, 1990-2030 (2005 dollars per barrel)

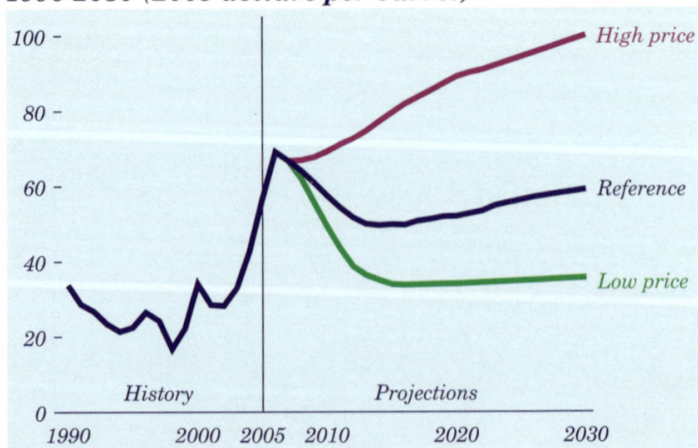

The prices in the reference case are high enough to trigger the entry into the market of some alternative energy supplies, including oil sands, ultra-heavy oils, GTL, CTL, and biomass-to-liquids, which are expected to become economically viable when oil prices are in the range of $30 to $50 per barrel. The same price range also increases the likelihood of greater investment in unconventional oil production.

Several non-OPEC countries, including Russia, Azerbaijan, Kazakhstan, Brazil, and Canada, are expected to increase production over the projection period, pursuing projects that are economically attractive with oil prices at or somewhat below those in the reference case. In Russia, oil production has recovered from a low of 6.0 million barrels per day in 1996, reaching 9.6 million barrels per day in 2006 [*61*]. While the Russian government has sought to increase its control of oil exploration, development, and production and recent actions have resulted in a markedly less desirable climate for foreign investment in Russian petroleum—a development that does not bode well for higher levels of petroleum production in the future—higher world oil prices have allowed the government to invest in additional exploration and production (E&P), which suggests continued production growth. The recent investments are projected to add 1 to 2 million barrels per day to Russia's oil production by 2030.

The Caspian Sea nations of Azerbaijan and Kazakhstan control large deposits of oil and natural gas. Because the two countries are landlocked, however, there was little incentive to develop their resources until pipelines began to be built. With the opening of the BTC oil pipeline in 2006 between the Caspian and Mediterranean Seas, production in Azerbaijan's Caspian offshore is expected to rise quickly, to 1.2 million barrels per day in 2010 [*62*]. Azerbaijan's production already has begun to surge, rising by more than 40 percent from 2005 to 2006, with similar volume growth expected in 2007 [*63*]. Production is expected to decline slowly in the future, however, to 1.0 million barrels per day in 2030.

Kazakhstan produced 1.4 million barrels per day in 2005 [*64*]. Recent access to the BTC pipeline is expected to lower its total production and export costs. The Kazakh government has stated goals of producing 3.5 million barrels per day by 2015. Kazakhstan's geology and economics might support that production level; however, uncertainties with regard to regulatory and tax policy could slow the rate of production growth. In addition, its success in reaching the stated target depends on access to export pipelines and adequate investment. In the *AEO2007* reference case, Kazakhstan's production is projected to reach 3.3 million barrels per day in 2030.

Brazil produced 1.7 million barrels per day of crude oil in 2006. Its production is expected to continue growing, based on proven reserves of more than 11 billion barrels, clear government policy objectives to increase production, and an increasingly competitive production market following the 1999 reforms that began to allow foreign oil companies to compete with the national oil company, Petrobras [*65*]. More than one-half of the country's oil reserves are in deepwater fields, and Brazil has long been a leader in developing deepwater production technology. Total liquids production from Brazil is projected to reach 4.6 million barrels per day in 2030.

Canada's conventional oil production is projected to remain relatively constant at 2.0 million barrels per day through 2015, but oil sands production is projected to grow rapidly. In recent years, net growth in production from Canada's oil sands has averaged 150,000 barrels per day [*66*], and production is projected to reach 2.3 million barrels per day in 2015 and 3.7 million barrels per day in 2030.

The production outlook for the countries highlighted here informs the three EIA world oil price cases. Sustained higher oil prices support the development and production of oil from more remote, technically challenging, and unconventional resources. Oil prices are significantly affected by assumptions about the ultimate size of world resources. Smaller resource estimates strengthen OPEC producers' influence over prices and raise their profits; however, the resulting higher prices encourage more extensive development of non-OPEC oil supplies, limiting the extent of OPEC's influence on prices. Oil production around the world over the next 25 years will also depend on the stability of government regulations and tax policies, access to export pipelines and ships, and adequate investment.

The projections for world petroleum production in 2030 are 101.6, 117.3, and 128.1 million barrels per day in the *AEO2007* high price, reference, and low price cases. The projected market share of world petroleum liquids production from OPEC in 2030 is about 33 percent in the high price case, 41 percent in the reference case, and 43 percent in the low price case. Because assumed production costs rise from the low price case to the reference case to the high price case, the differences in net profits among the three cases are smaller than they might have been if the underlying supply curves for OPEC and non-OPEC producers had remained unchanged. In the absence of tighter resources and higher costs, an OPEC strategy that attempted to pursue the output path in the high price case would subject OPEC to the risk of losing market share to other producers, as well as to alternatives to oil. The *AEO2007* projections for world oil production are shown in Table 3. Further discussions of the three world oil price cases and their implications for energy markets appear in the "Market Trends" section.

Impacts of Rising Construction and Equipment Costs on Energy Industries

Costs related to the construction industry have been volatile in recent years. Some of the volatility may be related to higher energy prices. Prices for iron and steel, cement, and concrete—commodities used heavily in the construction of new energy projects—rose sharply from 2004 to 2006, and shortages have been reported. How such price fluctuations may affect the cost or pace of new development in the energy industries is not known with any certainty, and short-term changes in commodity prices are not accounted for in the 25-year projections in *AEO2007*. Most projects in the energy industries require long planning and construction lead times, which can lessen the impacts of short-term trends.

From the late 1970s through 2002, steel, cement, and concrete prices followed a general downward trend. Since then, however, iron and steel prices have increased by 9 percent from 2002 to 2003, 9 percent from 2003 to 2004, and 31 percent from 2004 to 2005. (Early data from 2006 indicate that iron and steel prices have started to decline, but the direction of future prices remains to be seen.) Cement and concrete prices, as well as the composite cost index for all construction commodities, have shown similar trends, although with smaller increases, from 2004 to 2005 and 2005 to 2006 (Figure 11).

The cost index for construction materials has shown an average annual increase of 7 percent over the past 3 years in real terms. Over the past 30 years, however, it has shown an average annual decrease of 0.5 percent, with decreases following periods of increases in the early 1970s and early 1990s. *AEO2007* assumes that, for the purposes of long-term planning in the energy industries, costs will revert to the stable or slightly declining trend of the past 30 years.

Oil and Natural Gas Industry

Exploration and Production Costs

The American Petroleum Institute publishes an annual survey, *Joint Association Survey of Drilling Costs* [*67*], which reports the cost of drilling oil and natural gas wells in the United States. As shown in Figure 12, the average real cost of drilling an onshore natural gas development well to a depth of 7,500 to 9,999 feet roughly doubled from 2003 to 2004 [*68*].

Offshore drilling costs largely reflect the cost of renting an offshore drilling rig. ODS-Petrodata, Inc., has reported that, in real dollar terms from August 2004 to August 2006, daily rental costs for offshore jack-up rigs drilling at water depths of 250 to 300 feet increased by about 225 percent, while fleet utilization

Table 3. OPEC and non-OPEC oil production in three AEO2007 world oil price cases, 2005-2030 (million barrels per day)

	Low price	Reference	High price
OPEC			
2005	34.0	34.0	34.0
2010	34.7	34.7	31.2
2015	39.3	37.5	29.1
2020	43.9	40.2	29.3
2025	49.2	43.7	31.4
2030	54.7	47.6	33.3
Non-OPEC			
2005	50.3	50.3	50.3
2010	57.5	56.3	55.6
2015	62.1	60.2	60.9
2020	66.2	63.1	64.1
2025	70.1	66.3	66.0
2030	73.4	69.7	68.3

Figure 11. Changes in construction commodity costs, 1973-2006 (constant dollar index, 1973=100; 1981=100 for cement costs)

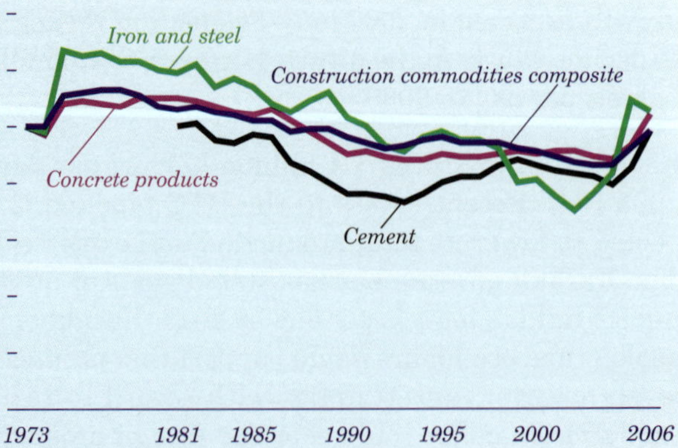

increased from about 80 percent to 89 percent; for semisubmersible rigs drilling at water depths of 2,001 to 5,000 feet, daily rental costs increased by approximately 340 percent, while fleet utilization increased from about 80 percent to just under 100 percent; and for floating rigs drilling at water depths of 5,001 feet or more, daily rental costs increased by approximately 266 percent, while fleet utilization increased from about 88 percent to 100 percent [*69*].

Petroleum Refinery Costs

Oil & Gas Journal uses Nelson-Farrar refinery construction cost indexes to track the overall cost of refinery construction. According to the Nelson-Farrar indexes, refinery construction costs increased overall by about 17 percent from 2002 to 2005 in real dollar terms. The escalation rate associated with petroleum refinery construction is lower than the rate for oil and natural gas drilling, because refinery costs in some categories have either declined or increased only slightly. Specifically, from 2002 to 2005, the following escalation rates for refinery construction were reported by *Oil & Gas Journal*: refinery composite index, 9 percent; pumps and compressors, 3 percent; electrical machinery, -10 percent; internal combustion engines, -5 percent; instruments, -3 percent; heat exchangers, 36 percent; materials, 22 percent; and construction labor, 5 percent [*70*].

In the aggregate, the large increases for heat exchangers and materials were largely offset by smaller increases or decreases for the other categories. More importantly, the 5-percent increase in labor costs is largely responsible for keeping the overall cost increase low, because labor costs account for about 60 percent of the overall cost of refinery construction.

Figure 12. Drilling costs for onshore natural gas development wells at depths of 7,500 to 9,999 feet, 1996-2004 (2004 dollars per well)

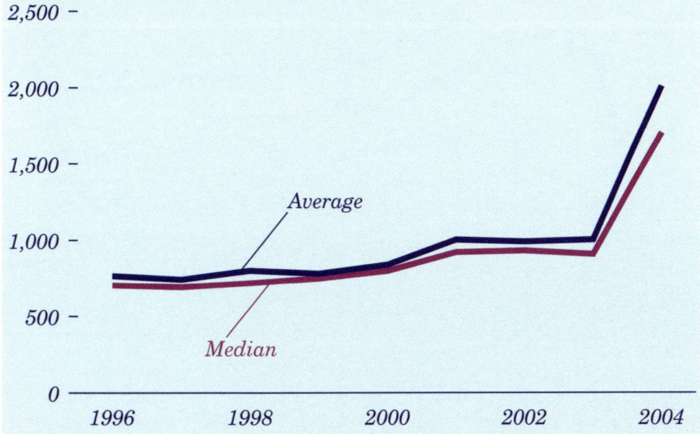

Discussion

Although the cost of steel and other commodities used in the oil and natural gas industry have posted significant cost increases over the past few years, the escalation of industry costs has not been caused by commodity cost increases alone, but also by higher crude oil and natural gas prices and the resulting increase in demand for exploration services (contract drilling, seismic data collection, well logging, fracturing, etc.). While iron and steel prices increased by 72 percent from May 2002 to June 2006 [*71*], onshore drilling costs increased by 100 percent and rental rates for offshore drilling rigs by 200 percent or more.

The growth in demand for services has occurred primarily in the E&P segment of the industry rather than refining sector. Higher crude oil and natural gas prices increase both producer cash flows and rates of return; greater potential profitability provides producers with the incentive to invest in and produce more oil and natural gas; and increased cash flow gives them more money to invest in more projects.

The increase in demand for services in the oil and natural gas industry is best illustrated by offshore drilling rig rates and fleet utilization. Similarly, the increase in demand for onshore drilling services is best illustrated by the growth in the number of onshore drilling rigs operating. Baker-Hughes, Inc., has reported that 1,656 onshore drilling rigs were in operation at the end of August 2006, compared with 738 at the end of August 2002 [*72*].

The refining sector has not experienced the same degree of cost escalation, largely because there has not been a significant increase in U.S. refining construction activity over the past few years. Consequently, cost increases in the petroleum refining sector largely mirror the increases associated with the various commodities used in refineries (steel, nickel, cobalt, etc.) rather than a significant increase in demand for refinery services and equipment.

Future cost changes in the E&P and refinery sectors of the oil and natural gas industry are expected to follow different patterns. Over the long term, new service capacity will be added to meet demand in the E&P sector; and if oil and natural gas prices stabilize, the demand—and consequently prices—for E&P services will decline. Conversely, if oil and natural gas prices increase in the future, it will take longer for E&P service capacity to catch up with the increased level of demand. In the refinery sector, construction costs are more likely to follow the path of construction commodity costs, barring a significant surge or reduction in demand for refinery equipment and construction services.

In NEMS, the real-world interaction between escalating petroleum E&P costs and the supply and demand for E&P services is captured in two ways. First, as oil and natural gas prices rise, E&P activities, such as the number of wells drilled, also increase. The increase in E&P activity, in turn, causes the cost of E&P activities to increase in the NEMS projections. Second, changes in E&P costs are addressed through annual econometric reestimations of equations related to oil and natural gas supply activities. The annual reestimations capture the latest trends in E&P costs and their impacts on E&P activity levels and outcomes. For example, for the *AEO2007* projections, the reestimations capture all the cost increases and outcomes for E&P activity that occurred through December 31, 2004. With regard to petroleum refining, the recent cost escalation for refining equipment resulting from higher commodity prices (including steel and concrete) is considered to be temporary and self-correcting over the long term, both through the addition of new commodity supplies and through a reduction in demand for those commodities. As a result, equipment costs for the petroleum refining sector are expected to rise at the overall rate of inflation over the long term.

Coal Industry

In the coal industry, both the mining and transportation sectors have been susceptible to the volatility of steel prices over the past few years. Higher prices for steel can make investments in machinery and equipment for coal mining more expensive; and coal transportation—predominantly by rail—depends on investments in freight cars, locomotives, and track, all of which require steel as a raw material.

The costs of rail equipment and, to a lesser extent, mining equipment and machinery followed the general pattern of declining steel prices from the mid-1970s through 2001 and 2002 (Figure 13). Although steel prices began to rise in 2003, rail equipment and mining machinery and equipment prices did not begin rising until 2005 and 2006, respectively. Although the early 2006 data suggest that steel prices have started to decline, there is no evidence yet of a decline in the equipment prices.

Coal Mining

The U.S. Census Bureau, in its Current Industrial Reports, combines surface mining equipment with construction machinery. In the construction machinery category, some subcategories provide better indicators than others of the price changes that have affected the surface mining industry. For example, the subcategory that includes draglines, excavators, and mining equipment has increased by 26 percent (average value in constant dollars) since 2002, while the number of units shipped has increased by 10 percent (Table 4). A smaller subcategory that includes draglines has increased by 33 percent in average value since 2002, with a 59-percent increase in quantity shipped. Larger hydraulically operated excavators show a different pattern, with a 10-percent decline in average value and a 57-percent increase in quantity shipped over the same time period, as does the subcategory that includes coal haulers, which did not show a significant increase in value between 2004 and 2005. For the subcategories with increases in average value, the largest increases occurred in 2004, coinciding with higher steel prices.

Both surface and underground mines rely on machinery made largely from steel to produce coal efficiently. Although specific costs typically are not publicly available, many of the major mining companies, including Peabody, CONSOL, and Massey, have indicated in their annual reports that they are susceptible to higher costs for machinery purchases as a result of increases in the cost of steel. Census Bureau data indicate that the mining industry as a whole (including coal mining) spent $597 million on underground mining machinery in 2005, as compared with $393 million in 2004 (constant 2005 dollars) [*73*]. In addition to

Figure 13. Changes in iron and steel, mining equipment and machinery, and railroad equipment costs, 1973-2006 (constant dollar index, 1973=100)

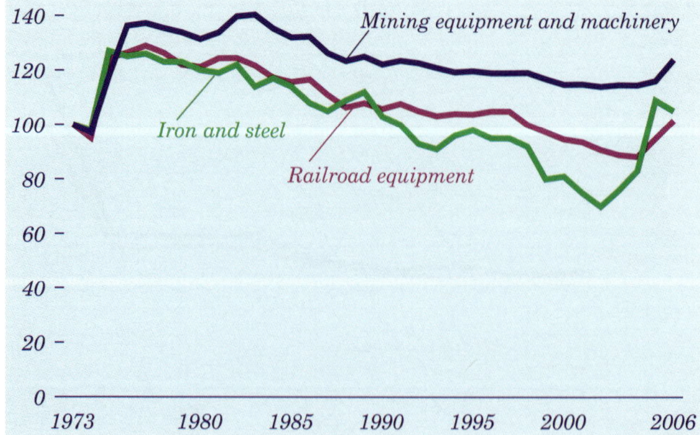

higher steel costs, the increase may also be due in part to the amount or mix of mining machinery purchased and in part to increases in other manufacturing costs.

Peabody listed the value of its mining and machinery assets at $1.2 billion in 2005, up from $910 million in 2004 and $759 million in 2003 (2005 dollars) [*74*]. The more recent annual increase, from 2004 to 2005, is larger than the earlier one, but the portion attributable to the effect of higher steel prices on the cost of newly acquired equipment is not publicly known. The company's operating costs, in constant dollars, rose by 8.4 percent from 2003 to 2005, from $11.23 per ton to $12.17 per ton of coal produced [*75*]. CONSOL cited both higher labor costs and higher commodity prices as the reasons for a 5.9-percent real increase in operating costs (to $30.06 per ton) in 2005 compared with 2004 [*76*]. For Massey, the average cash cost per ton of coal has risen to $35.62 per ton in 2005 from $26.58 per ton in 2001 (2005 dollars) [*77*].

Joy Global, a manufacturer of mining machinery [*78*], has mentioned in its annual report that some customers have delayed orders for manufacturing equipment in response to the short-term price volatility for steel and steel parts and that steel availability, in addition to prices, has been a problem in recent years. In general, the company has long-term contracts with steel suppliers, which help maintain steel availability, but those contracts also have surcharge provisions for increases in raw material costs. Caterpillar, Inc., another mining equipment manufacturer, has also been paying surcharges for steel.

As of February 2005, some steel prices paid by Joy Global were 100 percent higher than they had been 15 months earlier [*79*]. The company appears to have been able to pass through the higher steel prices to its customers (including coal producers), increasing its overall gross profit margins from 2004 to 2005.

Although the coal mining sector is hurt by higher costs for steel as an input factor in the production process, higher demand for steel and steel products also helps to boost metallurgical coal prices. Some coal companies are paying more for steel-based equipment, but at the same time their profit margins may be protected by their ability to sell their coal at higher prices.

The cost increases for coal mining equipment that occurred in 2006 are included in the *AEO2007* reference case. Thereafter, mine equipment costs are assumed to return to the long-term trend, increasing at the general rate of inflation.

Coal Transportation

Railroads are the primary mode for coal transportation in the United States, carrying about two-thirds of all coal shipments. The railroads use both steel and

Table 4. Changes in surface coal mining equipment costs, 2002-2005

Category		2002	2003	2004	2005
Power cranes, draglines, and excavators, including surface mining equipment, and attachments	*Million 2005 dollars*	*2,640.6*	*2,762.9*	*2,939.8*	*3,652.2*
	Quantity	*178,823*	*182,065*	*165,868*	*196,974*
	Index (2002=1.00)	*1.00*	*1.02*	*0.93*	*1.10*
	Average value (thousand dollars per unit)	*14.77*	*15.18*	*17.72*	*18.54*
	Constant dollar index (2002=100)	*1.00*	*1.03*	*1.20*	*1.26*
Excavators, hydraulic operated, more than 40 metric tons					
	Thousand 2005 dollars	*301,650*	*326,440*	*421,429*	*424,010*
	Quantity	*1,159*	*1,265*	*1,662*	*1,818*
	Index (2002=1.00)	*1.00*	*1.09*	*1.43*	*1.57*
	Average value (thousand dollars per unit)	*260.27*	*258.05*	*253.57*	*233.23*
	Constant dollar index (2002=1.00)	*1.00*	*0.99*	*0.97*	*0.90*
Excavators and draglines and some cranes not meeting other category classifications	*Thousand 2005 dollars*	*125,538*	*139,998*	*201,910*	*265,411*
	Quantity	*777*	*840*	*1,036*	*1,232*
	Index (2002=1.00)	*1.00*	*1.08*	*1.33*	*1.59*
	Average value (thousand dollars per unit)	*161.57*	*166.66*	*194.89*	*215.43*
	Constant dollar index (2002=1.00)	*1.00*	*1.03*	*1.21*	*1.33*
Off-highway trucks, coal haulers, truck-type tractor chassis, trailers, and wagons	*Thousand 2005 dollars*	—	—	*208,596*	*265,506*
	Quantity	—	—	*3,054*	*3,845*
	Index (2004=1.00)	—	—	*1.00*	*1.26*
	Average value (thousand dollars per unit)	—	—	*68.30*	*69.05*
	Constant dollar index (2004=1.00)	—	—	*1.00*	*1.01*

concrete to keep pace with the increased traffic demands placed on their network. (Concrete is used to provide a foundation for rail beds and, increasingly, is being used to make ties for tracks that carry heavier loads.) Consistent with the recent increase in steel prices, BNSF Railway Company, one of the largest coal haulers in the United States, has cited a $70 million increase in material costs associated with locomotive, freight car, and track structure in 2005 [*80*]. Freight cars and locomotive orders and new track installation often represent long-term decisions by railroads. BNSF, for instance, has contracted to take delivery of 845 locomotives by 2009. As of 2005, it had acquired 405 of the total [*81*]. Depending on the terms of those contracts, BNSF may or may not be susceptible to variation in steel prices.

For new freight car acquisitions, aluminum cars, lighter than steel cars and thus capable of carrying larger volumes of coal, tend to be preferred. The construction of aluminum cars still depends on some steel components, however, because more than 50 percent of the weight of a 42,000-pound aluminum car is made up of steel [*82*].

In 2005, more than 40,000 new freight cars of all types were acquired, representing an investment of roughly $3 billion. Some industry experts project that an additional 40,000 new freight cars per year is the minimum level that will be required to replace retired cars and maintain current capacity [*83*]. The average cost of all freight cars, including coal cars, ordered from Freight Car America was $68,000 both in 2004 and in 2005, as compared with $60,000 in 2003 (2005 dollars) [*84*]. In addition to reflecting the increase in steel prices in 2004 and 2005, the averages may vary according to the mix of cars delivered; however, 93 percent of the cars sold by Freight Car America in 2005 are used for coal transportation. Freight Car America has also indicated in its annual report that raw steel prices increased by 155 percent from October 2003 to December 2005, and that the company has successfully passed the increase on to purchasers for 96 percent of its car deliveries [*85*].

The railroads have already added a record number of locomotives to their fleets in recent years. In 2004, Class I railroads purchased or leased 1,121 new locomotives—91 percent more than in 2003 and 21 percent more than the previous high since 1988. In 2005, Norfolk Southern (NS) added 102 locomotives to its fleet, bringing its total to 4,000. In the same year, Union Pacific (UP) had plans to add 315 new locomotives. In 2004, Kansas City Southern ordered 30 new locomotives that were capable of transporting 9.6 percent more 110-ton cars than the rest of its existing fleet [*86*]. In 2006, BNSF has plans to add 310 locomotives to its fleet, at an estimated cost of $550 million [*87*]. Each new piece of equipment can have a much larger marginal impact on a railroad's capacity than its older existing equipment. Over time, the added economic benefit of more efficient equipment capable of moving heavier, longer train sets is likely to outweigh the recent increase in steel costs.

Finally, with increasingly heavy loads of coal being moved, the repair and maintenance cycle for existing railroad infrastructure becomes shorter, and the maintenance is more likely to be affected by short-term volatility in steel (and labor) prices. In 2004, for example, the seven Class I railroads spent $403 million (constant 2005 dollars) on rail and other materials for repair and maintenance of existing track [*88*]. In addition, over the next few years, the major railroads have plans to expand their network by adding multiple track systems and sidings. New track must be laid to handle higher freight volumes, and with heavier loads, more steel will be needed. For instance, track weighing 131 pounds per yard might be needed, as compared with 90 to 110 pounds per yard for less heavily used track. BNSF laid 749, 695, and 711 miles of track in 2003, 2004, and 2005, and an additional 884 miles is planned for 2006 [*89*].

The *AEO2007* reference case assumes that railroad equipment costs will rise in real terms through 2009, then return to their long-term declining trend.

Electric Power Industry

The Handy-Whitman index for electric utility construction provides an average cost index for six regions in the United States, starting from 1973. A simple average of the regional indexes for construction of electricity generation plants is used in Figure 14 to show a national cost trend relative to the cost index for construction materials. Because equipment and materials generally represent two-thirds to three-quarters of total power plant construction costs, it is not surprising that the trends are similar.

The long-term trend for construction costs in the electric power industry shows declining costs from 1975 to around 2000, after which it is relatively flat in real terms. The two indexes diverge in the early 2000s, with electric power construction costs showing a flat to slightly increasing trend, while general construction costs continue to decline. The difference coincides with a construction boom in the electric power sector from 2000 to 2004, when annual capacity additions averaged 38 gigawatts per year—well above previous build patterns (Figure 15). Over those years there were shortages and price increases specific to construction in the electric power industry due to the pace of building. For the past 3 years, the Handy-Whitman index shows an average annual increase of 5 percent, slightly less than that for the overall construction cost index.

Currently, new construction in the electric power sector is slowing down, with generating capacity additions averaging 16 gigawatts per year from 2004 to 2006. The slowdown is more likely a response to the oversupply of available capacity than a response to higher commodity prices. It is typical for investment in the power industry to cycle through patterns of increased building and slower growth, responding to changes in the expectations for future demand and fuel prices, as well as changes in the industry, such as restructuring.

AEO2007 does not project significant increases in new generating capacity in the electric power sector until after 2015. A total of 258 gigawatts of new capacity is expected between 2006 and 2030, representing a total investment of approximately $412 billion (2005 dollars). If construction costs were 5 to 10 percent higher than assumed in the reference case, the total investment over the period could increase by $21 billion to $41 billion.

Energy Demand: Limits on the Response to Higher Energy Prices in the End-Use Sectors

Energy consumption in the end-use demand sectors—residential, commercial, industrial, and transportation—generally shows only limited change when energy prices increase. Several factors that limit the sensitivity of end-use energy demand to price signals are common across the end-use sectors. For example, because energy generally is consumed in long-lived capital equipment, short-run consumer responses to changes in energy prices are limited to reductions in the use of energy services or, in a few cases, fuel switching; and because energy services affect such critical lifestyle areas as personal comfort, medical services, and travel, end-use consumers often are willing to absorb price increases rather than cut back on energy use, especially when they are uncertain whether price increases will be long-lasting. Manufacturers, on the other hand, often are able to pass along higher energy costs, especially in cases where energy inputs are a relatively minor component of production costs. In economic terms, short-run energy demand typically is inelastic, and long-run energy demand is less inelastic or moderately elastic at best [*90*].

Beyond the short-run inelasticity of demand in the end-use sectors, several factors make the long-run demand response to changes in energy prices relatively modest, including:

- Infrastructure—such as the network of roads, rails, and airports—that is unlikely to be substantially altered even in the long term

Figure 14. Changes in construction commodity costs and electric utility construction costs, 1973-2006 (constant dollar index, 1973=100)

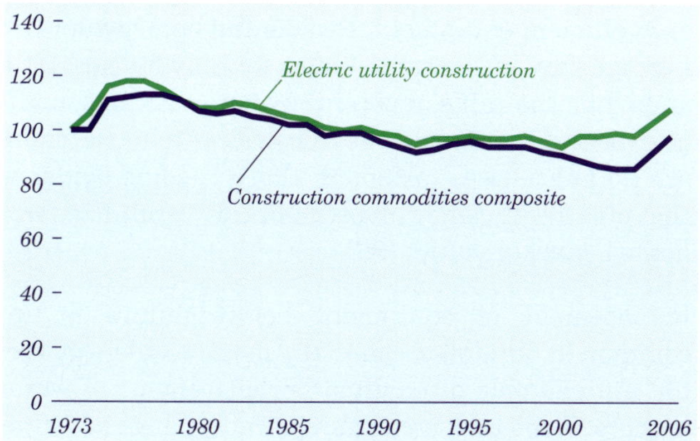

Figure 15. Additions to electricity generation capacity in the electric power sector, 1990-2030 (gigawatts, net summer capacity)

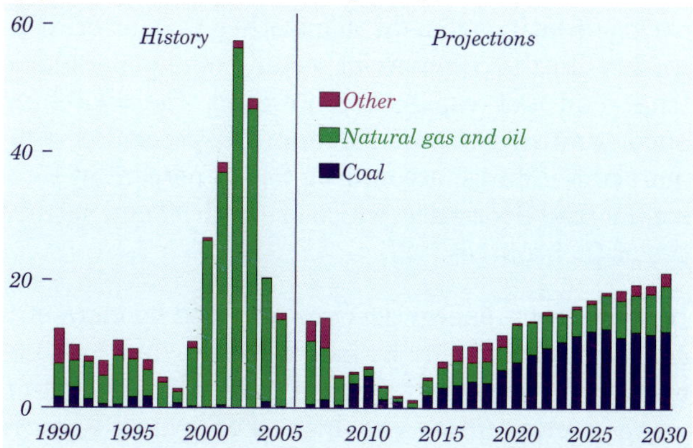

- General lack of fuel-switching capability in capital equipment
- Unattractive attributes of some energy-saving equipment, such as differences in quality or comfort and high cost
- Structural features of energy markets—including builder/owner versus buyer/renter incentives; incomplete information on energy-using equipment, such as consumption levels and potential savings; and inadequate price signals to consumers, resulting from rate design or other issues [*91*]

Uncertainty with regard to the value of potential energy savings and the opportunity costs of technology choices for long-lived equipment.

Buildings Sector

In the buildings sector, which includes residential and commercial end uses, building structures are long-lived assets that affect energy consumption through their overall design and "shell integrity" against unwanted heat transfers in or out of the building. A typical building may remain in the stock for 75 years. Beyond the structure itself, the energy-consuming equipment in a building typically lasts from 10 to 30 years. As a result, adjustments to the stock of buildings and equipment take many years, even if energy prices change dramatically. Because most previous disruptions in energy prices have been transitory, there is little evidence to indicate how quickly and how much the buildings sector could respond to a decades-long trend of increasing energy prices.

Limited capability for fuel switching is the rule rather than exception for equipment in buildings. In the residential sector, consumers have some limited choices between electricity and other fuels for a given energy service. For example, the thermostat on a natural gas water heater can be adjusted to reduce the use of the electric heating element in a clothes washer or dishwasher. In the commercial sector, some boilers have true dual-fuel capability; however, fuel-switching opportunities are available for only 3 percent of commercial buildings, accounting for 16 percent of total commercial floorspace, which use both oil and natural gas as fuel sources [*92*].

In some cases, energy services provided by more efficient equipment may be less desirable, and consumers may be slow to adopt the more efficient option when energy prices are high. For example, early versions of compact fluorescent lights (CFLs) had several quality issues, including bulky sizes that did not fit standard fixtures, poor light quality (flickering, poor color rendering, low light levels), and premature failures that caused life-cycle energy savings to be less than advertised [*93*]. Today's CFLs typically perform much better than the early models, and they are much less expensive. Even with those gains, however, some of their features remain less desirable than those of incandescent lights. CFLs typically have a warmup period, requiring several seconds to reach full output, and they cannot be dimmed. Other examples include lower outlet air temperatures for heat pumps than for other heating equipment and slower recovery times for heat pump water heaters.

Structural features of energy markets also contribute to the limited demand response. For example, investment decisions often are made by home builders, landlords, and property managers rather than the energy service consumers. In such cases, the decision-makers may prefer to purchase and install less costly, less efficient equipment, because they will not pay the future energy bills. Builders may choose less efficient equipment or offer fewer options to buyers in order to reduce design costs and increase profitability, even though consumers might be willing to pay higher home purchase prices or higher rents if they could lower their energy bills over the long term. A related issue arises from the inability of most consumers to evaluate the tradeoffs between capital cost and efficiency. Green building rating systems, such as the EPA's ENERGY STAR and DOE's Building America, do attempt to provide reliable information on the energy efficiency of buildings and potential energy savings [*94*].

In addition, because building equipment generally is expected to last for more than 10 years, many tenants will move before their cumulative energy savings can make up for the added expense of installing energy-efficient equipment. Residential homeowners on average stay in the same house for only 8 years [*95*], and while the value of potential energy savings might be expected to increase the sale price of a house, there are no guarantees (although there is some evidence that energy efficiency investments are capitalized in a home's market value) [*96*].

Replacement of equipment before failure is uncommon in buildings, especially in the residential sector. An example often cited is replacement of water heaters. Typically, a consumer waits until the water heater completely fails before replacing it. Because the failure creates considerable inconvenience, the consumer is likely to buy a new water heater as quickly as possible, without comparing price and efficiency tradeoffs before making a purchase decision. In the commercial sector, an exception is lighting retrofits, which often are made before the existing equipment wears out.

The potential for disruption of operations during equipment replacement can also affect decisions by purchasers, especially in the commercial sector, where energy costs are only a small fraction of business expenses for a typical commercial establishment. Efficiency investments may not be seen as cost-effective if the cost of the disruption outweighs potential savings, as is often the case with retrofits to improve the efficiency of building shells.

Demand response can also be attenuated by price signals that are incomplete or do not represent marginal costs. For example, because residential renters often pay electric bills but not natural gas bills, they may see the costs of air conditioning (electric) but not heating (natural gas, except for the electricity that powers the fan in a forced-air furnace). In commercial buildings, energy consumption choices (turning off computers or lights, for example) often are made by office workers who see no cost implications. Residential consumers, who typically see only monthly electric bills based on average costs, have no incentive to reduce their use of air conditioning on peak days. Under nonseasonal time-of-use rates, they would pay the higher marginal cost; but nonseasonal time-of-use rates currently are available in only about 5 percent of the residential market. For commercial customers, who tend to be larger consumers of electricity, the additional cost of more sophisticated demand metering or nonseasonal time-of-use metering is less significant, and their rates more often approximate the marginal cost of the electricity they use.

Industrial Sector

The industrial sector is more responsive to price changes for all inputs; however, the speed at which operational changes can be introduced to mitigate the cost impacts of rising energy prices is limited. Limitations arise from the fuel mix required by the existing capital stock (for example, it is not feasible in general to operate a natural-gas-fired boiler using coal), slow stock turnover, and falling capital investment rates. In addition, a strategy to reduce the demand for energy services by reducing production rates could prove to be more costly than the value of the energy savings if the reduction in output increased the probability of losing market share, reduced overall profitability, or led to contractual penalties.

Over a longer period, existing equipment could be scrapped and replaced with new equipment that uses different fuels or uses the same fuel more efficiently. The investments required to implement such changes would, however, compete with other uses of the funds available. Given the inherent uncertainty of energy prices, firms may be less than eager to invest in such measures as alternate fuel capability. Because most energy prices rise and fall together, dual-fuel investments may not be expected to have attractive paybacks. If high energy prices were sustained, however, companies might find previously neglected opportunities to reduce energy losses resulting from poor maintenance or other housekeeping items. Further, firms might find low-cost or no-cost options for reducing energy expenditures while maintaining the same level of energy services [*97*]. Successful examples include motor system optimization and steam line insulation, with implementation costs recovered in less than 1 year [*98*].

Energy costs account for only 2.8 percent of annual operating costs for U.S. manufacturing [*99*]. As a result, energy-saving investments may be less important than other factor-saving investments. Indeed, if energy prices rose substantially, corporate cash flow and the financial capital available for such investments could be reduced.

According to EIA's 2002 Manufacturing Energy Consumption Survey (MECS), more than 90 percent of petroleum consumption in the manufacturing sector is in the form of feedstocks [*100*]. In 2002, the sector's petroleum consumption for energy totaled only 450 trillion Btu, of which 140 trillion Btu was reported as switchable. Consumption of natural gas in the manufacturing sector totaled 6.5 quadrillion Btu in 2002, about 10 percent of which was used for feedstock. The 2002 MECS data indicate that 18 percent of the natural gas used for energy could be switched to another fuel, primarily petroleum. If all such switching did take place, the sector's petroleum consumption for energy would more than triple, increasing by 1 quadrillion Btu.

In summary, the manufacturing sector does respond to higher factor input prices, including energy prices, but energy expenditures do not constitute a large portion of most manufacturers' operating costs. Over time, however, the overall energy intensity of manufacturing does tend to decline in response to higher energy prices [*101*].

Transportation Sector

In the transportation sector, when consumers seek out energy-saving products and other cost-effective ways to service their travel needs, the energy cost savings are weighed against the perceived value of other factors considered in the decisionmaking process. Those factors include—but are not limited to—mobility, safety, comfort, quality, reliability, emissions, and capital cost.

The transportation sector is served primarily by four modes of travel: highway, air, rail, and water. Most of the energy consumed in the transportation sector is for highway vehicle travel, which accounts for approximately 85 percent of total consumption, followed by air (9 percent) and rail and water (6 percent combined). Energy consumption in the transportation sector consists almost exclusively (98 percent) of petroleum fuels. Thus, when there are appreciable increases in fuel prices, opportunities for reducing fuel expenditures through fuel switching are limited. As a result, savings can be realized only through reductions in travel demand, mode switching, improvements in system efficiency, and/or improvements in vehicle fuel efficiency.

The amount of efficiency improvement that could potentially be achieved varies greatly across modes and is limited by infrastructure constraints, vehicle lifetime and use patterns, and vehicle design criteria. For example, rail is a very energy-efficient way to move freight, about 11.5 times more energy-efficient on a Btu per ton-mile basis than heavy trucks. Opportunities for efficiency improvement in the rail mode are minimal, limited primarily to increases in system efficiency through higher equipment utilization and more efficient equipment operation—for example, by using unit and shuttle trains and by reducing locomotive idling. Limits are imposed by very long equipment lives, available infrastructure, and vehicle duty cycles. Similarly, waterborne travel is very efficient, and opportunities for energy savings are limited to improvements in system efficiency.

Air travel is serviced by a very competitive industry with significant investments in long-lived capital stock that operates in a constrained infrastructure. Immediate improvements in fuel efficiency can be gained through increased utilization of available infrastructure and increased load factors (ratio of passengers to available seats), but the desire of each company to maintain or increase market share limits opportunities for market players to act.

Long-term efficiency gains in air travel are realized through the adoption of technologies that improve either infrastructure efficiency (increased aircraft throughput at gates) or aircraft fuel efficiency (improved engine efficiency and lightweight materials); however, efficiency losses that result from changes in market structure to meet continued demand for increased flight availability and convenience generally cancel out efficiency gains. For example, the amount of air travel serviced by regional jets, which are about 40 percent less efficient than narrow-body jets, continues to increase as consumers look for improved destination and flight availability. As the share of the market served by regional jets increases, the overall fuel efficiency of the active aircraft stock is reduced, regardless of gains in the efficiency of larger aircraft.

Unlike the other transportation modes, highway vehicles have a relatively short life. The average age of the existing passenger car fleet is 9 years, and the average age of trucks (light and heavy) is 8 years, reflecting, in part, the shift toward light trucks for personal transportation over the past decade. In addition, the car stock turns over at a rate of about 6 percent per year. Heavy truck stocks turn over at a much slower rate, approximately 4 percent per year. Those slow stock replacement rates, coupled with consumer attitudes toward fuel economy improvement relative to other, more highly desired vehicle attributes, make it difficult to realize short-term increases in fuel economy for the vehicle stock as a whole.

Further limiting increases in vehicle fuel economy is the scarcity of cost-effective alternatives within the vehicle categories preferred by consumers. Whether the consumer rates the desirability of a vehicle purchase by quality, safety, seating capacity, storage capacity, towing capacity, luxury, or performance, once the criteria are established they limit the vehicle types considered. For example, someone shopping for a van or sport utility vehicle is unlikely to view a compact as a viable alternative.

In addition to efficiency improvements made within a mode, transportation efficiency can be improved by switching to more efficient modes of travel. For example, passenger and freight travel can be served by a variety of travel modes (highway, air, and rail), with mode selection determined by cost of service, access, convenience, mobility afforded, and time budgets. When energy prices increase, consumers seeking reductions in travel costs examine the expected savings associated with alternative mode choices in relation to the values placed on other considerations. For most consumers, alternative mode choices are limited, providing little opportunity for cost reductions. For others, the cost savings that would result from the choice of an alternative mode of travel are likely to be outweighed by the value placed on travel time, convenience, and mobility.

Miscellaneous Electricity Services in the Buildings Sector

Residential and commercial electricity consumption for miscellaneous services has grown significantly in recent years and currently accounts for more electricity use than any single major end-use service in either sector (including space heating, space cooling, water heating, and lighting). In the residential sector, a proliferation of consumer electronics and information technology equipment has driven much of the growth. In the commercial sector, telecommunications and network equipment and new advances in medical imaging have contributed to recent growth in miscellaneous electricity use [*102*].

Until recently, energy consumption for most miscellaneous electricity uses has not been well quantified. A September 2006 report prepared for EIA by TIAX LLC [*103*] provides much-needed information about many miscellaneous electricity services. For the report, TIAX developed estimates of current and future electricity consumption for the 10 largest miscellaneous electricity loads in the residential sector and for 10 key contributors to miscellaneous electricity use in the commercial sector, based on current usage and technology trends. The information has allowed EIA to disaggregate components of the "other" electricity consumption category and refine the *AEO2007* projections for the buildings sector. Based on the conclusions of the TIAX study, which allows a finer breakout of smaller electric uses in the buildings sector, the projected growth rate for miscellaneous electricity use in the *AEO2007* reference case is lower than was projected in the *AEO2006* reference case.

Residential Sector

The 10 miscellaneous electricity uses evaluated by TIAX account for about 40 percent of the comparable miscellaneous electricity use in 2005 (11 percent of total residential electricity use). Televisions (TVs), which were accounted for separately in previous *AEO*s, account for one-third of residential miscellaneous electricity use in 2005 in the TIAX study, and TVs and set-top boxes are projected to account for 80 percent of the growth in electricity use for the 10 miscellaneous loads from 2005 to 2030. It should be noted that considerable uncertainty surrounds the projections, in that technological change and innovation, as well as consumer preferences, can lead to rapid changes in the market for these products. Table 5 summarizes electricity use in 2005, 2015, and 2030 for the 10 residential loads included in the study.

As shown in Table 5, electricity use for TVs and set-top boxes nearly doubles from 2005 to 2030. This projection is based on factors such as number of TVs per house, screen size, technology type, satellite/cable penetration, and the transition away from analog to digital broadcasts. For most TVs in the current stock, the transition to digital broadcasts will require a set-top box to decode the signal, as reflected in the sharp increase of electricity use for set-top boxes from 2005 to 2015. After 2015, when newer TVs are expected to have the decoder built in, the rate of increase slows. Continued penetration of satellite and cable systems, as well as multi-function digital video recorders (DVRs) contributes to the increase in set-top boxes over the projection period.

There are many uncertainties that could affect future growth in electricity use for TVs. Although it is certain that screen sizes have increased over time in the past, and likely that they will continue to increase, it is far less certain which technology will come to

Table 5. Miscellaneous electricity uses in the residential sector, 2005, 2015, and 2030 (billion kilowatthours)

Electricity use	*2005*	*2015*	*2030*
Coffee makers	*4.0*	*4.7*	*5.5*
Home audio	*11.8*	*12.6*	*14.0*
Ceiling fans	*16.8*	*20.1*	*23.5*
Microwave ovens	*14.3*	*16.3*	*19.0*
Security systems	*1.9*	*1.8*	*2.4*
Spas	*8.3*	*9.6*	*12.7*
Set-top boxes	*17.1*	*30.0*	*32.7*
Color TVs	*52.1*	*72.9*	*92.5*
Hand-held rechargeable devices	*9.8*	*9.0*	*10.6*
DVRs/VCRs	*15.6*	*12.0*	*9.8*
Total, miscellaneous uses studied	***151.7***	***188.9***	***222.7***
Other miscellaneous uses	*232.5*	*325.2*	*432.7*
Total miscellaneous	***384.2***	***514.1***	***655.4***
Total residential sector electricity use	*1,364.8*	*1,591.2*	*1,896.5*

dominate the market. Plasma, liquid crystal display, and digital light processing screen technologies all have footholds in the current market for TVs, and they vary in electricity use. Moreover, future technologies, such as carbon nanotube displays, may use significantly less power than today's technologies, and TVs with point-of-deployment slots could make set-top boxes obsolete.

The projections in Table 5 assume that all TVs will meet the current ENERGY STAR requirements for off power (less than 1 watt); however, overall electricity use for TVs is largely insensitive to that assumption, because hours of use and screen size predominantly determine their electricity use. As shown in Table 6, bigger TVs with high-definition screens that require more energy per unit are projected to double in market share from 2005 to 2015, resulting in a 24-percent increase in active power draw per set, on average.

The eight other devices listed in Table 5 contribute little (about 20 percent) to the projected growth in total miscellaneous electricity use for the residential sector. Their functions are diverse, ranging from common appliances (microwave ovens) to less common products (spas). Their annual electricity consumption also varies widely, from 74 kilowatthours per year for security systems to more than 2,500 kilowatthours per year for spas.

Of the eight other devices, electricity use for ceiling fans (not including attached lights) is projected to increase the most through 2030, as newly constructed homes tend to have more ceiling fans installed, and more new homes are built in warmer areas where ceiling fans are used more intensively. Microwave ovens show a slight increase in household saturation, from 96 percent in 2005 to 98 percent in 2030, but energy use will grow faster as the number of households increases. For spas, electricity use per unit is expected

Table 6. Electricity use and market share for televisions by type, 2005 and 2015

Television type	Screen size (inches)	Active power draw (watts)	Market share (percent) 2005	Market share (percent) 2015
Analog	<40	86	69	10
	>40	156	16	2
Digital, standard definition	<40	96	<1	34
	>40	166	<1	<1
Digital, enhanced/high definition	<40	150	8	34
	>40	234	8	19

to decrease as efficiency standards tighten [104], but more units are expected to be installed, leading to an overall increase in electricity consumption. Hand-held rechargeable devices (mobile phones, cordless phones, hand-held power tools, and others) also are projected to use less electricity per unit, again, in response to tighter efficiency standards.

Commercial Sector

The 10 commercial uses evaluated in the TIAX study currently account for 137 billion kilowatthours of electricity demand (about 470 trillion Btu), or approximately 37 percent of miscellaneous electricity use in the commercial sector (Table 7). Two well-established areas of commercial electricity use, distribution transformers used to decrease the voltage of electricity received from suppliers to usable levels and water services (purification, distribution, and wastewater treatment) account for a large share of the electricity consumption evaluated in the study. Although those two uses are expected to continue accounting for a significant amount of commercial electricity use, neither shows rapid growth in the projections. EPACT2005 includes efficiency standards to limit electricity losses from low-voltage dry-type distribution transformers—the type most prevalent in the commercial sector—which should limit their contribution to growth in commercial electricity use. Trends in water conservation and wastewater reuse are expected to offset the increasing energy intensity of treatment, resulting in total projected growth in electricity use for public water services of more than 15 percent from 2005 to 2030—slightly less than the growth implied by the 0.8-percent average annual

Table 7. Miscellaneous electricity uses in the commercial sector, 2005, 2015, and 2030 (billion kilowatthours)

Electricity use	2005	2015	2030
Coffee makers	2.7	3.0	3.5
Distribution transformers	54.5	54.6	54.9
Non-road electric vehicles	4.0	5.1	7.1
Magnetic resonance imaging (MRI)	0.6	1.9	4.5
Computed tomography (CT) scanners	0.9	1.8	2.8
X-ray machines	4.0	6.8	12.0
Elevators	4.4	4.7	5.5
Escalators	0.7	0.8	1.0
Water supply: distribution	40.0	42.0	47.0
Water supply: purification	1.1	1.2	1.3
Wastewater treatment	24.5	25.3	27.2
Total, miscellaneous uses studied	**137.4**	**147.2**	**166.8**
Other miscellaneous uses	229.5	357.9	601.6
Total miscellaneous	**366.9**	**505.1**	**768.4**
Total commercial sector electricity use	1,266.7	1,548.2	2,061.6

rate of population growth projected in the *AEO2007* reference case.

Growth rates in electricity use for the remaining commercial uses included in the TIAX study are governed by the specific market segments serviced and by technology advances. The electricity requirements for medical imaging equipment—magnetic resonance imaging systems (MRIs), computed tomography (CT) scanners, and fixed-location x-ray machines—are expected to grow more quickly than consumption for the other commercial services studied. MRIs and CT scanners are relatively new technologies. They are expected to continue penetrating the healthcare arena, and the technology is expected to advance, leading to future increases in their total electricity use. Although x-ray machines have been in use for many years, the move toward digital x-ray systems and steady growth in the healthcare sector are expected to increase their electricity use as well.

Electricity use for non-road electric vehicles, including lift trucks, forklifts, golf carts, and floor burnishers, is projected to grow slightly faster than commercial floorspace in the *AEO2007* reference case, led by growing sales of electric golf carts. Commercial-style coffee makers are expected to grow with the food service and office segments, reflecting the two major markets for commercial coffee services. Electricity consumption for vertical transport (elevators and escalators) is expected to follow growth in the commercial sector, tempered by the expectation that increasing numbers of elevators will have the capability to enter standby mode, turning off lights and ventilation, for up to 12 hours per night.

Industrial Sector Energy Demand: Revisions for Non-Energy-Intensive Manufacturing

For the industrial sector, EIA's analysis and projection efforts generally have focused on the energy-intensive industries—food, bulk chemicals, refining, glass, cement, steel, and aluminum—where energy cost averages 4.8 percent of annual operating cost. Detailed process flows and energy intensity indicators have been developed for narrowly defined industry groups in the energy-intensive manufacturing sector. The non-energy-intensive manufacturing industries, where energy cost averages 1.9 percent of annual operating cost, previously have received somewhat less attention, however. In *AEO2006*, energy demand projections were provided for two broadly aggregated industry groups in the non-energy-intensive manufacturing sector: metal-based durables and other non-energy-intensive. In the *AEO2006* projections, the two groups accounted for more than 50 percent of the projected increase in industrial natural gas consumption from 2004 to 2030.

With the non-energy-intensive industries making up such a significant share of industrial natural gas demand, a more detailed review of the individual industries that made up the two groups has been conducted. The review showed that aggregation within those groups created a bias that contributed strongly to the projected increase in their natural gas use in *AEO2006*. The least energy-intensive component (computers and electronics) had the highest projected growth rate for value of shipments, whereas the more energy-intensive components had lower growth projections. To address the disparity, the *AEO2007* projections are based on more narrowly defined subgroups in the non-energy-intensive manufacturing sector, as shown in Table 8.

Among the non-energy-intensive industry subgroups analyzed for *AEO2007*, the computers and electronics group has the lowest energy intensity in the metal-based durables manufacturing sector (Figure 16) and the highest projected growth rate (Figure 17). Conversely, fabricated metals has the highest energy intensity and the lowest projected growth rate in value of shipments. Consequently, although the projected growth in value of shipments for metal-based durables as a whole is higher in *AEO2007* than it was in *AEO2006*, because of the disaggregation, its delivered energy consumption in 2030 is 15 percent lower in *AEO2007* than in *AEO2006* (Figure 18), and its

Table 8. Revised subgroups for the non-energy-intensive manufacturing industries in AEO2007: energy demand and value of shipments, 2002

Manufacturing group and subgroups	*NAICS code*	*Energy demand (trillion Btu)*	*Value of shipments (billion 2000 dollars)*
Metal-based durables			
Fabricated metals	*332*	*386*	*244.2*
Machinery	*333*	*174*	*250.3*
Computers and electronics	*334*	*211*	*438.9*
Transportation equipment	*336*	*391*	*641.1*
Electrical equipment	*335*	*169*	*91.2*
Total		***1,331***	***1,665.7***
Other non-energy-intensive			
Wood products	*321*	*361*	*91.5*
Plastics and rubber products	*326*	*344*	*172.7*
Balance of manufacturing	*NA*	*1,876*	*918.9*
Total		***2,581***	***1,183.1***

natural gas consumption in 2030 is nearly 200 trillion Btu (19 percent) lower.

In the "other non-energy-intensive" sector of the non-energy-intensive manufacturing industries, data limitations and the lack of a dominant energy user make it more difficult to disaggregate industry subgroups. Based on EIA's 2002 MECS data, however, two specific industries—wood products (North American Industry Classification System [NAICS] 321) and plastics manufacturing (NAICS 326)—have been separated in the *AEO2007* projections, with the remainder of the other non-energy-intensive sector treated as a third subgroup. Wood products is of interest because that industry derives 58 percent of the energy it consumes (209 trillion Btu out of a total 361 trillion Btu in 2002) from biomass in the form of wood waste and residue. In the plastics manufacturing industry, which produces goods by processing plastic materials (it does not produce the plastic), one-half of the energy consumed (182 trillion Btu out of a total 344 trillion Btu in 2002) is in the form of electricity. Together, the two industries account for 4 percent of the total energy demand for all manufacturing (about 700 trillion Btu) and 7 percent of the value of shipments for all manufacturing.

Figure 16. Energy intensity of industry subgroups in the metal-based durables group of non-energy-intensive manufacturing industries, 2002 (thousand Btu per 2000 dollar value of shipments)

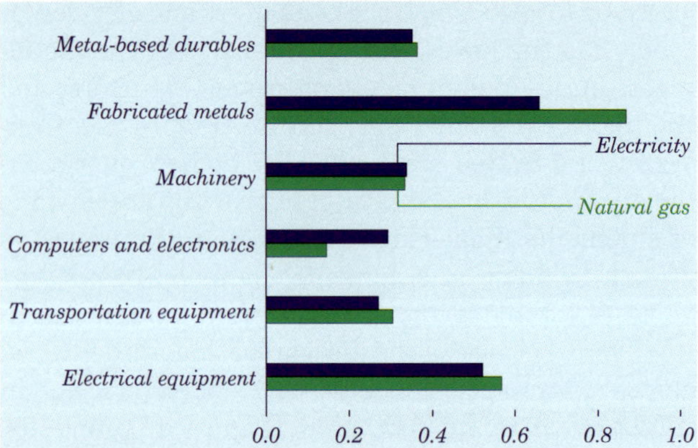

In addition to the disaggregation described above, EIA has also reexamined the use of steam as an energy source in the non-energy-intensive manufacturing industries. For the other non-energy-intensive group, it was found that steam is used primarily for space heating in buildings rather than in manufacturing processes. As a result, *AEO2007* projects slower growth in its demand for steam than was projected in *AEO2006*. In combination, the two revisions described here result in a significantly lower projection of energy demand for non-energy-intensive manufacturing in 2030 in the *AEO2007* reference case, about 20 percent lower than was projected in *AEO2006* (Figure 19).

Loan Guarantees and the Economics of Electricity Generating Technologies

The loan guarantee program authorized in Title XVII of EPACT2005 is not included in *AEO2007*, because the Federal Credit Reform Act of 1990 requires congressional authorization of loan guarantees in an appropriations act before a Federal agency can make a binding loan guarantee agreement. As of October 2006, Congress had not provided the legislation necessary for DOE to implement the loan guarantee program (see "Legislation and Regulations"). In August

Figure 17. Average annual growth rates of value of shipments for metal-based durables industries in the AEO2006 and AEO2007 reference case projections, 2005-2030 (percent per year)

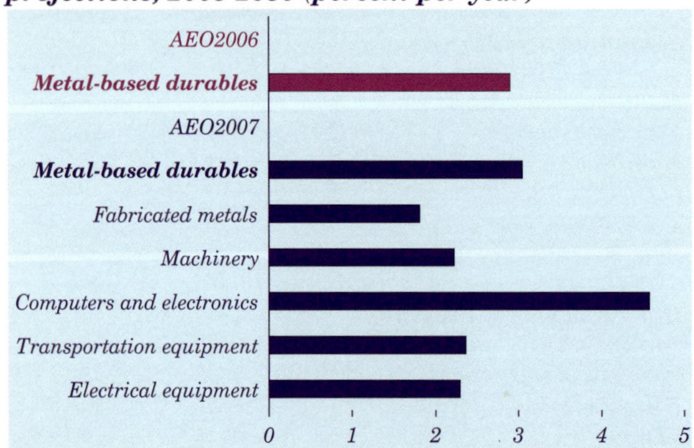

Figure 18. Average annual increases in energy demand for metal-based durables industries in the AEO2006 and AEO2007 reference case projections, 2005-2030 (percent per year)

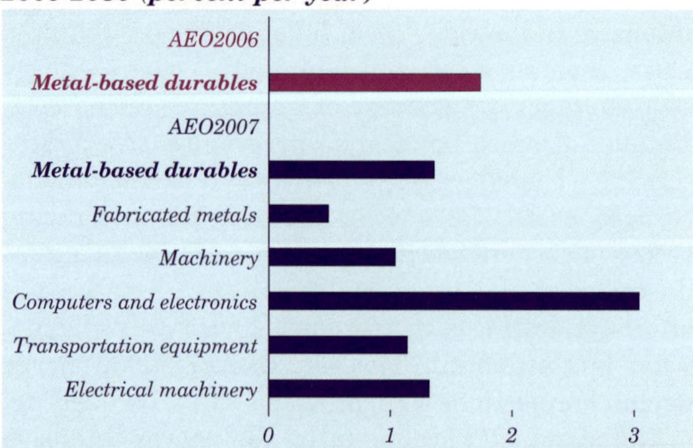

2006, however, DOE invited firms to submit "pre applications" for the first $2 billion in potential loan guarantees.

The EPACT2005 loan guarantee program could provide incentives for a wide array of new energy technologies. Technologies potentially eligible for loan guarantees include renewable energy systems, advanced fossil energy technologies, hydrogen fuel cell technologies, advanced nuclear energy facilities, CCS technologies, efficient generation, transmission, and distribution technologies for electric power, efficient end-use technologies, production facilities for fuel-efficient vehicles, pollution control technologies, and new refineries.

In the electric power sector, the loan guarantee program could substantially affect the economics of new power plants, for three reasons. First, Federal loan guarantees would allow lenders to be reimbursed in cases of default, but only for certain electric power sector technologies. Consequently, they would be willing to provide loans for power plant construction at lower interest rates, which would reduce borrowing costs. For example, a number of private companies guarantee loans made by State and local governments. Such insured loans typically are rated AAA (very low risk) and therefore have relatively low yields. Indeed, municipalities purchase such insurance because the decrease in interest rate is greater than the insurance premiums.

Second, firms typically finance construction projects by using a capital structure that consists of a mix of debt (loans) and equity (funds supplied from the owners of the firm). Debt financing usually is less

Figure 19. Annual delivered energy demand for the non-energy-intensive manufacturing industry groups in the AEO2006 and AEO2007 reference case projections, 2005-2030 (trillion Btu)

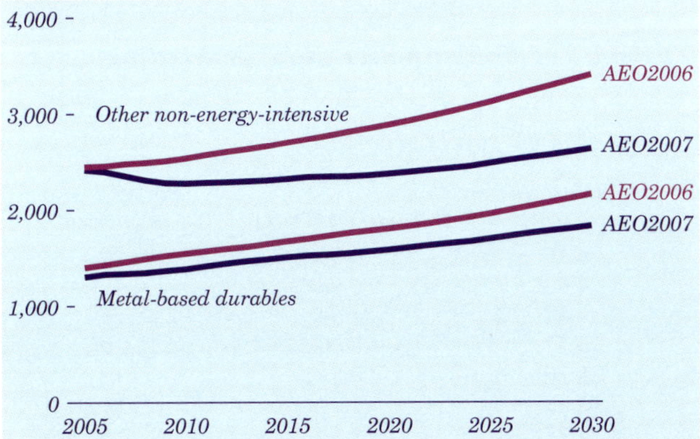

expensive than equity financing, and up to some point, the average cost of capital (the weighted average cost of debt and equity financing) can be reduced by substituting debt for equity financing. (The substitution of debt for equity is called leveraging.) After that point, however, projects financed with large amounts of debt can be very risky, and additional debt financing can increase the average cost of capital rather than lower it. Thus, there are constraints on the use of leverage. In many industries, capital structures tend to include 40 to 60 percent debt. With loan guarantees, however, the risks of highly leveraged projects are shifted to the guarantor, and more leveraging can be used to reduce the average cost of capital for construction projects.

Federal loan guarantees also can allow potential sponsors to participate in one or more major projects while avoiding the risk of possible failure, which might be caused by factors such as construction cost overruns or lower than expected electricity prices and, potentially, could threaten the financial viability of the sponsoring firm. To avoid this problem, beginning in the 1990s, many firms used project financing to build electric power plants, including a number of merchant natural-gas-fired plants that were built in the late 1990s and early 2000s.

Under project financing, a power plant under construction is treated as if it were owned by a separate entity whose sole asset is that new power plant. Thus, the loan is secured only by the new plant. This is also referred to as non-recourse financing. Because lenders for the plant's construction have claims only on the power plant in case of default, the project's risk is quarantined. That is, the lenders have no claims on the firm's other assets in case of default, and the project's failure will have only limited effect on the firm's creditworthiness and overall financial health.

From the firm's perspective, there are clear advantages to using project financing. From the lender's perspective, however, project (non-recourse) financing can be very risky, especially if the project is highly leveraged. If the project fails and the firm defaults on its loans, the power plant will be sold; but if market electricity prices and thus the value of the asset are depressed at the time of the sale, the lender may not be able to recover all its costs. In addition, the administrative costs associated with bond default can be substantial. Consequently, given the inherent risk of large-scale projects, it could be very difficult to obtain project financing for a multi-billion-dollar power plant at a cost that would allow the project to remain economical. Federal loan guarantees would thus provide an incentive program for potential lenders.

To examine the potential impacts of DOE's loan guarantee program on the economics of various capital-intensive electricity generating technologies, the levelized costs of electricity generation from newly built power plants financed with and without loan guarantees were computed, using plant cost and performance assumptions from the *AEO2007* reference case. In the case without guarantees, financial assumptions from the reference case were also used, including average equity financing costs of about 14 percent over the 2006-2030 period, average debt financing costs of about 8.0 percent, capital structures consisting of 55 percent equity and 45 percent debt, and a capital recovery period of 20 years. In the case with loan guarantees, capital structures of 20 percent equity and 80 percent debt were assumed.

The capital structure assumption in the loan guarantees case is typical of the financing for construction projects for some merchant natural-gas-fired power plant that have been built by companies with long-term power purchase contracts. In addition, DOE has stated that its loan guarantees under the new program will cover no more than 80 percent of the debt for any project. It was assumed that the yields on such guaranteed debt would be halfway between risk-free 10-year Treasury bonds and very low but not riskless AAA corporate bonds. Based on average yields over the past 25 years, this assumption implies that, with the loan guarantees, the cost of the insured portion of the debt would fall by about 1.5 percentage points, to about 6.5 percent on average over the 2006-2030 period.

The uninsured portion of the debt (20 percent of 80 percent) would be relatively risky, however, and probably would be rated below investment grade. Thus, it was assumed that the cost of the uninsured debt would be at the lower end of the yields to high-yield (fairly risky) corporate bonds, or about 1.5 percentage points higher than the 8.0 percent assumed in the case without guarantees. In total, the cost of debt averaged over the insured and uninsured portions of project debt financing in the case with loan guarantees would be 7.1 percent—about 0.9 percentage point below the 8.0 percent assumed in the case without loan guarantees.

Projections from the two alternative cases are shown in Table 9 for the levelized costs of generating electricity from various technologies at power plants becoming operational in 2015. The results show that loan guarantees would significantly lower the levelized costs for eligible generating technologies. (Conventional coal-fired and combined-cycle natural-gas-fired plants do not qualify for the loan guarantee program.) In addition, because the loan guarantee program reduces financing costs, the greater a technology's capital intensity, the greater would be the percentage reduction in total generation costs. For a (capital-intensive) new nuclear power plant or wind farm that received a loan guarantee, the levelized cost of its electricity production is reduced by about 25 percent under the assumptions outlined above.

Impacts of Increased Access to Oil and Natural Gas Resources in the Lower 48 Federal Outer Continental Shelf

The OCS is estimated to contain substantial resources of crude oil and natural gas; however, some areas of the OCS are subject to drilling restrictions. With energy prices rising over the past several years, there has been increased interest in the development of more domestic oil and natural gas supply, including OCS resources. In the past, Federal efforts to encourage exploration and development activities in the deep waters of the OCS have been limited primarily to

Table 9. Effects of DOE's loan guarantee program on the economics of electric power plant generating technologies, 2015 (2005 cents per kilowatthour)

	Levelized cost of generation			
Technology	*Without loan guarantee*	*With loan guarantee*	*Cost reduction*	*Percent cost reduction*
Pulverized coal	5.36	5.36	0.00	0
Integrated coal gasification combined cycle (IGCC)	5.61	4.66	0.95	17
IGCC with carbon sequestration	7.37	6.03	1.34	18
Advanced combined cycle	5.53	5.53	0.00	0
Advanced combined cycle with carbon sequestration	7.59	6.70	0.89	12
Wind	6.80	5.06	1.75	26
Nuclear	6.33	4.78	1.55	25

regulations that would reduce royalty payments by lease holders. More recently, the States of Alaska and Virginia have asked the Federal Government to consider leasing in areas off their coastlines that are off limits as a result of actions by the President or Congress. In response, the Minerals Management Service (MMS) of the U.S. Department of the Interior has included in its proposed 5-year leasing plan for 2007-2012 sales of one lease in the Mid-Atlantic area off the coastline of Virginia and two leases in the North Aleutian Basin area of Alaska. Development in both areas still would require lifting of the current ban on drilling.

For *AEO2007*, an OCS access case was prepared to examine the potential impacts of the lifting of Federal restrictions on access to the OCS in the Pacific, the Atlantic, and the eastern Gulf of Mexico. Currently, except for a relatively small tract in the eastern Gulf, resources in those areas are legally off limits to exploration and development. Mean estimates from the MMS indicate that technically recoverable resources currently off limits in the lower 48 OCS total 18 billion barrels of crude oil and 77 trillion cubic feet of natural gas (Table 10).

Although existing moratoria on leasing in the OCS will expire in 2012, the *AEO2007* reference case assumes that they will be reinstated, as they have in the past. Current restrictions are therefore assumed to prevail for the remainder of the projection period, with no exploration or development allowed in areas currently unavailable to leasing. The OCS access case assumes that the current moratoria will not be

Table 10. Technically recoverable undiscovered oil and natural gas resources in the lower 48 Outer Continental Shelf as of January 1, 2003

OCS areas	*Crude oil (billion barrels)*	*Natural gas (trillion cubic feet)*
Available for leasing and development		
Eastern Gulf of Mexico	2.27	10.14
Central Gulf of Mexico	22.67	113.61
Western Gulf of Mexico	15.98	86.62
Total available	**40.92**	**210.37**
Unavailable for leasing and development		
Washington-Oregon	0.40	2.28
Northern California	2.08	3.58
Central California	2.31	2.41
Southern California	5.58	9.75
Eastern Gulf of Mexico	3.98	22.16
Atlantic	3.82	36.99
Total unavailable	**18.17**	**77.17**
Total Lower 48 OCS	**59.09**	**287.54**

reinstated, and that exploration and development of resources in those areas will begin in 2012.

Assumptions about exploration, development, and production of economical fields (drilling schedules, costs, platform selection, reserves-to-production ratios, etc.) in the OCS access case are based on data for fields in the western Gulf of Mexico that are of similar water depth and size. Exploration and development on the OCS in the Pacific, the Atlantic, and the eastern Gulf are assumed to proceed at rates similar to those seen in the early development of the Gulf region. In addition, it is assumed that local infrastructure issues and other potential non-Federal impediments will be resolved after Federal access restrictions have been lifted. With these assumptions, technically recoverable undiscovered resources in the lower 48 OCS increase to 59 billion barrels of oil and 288 trillion cubic feet of natural gas, as compared with the reference case levels of 41 billion barrels and 210 trillion cubic feet.

The projections in the OCS access case indicate that access to the Pacific, Atlantic, and eastern Gulf regions would not have a significant impact on domestic crude oil and natural gas production or prices before 2030. Leasing would begin no sooner than 2012, and production would not be expected to start before 2017. Total domestic production of crude oil from 2012 through 2030 in the OCS access case is projected to be 1.6 percent higher than in the reference case, and 3 percent higher in 2030 alone, at 5.6 million barrels per day. For the lower 48 OCS, annual crude oil production in 2030 is projected to be 7 percent higher—2.4 million barrels per day in the OCS access case compared with 2.2 million barrels per day in the reference case (Figure 20). Because oil prices are determined on the international market, however, any impact on average wellhead prices is expected to be insignificant.

Similarly, lower 48 natural gas production is not projected to increase substantially by 2030 as a result of increased access to the OCS. Cumulatively, lower 48 natural gas production from 2012 through 2030 is projected to be 1.8 percent higher in the OCS access case than in the reference case. Production levels in the OCS access case are projected at 19.0 trillion cubic feet in 2030, a 3-percent increase over the reference case projection of 18.4 trillion cubic feet. However, natural gas production from the lower 48 offshore in 2030 is projected to be 18 percent (590 billion cubic feet) higher in the OCS access case (Figure 21). In

2030, the OCS access case projects a decrease of $0.13 in the average wellhead price of natural gas (2005 dollars per thousand cubic feet), a decrease of 250 billion cubic feet in imports of liquefied natural gas, and an increase of 360 billion cubic feet in natural gas consumption relative to the reference case projections. In addition, despite the increase in production from previously restricted areas after 2012, total natural gas production from the lower 48 OCS is projected generally to decline after 2020.

Although a significant volume of undiscovered, technically recoverable oil and natural gas resources is added in the OCS access case, conversion of those resources to production would require both time and money. In addition, the average field size in the Pacific and Atlantic regions tends to be smaller than the average in the Gulf of Mexico, implying that a significant portion of the additional resource would not be economically attractive to develop at the reference case prices.

Alaska Natural Gas Pipeline Developments

The *AEO2007* reference case projects that an Alaska natural gas pipeline will go into operation in 2018, based on EIA's current understanding of the project's time line and economics. There is continuing debate, however, about the physical configuration and the ownership of the pipeline. In addition, the issue of Alaska's oil and natural gas production taxes has been raised, in the context of a current market environment characterized by rising construction costs and falling natural gas prices. If rates of return on investment by producers are reduced to unacceptable levels, or if the project faces significant delays, other sources of natural gas, such as unconventional natural gas production and LNG imports, could fulfill the demand that otherwise would be served by an Alaska pipeline.

The primary Alaska North Slope oil and natural gas producers—BP, ExxonMobil, and ConocoPhillips—became interested in building an Alaska natural gas pipeline after natural gas prices began to increase substantially during 2000. In May 2002, they released a report on the expected costs of building a pipeline along two different routes. Since then, construction of a pipeline has been stalled by differences of opinion within Alaska regarding the ultimate destination of the pipeline and the level of taxation applied to the State's oil and natural gas production. Recent increases in construction costs and trends in natural gas prices are important factors that will determine the economic viability of the pipeline.

Physical Configuration of the Pipeline

There are three different visions for the physical configuration of the Alaska natural gas pipeline. One vision—the southern route—supports the construction of a pipeline that would serve lower 48 natural gas markets exclusively, following the TransAlaska Pipeline System to Fairbanks and then the Alaska Highway into Canada. A second vision—the northern route—as proposed by the North Slope producers, advocates a pipeline route going east along the Alaska's north coast to the Mackenzie Delta in Canada and then proceeding south to the lower 48 States. In 2002, the producers estimated that the northern route would cost approximately $800 million less to build than the southern route, because it would be about 338 miles shorter and would traverse less mountainous terrain. In 2001, Alaska enacted legislation to foreclose the northern route. A third view—the south

Figure 20. Lower 48 offshore crude oil production in two cases, 1990-2030 (million barrels per day)

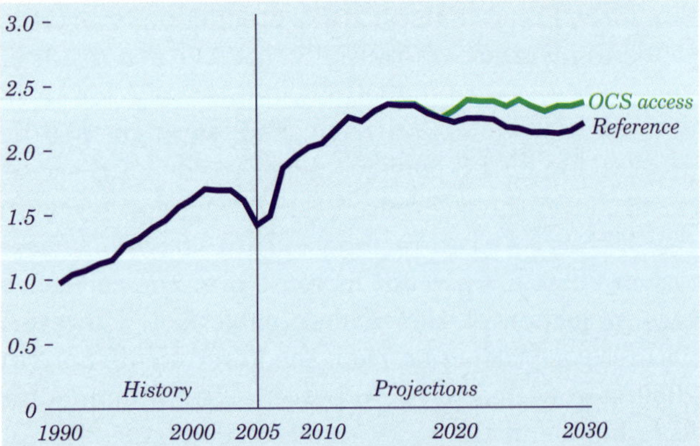

Figure 21. Lower 48 offshore natural gas production in two cases, 1990-2030 (trillion cubic feet)

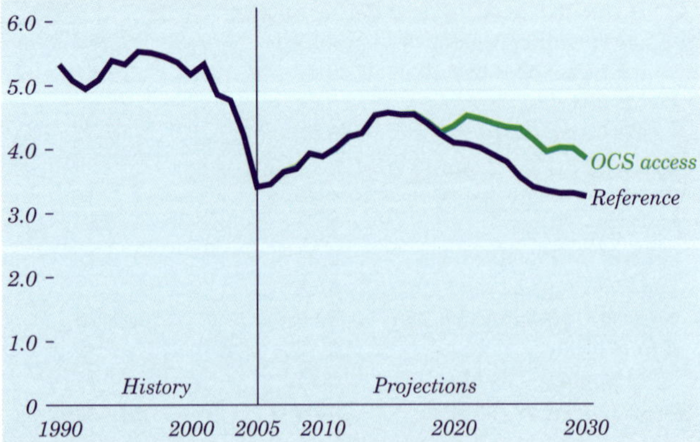

central design—supports the construction of a pipeline that would transport natural gas to south central Alaska, both to serve local consumers and to provide LNG to overseas consumers.

The three pipeline proposals are based on fundamentally different priorities. The northern and southern routes are premised on the notion that an Alaska natural gas pipeline would be economically feasible only if it captured the greatest possible economies of scale (the greatest pipeline throughput), thereby ensuring the highest possible wellhead price for North Slope natural gas and the greatest State royalty collection. The south central design is premised largely on the idea that, because natural gas reserves in the Cook Inlet region are declining, North Slope production should be transported to south central Alaska to ensure the future availability of natural gas to that region's consumers.

Production Taxes

The Alaska Stranded Natural Gas Development Act was signed in 1998 to make a natural gas pipeline project in Alaska commercially feasible. When the Act was passed, lower 48 wellhead natural gas prices averaged $1.96 per thousand cubic feet. Since then, as lower 48 prices have increased, the political climate in Alaska has changed from one in which financial incentives were thought to be crucial to the construction of a pipeline to one in which some interests believe that State taxes on oil and natural gas production are not high enough.

In May 2006, a draft stranded gas contract was made publicly available. In the draft, the North Slope producers and the State agreed to a 20-percent production tax with a 20-percent tax credit for future investments in Alaska's oil and natural gas development. The terms and conditions were negotiated to remain in effect for the next 30 years. After the release of the draft contact, opponents argued that the contract's production tax rate was too low and the investment credits too large.

In August 2006, the Alaska legislature in a special session passed an oil and natural gas production tax, which raised the oil production tax from the negotiated 20 percent up to 22.5 percent. The legislation, which was signed into law that same month, also reduced the level of investment tax credits that North Slope producers could use to offset their production tax liabilities.

At a minimum, the discrepancy between the provisions in the August 2006 law and the draft standard gas contract will necessitate renegotiation between the producers and the State. The governor who negotiated the draft contract and signed the August 2006 law was defeated in his bid for reelection. The pipeline was a major issue in the campaign, and the new governor may not want to use the existing draft contract as the starting point for negotiation.

Other Issues

Until the State of Alaska and the North Slope producers come to some agreement on an Alaska natural gas pipeline, a number of other issues will remain unresolved. One issue is whether the State should be an equity investor and owner of the pipeline [*105*]. Another involves the issuing of environmental permits for the pipeline route, a process that has been contentious for other pipeline projects, sometimes resulting in significant delays.

A third issue is who will construct, own, and operate the portion of an Alaska natural gas pipeline that runs through Canada. TransCanada Pipelines maintains that it has the legislated right to be the owner and operator of the Canadian portion, as specified in Canada's Northern Pipeline Act of 1978 [*106*]. Finally, the pipeline's regulatory framework could prove contentious. For the portion located within the confines of the State, Alaska's Regulatory Commission will have jurisdiction over rates and tariffs, including the terms and conditions associated with third-party access to the pipeline. These other issues will not be fully addressed until after all the issues between the State and the North Slope producers have been resolved, and it is not clear how contentious the issues will be or how quickly they can be settled.

Construction Costs and Natural Gas Prices

In May 2002, the three primary Alaska North Slope producers estimated the cost of construction for a proposed southern route pipeline to the Chicago area and its associated facilities at approximately $19.4 billion [*107*]. On the basis of that capital cost, they estimated a pipeline transportation tariff of $2.39 per thousand cubic feet for natural gas moving from the North Slope to Chicago. From May 2002 to June 2006, however, iron and steel prices increased by 72 percent [*108*]. Although it has been estimated that only 25 percent of the total pipeline cost would be associated with steel pipe, construction costs have been increasing across the board, as equipment, labor, and contractor costs have also risen.

A Federal law enacted in 2004 permits the Secretary of Energy to issue Federal loan guarantees for the construction of an Alaska natural gas pipeline. The guarantees would be limited to 80 percent of the pipeline's total cost, up to a maximum of $18 billion. Because the Federal loan guarantees would lower the risk associated with recovery of the project's capital costs, pipeline sponsors would be able to secure debt financing at a lower interest rate than they could in the absence of such guarantees, and the pipeline's financial viability would be enhanced.

Recent increases in natural gas prices, which began in 2000, have also improved the economic outlook for an Alaska natural gas pipeline. Lower 48 wellhead prices, which averaged $2.19 per thousand cubic feet in 1999, rose to an average of $7.51 per thousand cubic feet in 2005. Although prices have declined since then, the *AEO2007* reference case price projections are at a level at which an Alaska natural gas pipeline would remain economically viable if other issues surrounding the project could be resolved in a manner that met the needs of all parties. The parties would have to agree on a division of the projected benefits before the pipeline could be built.

Coal Transportation Issues

Most of the coal delivered to U.S. consumers is transported by railroads, which accounted for 64 percent of total domestic coal shipments in 2004 [*109*]. Trucks transported approximately 12 percent of the coal consumed in the United States in 2004, mainly in short hauls from mines in the East to nearby coal-fired electricity and industrial plants. A number of minemouth power plants in the West also use trucks to haul coal from adjacent mining operations. Other significant modes of coal transportation in 2004 included conveyor belt and slurry pipeline (12 percent) and water transport on inland waterways, the Great Lakes, and tidewater areas (9 percent) [*110*].

Rail is particularly important for long-haul shipments of coal, such as the transport of subbituminous coal from mines in Wyoming to power plants in the eastern United States. In 2004, rail was the primary mode of transportation for 98 percent of the coal shipped from Wyoming to customers in other States.

Rail Transportation Rates

When the railroad industry was deregulated in the early 1980s, consumers benefited from a long period of declining coal transportation rates. For coal shipments to electric utilities, rates in constant dollars per ton fell by 42 percent from 1984 to 2001 [*111*]. More recently, railroads have been raising base transportation rates and implementing fuel surcharge programs. There are also concerns that railroads are failing to meet their common carrier obligation with regard to reliability of service [*112*].

The national average rate for coal transportation in 2005 was approximately 6 percent higher (in constant dollars) than in 2004 [*113*]; and according to BNSF, average revenue per car in the first 6 months of 2006 was 7 percent higher than in the same period of 2005 as a result of contract rate escalations, fuel surcharges, and increases in hauling distances [*114*]. Recent increases in rates have caused shippers to question their fairness and to raise the possibility that the railroads may be exercising market power. Since deregulation, four railroads have dominated rail transportation of coal: CSX Transportation (CSX) and NS in the East and UP and BNSF in the West.

The concentration of coal freight business among a few carriers has led to claims of pricing power, in particular from coal shippers that have no alternative to relying on a single railroad. In 2004, when both UP and BNSF made their rates public by posting them on their web sites, some called it price collusion, in that the two companies could see each other's rates and, potentially, harmonize them. In February 2005, the U.S. Department of Justice initiated an investigation of their pricing activities. In October 2006, while not drawing any conclusions, the Government Accountability Office recommended that the state of competition in the freight railroad industry be analyzed [*115*].

The U.S. Department of Transportation's Surface Transportation Board (STB) has also been asked to review the reasonableness of rates imposed on some captive customers. Typically, for a rate case to be brought before the STB, there must be evidence suggesting not only that the railroads charge more than 180 percent of their variable cost to the captive shipper but also that construction of a new rail line to serve the captive customer's needs would be more economical than the prices currently charged. In cases decided from 2004 through June 2006, one showed an unreasonable rate, three were settled voluntarily, and two were decided in favor of the railroads [*116*]. Because concerns have been raised about the cost and time involved in preparing rate cases, the STB instituted a series of rulemakings in 2006 to improve the process by modifying its methods and procedures for large rail rate disputes and revising its simplified guidelines for smaller rate disputes.

A number of factors, including railroad profitability, the need for more investment, and increased fuel expenses in recent years, may be contributing to the recent increase in coal transportation rates. One motive for price increases by the railroads is to improve their rate of return on investment. The STB identifies a railroad as "revenue adequate" if its return on investment exceeds the industry's average cost of capital, as estimated by the STB. By this standard, only NS was considered revenue adequate in 2004 and 2005, whereas none of the railroads was considered revenue adequate in 2003 [*117*].

The railroads have argued that, after deregulation, savings resulting from consolidation of redundant infrastructure were passed on to their customers, but that such savings are no longer attainable. Instead, they typically state that higher prices are needed to add infrastructure in order to keep pace with demand. Most recently, each of the railroads has instituted a fuel surcharge program in response to rising fuel prices. The surcharge programs have been cited by many of the railroads as a success, and they have contributed to record-breaking profits. UP, for instance, reported profits for the fourth quarter of 2005 that were triple those of the fourth quarter of 2004 [*118*]. Some rail customers in the coal industry have in turn claimed that the railroads are "double dipping," recovering more through the surcharges than they spend on fuel.

The railroads have maintained that their fuel surcharge programs are transparent, but most customers appear to disagree. Each of the railroads has implemented its program differently, choosing different fuel price targets and thresholds that trigger the surcharge. For instance, BNSF and UP use EIA's on-highway diesel price as the basis for determining whether a fuel surcharge will be implemented, whereas NS and CSX use the WTI crude oil price. As of July 1, 2006, NS was applying a surcharge when the monthly WTI average price exceeded $64 per barrel [*119*]. CSX begins its price adjustments when the WTI price reaches $23.01 per barrel [*120*].

The STB has stated that the surcharge programs, while not unreasonable, were implemented in an unreasonable manner that lacked transparency. It simultaneously recommended the use of a program that would be linked more tightly to actual fuel usage and would require all carriers to use the same fuel index [*121*]. The response from the railroads has been mixed, with BNSF stating that the STB lacks authority to make a ruling unless a formal shipper's complaint is brought forward [*122*] and CSX expressing a willingness to comply "under future guidance from the STB" [*123*].

Wyoming Powder River Basin

One of the most important U.S. coal-producing areas is Wyoming's Powder River Basin. Almost all the coal produced there is carried out by rail, and disruptions in the rail transportation network can have significant effects on the flow of coal from the region. Key factors that can lead to disruptions include the need to perform major maintenance on important segments of a rail corridor and the development of bottlenecks due to unforeseen growth in the demand for rail transportation services. The problems that arose in the Powder River Basin in 2005 and 2006 illustrate the potential impact of these factors.

In May 2005, adverse weather conditions and accumulated coal dust in the roadbed of the Joint Line railroad combined to create track instability that contributed to two train derailments. The Joint Line Railroad, a 103-mile stretch of dedicated coal railway, is jointly owned and operated by BNSF and UP. It serves 8 of the 14 active coal mines in Wyoming's Powder River Basin and is one of the most heavily used sections of rail line in the world.

During 2005 and 2006, coal shippers expressed their concerns about operating conditions on the Joint Line in testimony before both houses of Congress and the FERC. Some power plant operators indicated that inadequate shipments of coal from the Powder River Basin had forced them to draw down their on-site stockpiles of coal to unprecedented levels in early to mid-2006. Others said they were forced to dispatch more expensive generating capacity, purchase electricity from other generators to meet customer demand, or buy high-priced coal on the spot market or from offshore suppliers. In testimony before the U.S. Senate in May 2006, EIA indicated that monthly data reported by electric power plants did show a drop in inventories of subbituminous coal (most of which comes from Wyoming) from mid-2005 through early 2006, consistent with press reports that generators relying on subbituminous coal were taking steps to conserve coal supplies [*124*].

A study recently produced for the U.S. Bureau of Land Management found that capacity utilization of the Joint Line in 2003 exceeded 88 percent, as compared with 22 percent for the BNSF rail line that served five active Wyoming mines north of the Joint Line in 2003 (Wyodak, Dry Fork, Rawhide, Eagle Butte, and Buckskin). The combined output of those mines has increased significantly, from 55 million tons in 2003 to 65 million tons in 2005, and is likely to surpass 70 million tons in 2006. As a result, utilization of the BNSF line is now slightly higher than it was in 2003. The mines served by the Joint Line produced and shipped 325 million tons of coal in 2005, accounting for 29 percent of the year's total U.S. coal production. Joint Line shipments for the year were 3 million tons higher than in 2004 but still 20 million tons less than had been planned [*125*].

BNSF and UP have completed maintenance work related to the 2005 train derailments and have embarked on major upgrades to increase haulage capacity on the Joint Line; however, demand in 2006 was expected to exceed the capability of the railroads and mines to supply coal from the area to the market. In mid-2006, a representative from BNSF indicated that the potential demand for Powder River Basin coal for the year probably would exceed supply by 20 to 25 million tons [*126*]. Through August 2006, coal shipments on the Joint Line were 9 percent higher than in the same period of 2005, corresponding to an annualized increase of approximately 25 million tons.

Beyond 2006, investments in new track and rail equipment for the Joint Line indicate an improved outlook for shipping capacity. Recently announced plans for investments in 2005 through 2007, totaling about $200 million, will add nearly 80 miles of third and fourth mainline track to the Joint Line, increasing annual shipping capacity to almost 420 million tons [*127*]. In a recent study for BNSF and UP, the consulting firm CANAC identified investments that could further increase the Joint Line's capacity to approximately 500 million tons by 2012 [*128*]. The potential increase in shipments was arrived at through discussions with individual mine operators along the Joint Line. According to the study, an additional 80 million tons of shipping capacity after 2007 would require the construction of 12 new loading spots at mines and 45 additional miles of mainline track. Also key to meeting the target of 500 million tons is the expectation that railroads will be able to move gradually to longer trains over the next few years, from current lengths of 125 to 130 cars to approximately 150 cars [*129*].

The authors of the CANAC report indicated that the timing of investments will depend on the market for Powder River Basin coal in coming years and could deviate from the schedule outlined. Although production from mines on the Joint Line were not explicitly modeled by EIA, the projected growth of coal production from Wyoming's Powder River Basin in the *AEO2007* reference case is not inconsistent with the expansion potential identified in the CANAC report. In all the cases modeled for *AEO2007*, the projected increase in annual coal production from active mines in Wyoming's Powder River Basin is less than 175 million tons (the sum of Joint Line expansion projects identified in the report) until after 2019.

Another potential investment under consideration is an expansion of the Dakota Minnesota & Eastern Railroad (DM&E) westward to the Powder River Basin. The project would include 280 miles of new construction and provide an alternative rail option for Wyoming coal. It would provide access to the mines currently active south of Gillette, Wyoming, and would be independent of the existing Joint Line [*130*]. The extension would provide enough rail capacity for the transport of 100 million tons of coal annually according to DM&E, which is seeking a loan from the Federal Railroad Administration to support it.

Coal Production and Consumption Projections in AEO2007

In the *AEO2007* reference case, coal remains the primary fuel for electricity generation through 2030. Coal production is projected to increase significantly, particularly in the Powder River Basin. From 2005 to 2030, production in the Wyoming Powder River Basin is projected to grow by 289 million tons, but the projected annual increases do not exceed 30 million tons. The resulting increase in coal transport requirements is not beyond the level of expansion projects currently being discussed.

The Rocky Mountain, Central West, and East North Central regions are projected to show the largest increases in coal demand, by about 100 million tons each, from 2005 to 2030. The majority of the coal delivered to the Rocky Mountain region is projected to continue to come from Colorado and Utah. In addition, most of the growth in the region is projected to come from new plants that are likely to be built as close as possible to supply sources, potentially reducing the need for extensive new development of rail infrastructure. At a minimum, new plants will be located only after careful consideration of transportation options, to reduce the potential for rail bottlenecks. For the Central West region, 42 percent of the increase in coal demand is projected to be supplied by Wyoming Powder River Basin coal; however, the largest supply increase (meeting 55 percent of the region's total increase in demand) is projected to come from the Dakota lignite supply region, to provide feedstocks for new CTL plants that are likely to be situated as close to their supply sources as possible.

In the East North Central region, most of the coal supply to meet the projected growth in consumption (120 million tons from 2005 to 2030) is expected to come from the Wyoming Powder River Basin. The increase in the region's demand for coal could lead to congestion on heavily traveled rail lines, such as those surrounding the Chicago area, where coal and other bulk commodities already make heavy use of the system. The strongest growth in the region's coal consumption is projected to occur between 2020 and 2025, when deliveries from Wyoming's Powder River Basin are projected to grow by 43 million tons, with the largest single-year increase being 12 million tons.

Biofuels in the U.S. Transportation Sector

Sustained high world oil prices and the passage of the EPACT2005 have encouraged the use of agriculture-based ethanol and biodiesel in the transportation sector; however, both the continued growth of the biofuels industry and the long-term market potential for biofuels depend on the resolution of critical issues that influence the supply of and demand for biofuels. For each of the major biofuels—corn-based ethanol, cellulosic ethanol, and biodiesel—resolution of technical, economic, and regulatory issues remains critical to further development of biofuels in the United States.

In the transportation sector, ethanol is the most widely used liquid biofuel in the world. In the United States, nearly all ethanol is blended into gasoline at up to 10 percent by volume to produce a fuel called E10 or "gasohol." In 2005, total U.S. ethanol production was 3.9 billion gallons, or 2.9 percent of the total gasoline pool. Preliminary data for 2006 indicate that ethanol use rose to 5.4 billion gallons. Biodiesel production was 91 million gallons, or 0.21 percent of the U.S. distillate fuel oil market, including diesel, in 2005 (Table 11). All cars and light trucks built for the U.S. market since the late 1970s can run on the ethanol blend E10. Automakers also produce a limited number of FFVs for the U.S. market that can run on any blend of gasoline and ethanol up to 85 percent ethanol by volume (E85). Because auto manufacturers have been able to use FFV sales to offset CAFE requirements, more than 5 million FFVs were produced for the U.S. market from 1992 through 2005. E10 fuel is widely available in many States. E85 has limited availability, at stations clustered mostly in the midwestern States.

In the *AEO2007* reference case, ethanol use increases rapidly from current levels. Ethanol blended into gasoline is projected to account for 4.3 percent of the total gasoline pool by volume in 2007, 7.5 percent in 2012, and 7.6 percent in 2030. As a result, gasoline demand increases more rapidly in terms of fuel volume (but not in terms of energy content) than it would in the absence of ethanol blending. Overall, gasoline consumption is projected to increase by 32 percent on an energy basis, and by 34 percent on a volume basis, from 2007 to 2030.

Ethanol can be produced from any feedstock that contains plentiful natural sugars or starch that can be readily converted to sugar. Popular feedstocks include sugar cane (Brazil), sugar beets (Europe), and maize/corn (United States). Ethanol is produced by fermenting sugars. Corn grain is processed to remove the sugar in wet and dry mills (by crushing, soaking, and/or chemical treatment), the sugar is fermented, and the resulting mix is distilled and purified to obtain anhydrous ethanol. Major byproducts from the ethanol production process include dried distillers'

Table 11. U.S. motor fuels consumption, 2000-2005 (million gallons per year)

	Gasoline	*Ethanol*	*Percent of gasoline pool*
2000	*128,662*	*1,630*	*1.27*
2001	*129,312*	*1,770*	*1.37*
2002	*132,782*	*2,130*	*1.60*
2003	*134,089*	*2,800*	*2.09*
2004	*137,022*	*3,400*	*2.48*
2005	*136,949*	*3,904*	*2.85*
	Diesel	***Biodiesel***	***Percent of diesel fuel pool***
2000	*37,238*	*—*	*—*
2001	*38,155*	*9*	*0.02*
2002	*38,881*	*11*	*0.03*
2003	*40,856*	*18*	*0.04*
2004	*42,773*	*28*	*0.07*
2005	*43,180*	*91*	*0.21*

grains and solubles (DDGS), which can be used as animal feed. On a smaller scale, corn gluten meal, gluten feed, corn oil, CO_2, and sweeteners are also byproducts of the ethanol production process used in the United States.

With additional processing, plants and other biomass residues (including urban wood waste, forestry residue, paper and pulp liquors, and agricultural residue) can be processed into fermentable sugars. Such potentially low-cost resources could be exploited to yield significant quantities of fuel-quality ethanol, generically termed "cellulosic ethanol." Cellulose and hemicellulose in biomass can be broken down into fermentable sugars by either acid or enzymatic hydrolysis. The main byproduct, lignin, can be burned for steam or power generation. Alternatively, biomass can be converted to synthesis gas (hydrogen and carbon monoxide) and made into ethanol by the Fischer-Tropsch process or by using specialized microbes.

Capital costs for a first-of-a-kind cellulosic ethanol plant with a capacity of 50 million gallon per year are estimated by one leading producer to be $375 million (2005 dollars) [*131*], as compared with $67 million for a corn-based plant of similar size, and investment risk is high for a large-scale cellulosic ethanol production facility. Other studies have provided lower cost estimates. A detailed study by the National Renewable Energy Laboratory in 2002 estimated total capital costs for a cellulosic ethanol plant with a capacity of 69.3 million gallons per year at $200 million [*132*]. The study concluded that the costs (including capital and operating costs) remained too high in 2002 for a company to begin construction of a first-of-its-kind plant without significant short-term advantages, such as low costs for feedstocks, waste treatment, or energy.

If future oil prices follow a path close to that in the *AEO2007* reference case, significant reductions in the capital cost and operating costs of a cellulosic ethanol plant will be needed for cellulosic ethanol to be economically competitive with petroleum-based fuels. The extent to which costs can be reduced through a combination of advances in the production process for cellulosic ethanol and learning as plants are constructed in series will be important to the future competitiveness of cellulosic ethanol. World oil price developments also will play a central role.

Currently, no large-scale cellulosic ethanol production facilities are operating or under construction. EPACT2005 provides financial incentives that in the *AEO2007* reference case are projected to bring the first cellulosic ethanol production facilities on line between 2010 and 2015, with a total capacity of 250 million gallons per year. Cellulosic ethanol currently is not cost-competitive with gasoline or corn-based ethanol, but considerable R&D by the National Renewable Energy Laboratory and its partners has significantly reduced the estimated cost of enzyme production. Although technological breakthroughs are inherently unpredictable, further significant successes in R&D could make cellulosic ethanol a viable economic option for expanded ethanol production in the future.

Biodiesel is a renewable-based diesel substitute used in Europe with early commercial market development in the United States. Biodiesel is composed of mono-alkyl esters of long-chain fatty acids derived from vegetable oils or animal fats [*133*]. It is similar to distillate fuel oil (diesel fuel) and can be used in the same applications, but it has different chemical, handling, and combustion characteristics. Biodiesel can be blended with petroleum diesel in any fraction and used in compression-ignition engines, so long as the fuel system that uses it is constructed of materials that are compatible with the blend. The high lubricity of biodiesel helps to offset the impact of adopting low-sulfur diesel.

Common blends of biodiesel are 2 percent, 5 percent, and 20 percent (B2, B5, and B20). Individual engine manufacturers determine which blends are warranted for use in their engines, but generally B5 blends are permissible and some manufacturers support B20 blends. Blends of biodiesel are distributed at stations throughout the United States. Some States have mandated levels of biodiesel use when in-State production reaches prescribed levels.

Predominant feedstocks for biodiesel production are soybean oil in the United States, rapeseed and sunflower oil in Europe, and palm oil in Malaysia. Biodiesel also can be produced from a variety of other feedstocks, including vegetable oils, tallow and animal fats, and restaurant waste and trap grease. To produce biodiesel, raw vegetable oil is chemically treated in a process called transesterification. The properties of the biodiesel (cloud point, pour point, and cetane number) depend on the type of feedstock used. Crude glycerin, a major byproduct of the reaction, usually is sold to the pharmaceutical, food, and cosmetic industries.

Energy Content and Fuel Volume

On a volumetric basis, ethanol and biodiesel have lower energy contents than do gasoline and distillate fuel oil, respectively. Table 12 compares the energy contents of various fuels on the basis of Btu per gallon and gallons of gasoline equivalent. The table shows both the low heating value (the amount of heat released by the fuel, ignoring the latent heat of vaporization of water) and the high heating value (the amount of heat released by the fuel, including the latent heat of vaporization of water). The lower energy content of ethanol and biodiesel generally results in a commensurate reduction in miles per gallon when they are used in engines designed to run on gasoline or diesel. Small-percentage blends of ethanol and biodiesel (E10, B2, and B5) result in smaller losses of fuel economy than do biofuel-rich blends (E85 and B20).

Today, most fuel ethanol is used in gasoline blends, where it accounts for as much as 10 percent of each gallon of fuel—a level that all cars can accommodate. In higher blends, ethanol can make up as much as 85 percent of each gallon of fuel by volume. In the future, increased use of ethanol as a transportation fuel will raise the issue of fuel volume versus energy content. Ethanol contains less energy per gallon than does conventional gasoline. A gallon of ethanol has only two-thirds the energy of a gallon of conventional gasoline, and the number of miles traveled by a given vehicle per gallon of fuel is directly proportional to the energy contained in the fuel.

E10 (10 percent ethanol) has 3.3 percent less energy content per gallon than conventional gasoline. E85 (which currently averages 74 percent ethanol by volume) has 24.7 percent less energy per gallon than conventional gasoline. *AEO2007* assumes that engine thermal efficiency remains the same whether the vehicle burns conventional gasoline, E10, or E85. This means that 1.03 gallons of E10 or 1.33 gallons of E85 are needed for a vehicle to cover the same distance that it would with a gallon of conventional gasoline. Although the difference is not expected to have a significant effect on purchases of E10, *AEO2007* assumes that motorists whose vehicles are able to run on E85 or conventional gasoline will compare the two fuels on the basis of price per unit of energy.

The issue of gasoline energy content first arose in the early 1990s with the introduction of oxygenated gasoline made by blending conventional gasoline with 15 percent MTBE or 7.7 percent ethanol by volume. When oxygenated gasoline was introduced, MTBE was the blending agent of choice. Since then, ethanol has steadily replaced MTBE in oxygenated and RFG blends. The fuel economy impact of switching from MTBE-blended gasoline to an ethanol blend is smaller than the impact of switching from conventional gasoline. For example, changing from 15 percent MTBE to 7.7 percent ethanol in blended gasoline results in a reduction in energy content of only 1.2 percent per gallon of fuel, and changing from 15 percent MTBE to 10 percent ethanol results in a reduction of 1.9 percent.

Current State of the Biofuels Industry

The nascent U.S. biofuel industry has recently begun a period of rapid growth. Over the past 6 years, biofuel production has been growing both in absolute terms and as a percentage of the gasoline and diesel fuel pools (see Table 11). High world oil prices, firm government support, growing environmental and energy security concerns, and the availability of low-cost corn and soybean feedstocks provide favorable market conditions for biofuels. Ethanol, in particular, has been buoyed by the need to replace the octane and clean-burning properties of MTBE, which has been removed from gasoline because of concerns about groundwater contamination. About 3.9 billion gallons of ethanol and 91 million gallons of biodiesel were produced in the United States in 2005. According to estimates based on the number of plants under construction, ethanol production capacity could rise to about 7.5 billion gallons and biodiesel capacity to about 1.1 billion gallons by 2008, possibly resulting in excess capacity in the near term (Figure 22).

Table 12. Energy content of biofuels

Fuel	Btu per gallon (low heating value)	Btu per gallon (high heating value)	Gallons of gasoline equivalent (high heating value)
Conventional gasoline	115,500	125,071	1.00
Fuel ethanol (E100)	76,000	84,262	0.67
E85 (74% blend on average)	—	94,872	0.76
Distillate fuel oil (diesel)	128,500	138,690	1.11
Biodiesel (B100)	118,296	128,520	0.95

The American Jobs Creation Act of 2004 established and extended blender's tax credits to reduce the final cost (in nominal terms) of pure ethanol by $0.51 per gallon, biodiesel made from virgin oil by $1.00 per gallon, and biodiesel made from waste grease by $0.50 per gallon [*134*]. The national RFS legislated in EPACT2005 provides biofuels with a reliable market of at most 7.5 billion gallons annually by 2012. Ethanol fuel is expected to fulfill most of the RFS requirement.

In the *AEO2007* reference case, ethanol demand is projected to exceed the applicable RFS requirements between now and 2012, because of the need for ethanol as a fuel oxygenate to meet Federal gasoline specifications and as an octane enhancer and because of the blender's tax credit. Ethanol consumption is projected to rise to 11.2 billion gallons, representing 7.5 percent of the gasoline pool, by volume, in 2012. Current and projected real oil prices far above those experienced during the 1990s, coupled with the availability of significant tax incentives and the RFS requirement have created a favorable market for biofuels. Accelerated investments in biofuel production facilities and rapid expansion of existing capacity underscore the attractiveness of biofuel investments.

Short-run production costs, which include feedstock costs, cash operating expenses, producer subsidies, and byproduct credits but exclude capital costs, transportation fees, tax credits, and fuel taxes, vary considerably according to plant size, design, and feedstock supply. Assuming corn prices of about $2 per bushel and excluding capital costs, corn-based ethanol can be produced by the dry-milling process for approximately $1.00 to $1.06 per gallon (2005 dollars) or $11.90 to $12.60 per million Btu [*135*, *136*]. Corn prices spiked to well above that level in 2006 because of tightness in the supply-demand balance for corn, caused by farmers' removing about 3 million acres from corn production and using it for soybean production instead.

Biodiesel can be produced from soybean oil for $1.80 to $2.40 per gallon ($15.20 to $20.30 per million Btu) and from yellow grease for $0.90 to $1.10 per gallon ($7.60 to $9.30 per million Btu) [*137*, *138*]. Feedstock costs for virgin soybean oil, which are dictated by commodity markets and vary between $0.20 and $0.30 per pound, constitute 70 to 78 percent of final production costs. Non-virgin feedstocks generally are cheaper, ranging from virtually no cost (for reclaimed restaurant trap grease) to 70 percent of the final production cost. For the production costs calculated above, virgin soybean oil was assumed to cost $0.26 per pound, and yellow grease was valued at 50 percent of the cost of an equivalent amount of soybean oil.

When the blender's tax credit for ethanol and biodiesel is subtracted from the wholesale prices (which include capital recovery and transportation fees), biofuels are price competitive with petroleum fuels on a volumetric basis [*139*]. Figure 23 compares the rack price of ethanol (including the blender's tax credit) with the price of unleaded gasoline. The "rack price" is defined as the wholesale price of ethanol fuel where title is transferred at the terminal.

Profitability in the biofuels industry depends heavily on the cost of feedstocks. For ethanol, corn feedstock made up nearly 57 percent of the total production cost in 2002 [*140*]. For biodiesel, soybean oil makes up 70 to 78 percent of the total production cost [*141*, *142*]. Fluctuations in the price of either feedstock can have dramatic effects on the production costs, and the industry assumes considerable market risk by relying on a limited array of feedstocks.

Figure 22. U.S. ethanol production and production capacity, 1999-2007 (billion gallons)

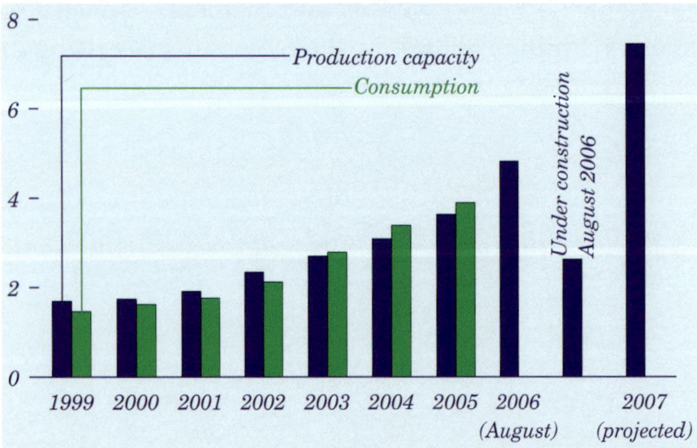

Figure 23. Average U.S. prices for ethanol and gasoline, 2003-2006 (nominal cents per gallon)

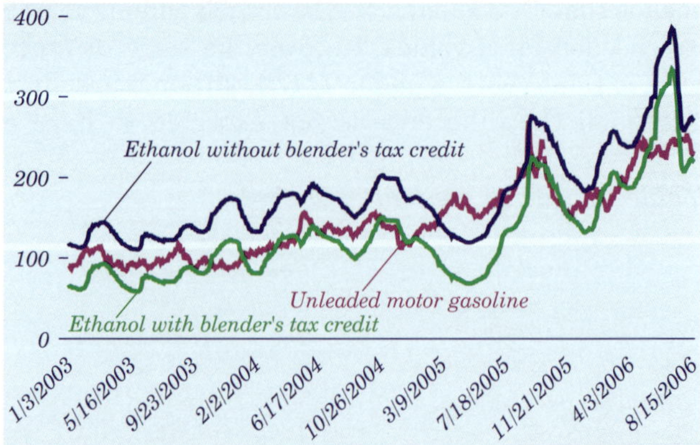

The U.S. ethanol industry relies almost exclusively on corn, consuming 20 percent of the available corn supply in 2006 [*143*]. At current production levels, corn—which is produced domestically in large volumes—is the most attractive feedstock for ethanol. As ethanol production increases, competition for corn supplies among the fuel, food, and export markets, along with a decline in the marginal value of ethanol co-products, is expected to make production more expensive [*144*].

Assuming the development of cost-effective production facilities, cellulosic biomass feedstocks like switchgrass, agricultural residues, and hybrid poplar trees could supply a growing ethanol industry with large quantities of less expensive raw materials. To differentiate the current use of corn with the future use of cellulosic biomass and the differences in production technology, corn is generally characterized as a "first generation" energy crop, whereas switchgrass and other cellulosic materials are "second generation" energy crops.

The U.S. biodiesel industry relies almost exclusively on soybean oil as a feedstock. Soybean oil has historically been a surplus product of the oilmeal crushing industry, available in large quantities at relatively low prices. At production levels nearing 300 to 600 million gallons of biodiesel per year (less than 2 percent of the diesel fuel pool), the marginal cost of using soybean oil as a feedstock rises to the point where other oilseeds—canola, rapeseed, sunflower, and cottonseed—become viable feedstocks [*145*]. There are no significant differences in processing for the numerous biodiesel feedstocks, and they cannot easily be grouped into first- and second-generation categories. The major differences among biodiesel feedstocks are regional availability, co-product value, and the composition of fatty acids in the refined vegetable oil.

Resource Utilization and Land Availability

Currently, corn and soybean feedstocks for biofuels are grown almost exclusively on prime agricultural land in the Midwest. Increases in the supply of biofuel feedstocks could come from a combination of three strategies: increasing the amount of land used as cropland, boosting the yields of existing energy crops, and replacing or supplementing corn with cellulosic biomass and soybeans with oilseeds more appropriate for biodiesel production. All three strategies may be required to overcome the constraints of currently available feedstocks and sustain biofuel production levels that could displace at least 10 percent of gasoline consumption.

According to the most recent Agricultural Census (2002), the amount of cropland available in the lower 48 United States is 434 million acres [*146*], or 23 percent of the total land area [*147*]. The total amount of cropland—defined as the sum of land used for crops, idle land, and pasture—has been declining for the past 50 years and, increasingly, is becoming concentrated in the Midwest. The trend is expected to continue as population pressure leads to permanent conversion of some agricultural lands to other uses. It is unlikely that additional cropland will be added in the United States to accommodate increases in the demand for biofuels. Instead, the cultivation of biofuels will compete with other agricultural uses, such as pastureland and idle land, much of which is in the Conservation Reserve Program (CRP) [*148*].

The potential use of CRP acreage to grow corn and soybeans is constrained by productivity, environmental, and contractual limitations. Nevertheless, there may be significant opportunities in the future to use some CRP acres to grow such "low-impact" energy crops as native grasses (switchgrass) and short-rotation trees (willows or poplars) to generate cellulosic biomass. Pilot programs are underway in Minnesota, Iowa, New York, and Pennsylvania to determine whether CRP acres can be used to grow energy crops while preserving the environmental mandate of the CRP.

Land Use and Productivity

With a limited supply of cropland available for biofuel feedstocks, increasing yield (bushels per acre) on an annual basis could significantly boost available supplies of corn and soybeans without requiring additional land. With more than 81 million acres devoted to corn and nearly 72 million acres devoted to soybeans (2005 U.S. planted acres), even small increases in annual yield could boost supplies significantly [*149*].

There have been large annual increases in yields of both corn and soybeans over the past 30 years. Corn yields increased from 86.4 bushels per acre in 1975 to 151.2 bushels per acre in 2006, and soybean yields increased from 28.9 bushels per acre to 43 bushels per acre over the same period [*150*]. If corn yields continue to increase at the same rate (approximately 1.8 bushels per acre per year), production could increase by more than 3.1 billion bushels (29 percent) by 2030 without requiring any additional acreage. Similarly, soybean production could increase by nearly 1.0 billion bushels per year by 2030 with no additional acreage requirement if yields continue to grow at the rate of 0.5 bushels per acre per year [*151*]. Improvements in biofuel collection and refining and bioengineering of corn and soybeans also could contribute to improved biofuel yields. Research on methods to increase the starch content of corn and the oil content of soybeans is also ongoing.

Crop Competition

A key uncertainty is the availability of sufficient land resources for large-scale expansion of the cultivation of biofuel crops, given the intense competition with conventional agricultural products for arable land. Competition will favor those crops most profitable for farmers, accounting for such factors as growing region, farming practice, and soil type. Currently, corn and soybeans are competitive energy crops, because they provide high value to farmers at prices low enough to allow the biofuel industry to produce a product competitive with petroleum fuels.

Cellulosic biomass from switchgrass, hybrid willow and poplar trees, agricultural residues, and other sources has significant supply potential, possibly up to 4 times the potential of corn [*152*]. Switchgrass and poplars could be grown on CRP lands, where corn cannot be grown economically, but they would not be competitive with corn until corn prices rose or the capital and non-feedstock production costs of cellulosic ethanol were significantly reduced. To expand beyond a production level of 15 to 20 billion gallons per year without seriously affecting food crop production and prices, the industry must make a transition to crops with higher yields per acre and grow crops in an environmentally permissible manner on CRP lands, while continuing to provide profits for producers.

Role of Co-products in Biofuel Economics

The value of co-products will play a significant role in determining which crops are most profitable for farmers to grow and biofuel producers to use. High prices for raw crop material are desirable for farmers but undesirable for biofuel producers. High prices for co-products, on the other hand, increase revenues for agricultural processors, sustain high prices for raw crop materials, and offset feedstock costs for biodiesel producers. Corn and soybeans not only provide starch and oil for biofuel production but also generate significant quantities of co-products, such as DDGS, gluten feed, gluten meal, corn oil, and soybean oil meal with high protein content (Table 13). As a result, corn grain and soybean oil can be offered at prices lower than those of other feedstocks, and currently they are the most competitive biofuel crops.

Co-products of the 3.9 billion gallons of ethanol produced in 2005 were significant, including 10 million short tons of DDGS, 473,000 short tons of corn gluten meal, 2.6 million short tons of corn gluten feed, and

Table 13. U.S. production and values of biofuel co-products

Biofuel feedstock	*Co-products*	*Volume produced (pounds per 100 pounds of feedstock)*	*Approximate value (dollars per pound)*
Ethanol			
Corn, wet mill	*Corn gluten feed*	*24.0*	*0.033*
	Corn gluten meal	*4.5*	*0.135*
	Corn oil	*2.9*	*0.260*
Corn, dry mill	*Dried distillers' grains and solubles*	*30.5*	*0.045*
Sugar	*Sugar stalks, bagasse*	*27.0*	—
Cellulosic ethanol			
Switchgrass *Hybrid poplar* *Forest residue* *Agricultural residue*	*Lignin*	*27.0*	—
Biodiesel			
Soybeans	*Meal (44-48% protein)*	*80-82*	*0.097*
Canola	*Meal (28-36% protein)*	*60-62*	*0.079*
Sunflower	*Meal (28% protein)*	*60-63*	*0.035*
Mustard	*Meal (28-36% protein)*	*60-62*	—
Cotton	*Meal (41% protein)*	*84-86*	*0.088*
	Crude glycerin	*10*	*0.050*

283,000 short tons of corn oil [*153*]. As biofuel production continues to expand to the level of 7.5 billion gallons per year mandated in EPACT2005, production of DDGS, used primarily as animal feed, will grow to more than 12 million short tons annually and may depress prices in the feed market.

Biodiesel production in 2005 was considerably less than ethanol production, at 90.8 million gallons. Because U.S. biodiesel production currently uses surplus soybean oil (generated as a co-product in the soybean meal industry), it has little effect on other markets for soybeans; however, annual production of 300 to 600 million gallons of biodiesel would begin to compete with food and feed markets for soybeans [*154*]. For every 100 pounds of biodiesel production, about 10 pounds of crude glycerin is generated as a co-product [*155*]. The glycerin generated by a 300 to 600 million gallon per year biodiesel industry could displace nearly one-half of the 692 million pounds of glycerin produced domestically in North America [*156*] and result in substantial oversupply.

Market Effects of Biofuel Growth

The feedstocks used to produce biofuels currently make up only 15 percent of available crop matter and are located at the end of a long agricultural supply chain. The markets for biofuels, biofuel co-products, and crop commodities are linked and susceptible to changes in the prices and availability of crops. Surging demand for biofuel feedstocks is likely to exert upward price pressure on corn and soybean commodities and influence export, food, and industrial feedstock markets, particularly in the short term.

Co-product production also increases with biofuel production. At higher levels of biofuel production in the future, co-products may be oversupplied, resulting in depressed prices for the co-products and lower revenues from their sale to offset fuel production costs. Finding new, high-value uses for co-products could ensure that market prices for co-products remain stable. To the extent that other energy crops, such as switchgrass and inedible oilseeds, could be grown on less productive land (like the CRP), upward pressure on the prices of corn, soybeans, and other high-value food crops could also be mitigated.

Some studies have suggested that up to 16 billion gallons of ethanol (slightly more than 10 percent of the total gasoline pool by volume) can be produced from corn in 2015 without adversely affecting the price of corn and upsetting domestic food, feed, and export markets [*157*]. A growing corn supply—the result of increasing yields and relatively slow growth in the demand for corn in the food, feed, and export markets—contributes to stable corn prices [*158*]. Between 33 and 38 percent of domestic corn production would be needed to produce 12 to 16 billion gallons of ethanol in 2015/2016, as compared with the 14.6 percent of domestic production that was used for ethanol feedstocks in 2005 [*159*].

Biofuel Distribution Infrastructure

Another issue that could limit the growth of the U.S. biofuels industry is development of the necessary infrastructure for collecting, processing, and distributing large volumes of biofuels. Currently, nearly all U.S. biofuel production facilities are located close to corn and soybean acreage in the Midwest, minimizing the transportation costs for bulky, unrefined materials. The facilities are far from the major biofuel consumption centers on the East and West Coasts. Further complicating matters is the fact that biodiesel and ethanol cannot be blended at the refinery and batched through existing pipelines. Ethanol can easily be contaminated by water, and biodiesel dissolves entrained residues in the pipelines. As a result, railroad cars and tanker trucks made from biofuel-compatible materials are needed to transport large volumes of biofuels to market.

Limited rail and truck capacity has complicated the delivery of ethanol, contributing to regional ethanol supply shortages and price spikes between April and June 2006. Feedstock and product transportation costs and concerns remain problematic for the biofuel industry and have led many biofuel producers to explore the prospect of locating near a dedicated feedstock supply or large demand center to minimize transportation costs and susceptibility to bottlenecks.

Distribution of biofuels to end-use markets is also hampered by a number of other factors. Although E10 is readily obtainable throughout the United States, there are limited numbers of fueling stations for biodiesel and E85 (Table 14). Further, some station owners may be averse to carrying B20 or E85, because

Table 14. Vehicle fueling stations in the United States as of July 2006

Fuel	*Number of stations*	*Percent of total*
All fuels	*169,000*	*100.0*
Biofuels	*1,767*	*1.0*
E85	*799*	*0.5*
Biodiesel	*968*	*0.5*

the unique physical properties of the blends may require costly retrofits to storage and dispensing equipment.

Recent EIA estimates for replacing one gasoline dispenser and retrofitting existing equipment to carry E85 at an existing fueling station range from $22,000 to $80,000 (2005 dollars), depending on the scale of the retrofit. Some newer fueling stations may be able to make smaller upgrades, with costs ranging between $2,000 and $3,000. Investment in an E85 pump that dispenses one-half the volume of an average unleaded gasoline pump (about 160,000 gallons per year) would require an increase in retail prices of 2 to 7 cents per gallon if the costs were to be recouped over a 15-year period. The costs would vary, depending on annual pump volumes and the extent of the station retrofit. The installation cost of E85-compatible equipment for a new station is nearly identical to the cost of standard gasoline-only equipment.

Independent station owners may also be uncomfortable with the relative novelty of biofuels and the murky regulatory environment that surrounds their use and distribution at retail locations. For gasoline outlets operated by major distributors, owners are more likely to be aware of the environmental regulations and more willing to seek appropriate permits when confronted with favorable biofuel economics. Awareness of various biofuels is limited, and station operators will need to post appropriate labels, placards, and warning signs to ensure that customers put the appropriate fuels in their vehicles. With the rapid growth and change in the biofuels industry, quality control programs are also critical to ensure that biofuels meet accepted quality specifications from the American Society for Testing and Materials for ethanol (ASTM D4806) and biodiesel (ASTM D6751).

Consumer Demand, Awareness, and Attitudes

Biofuel production capacity is expanding rapidly in response to heightened market demand resulting from high petroleum prices, favorable tax incentives, and consumer concerns over environmental and energy security issues. The market potential for biofuel blends (E10, B5, and B20) remains significantly larger than current production levels and will continue to absorb the biofuel supply for the foreseeable future (Table 15). Consumer behavior, however, will play an increasingly important role in determining demand for biofuels. Consumer attitudes about fuel prices, relative fuel performance, biofuel-capable vehicles, and the environment will affect the volume and type of biofuels sold.

Table 15. Potential U.S. market for biofuel blends, 2005 (billion gallons)

Fuel	Production	Motor fuel consumption	Blend	Current blend consumption
Ethanol	3.90	136.9	E10	13.70
Biodiesel	0.08	43.2	B2	0.86
			B5	2.16
			B20	8.64

Price, availability, and familiarity are the primary attributes by which many consumers judge the value of biofuels. Biofuel-rich blends, such as E85 and B20, are much less common in the United States than are petroleum-rich blends, such as gasohol (E10). Consistent with economic theories of adoption, consumers who are generally unfamiliar with biofuels have been hesitant to use them, even where they are available. On a gallon of gasoline equivalent basis, biofuels have historically been more expensive than gasoline and diesel. Because of high prices, low availability, and lack of familiarity, there has been little consumer demand for biofuels for many years. Current use of ethanol in E10 blends does not require any explicit consumer choice, because E10 and conventional gasoline have similar attributes and are rarely, if ever, offered as alternatives.

Availability of Biofuel Vehicles

The long-term market potential for biofuels will also depend on the availability of light-duty vehicles capable of using rich biofuel blends. For ethanol demand to grow beyond the market for E10, fuel containing up to 85 percent ethanol must be marketed and sold. Although the incremental cost for vehicle manufacturers to make some models E85-capable at the factory is low (about $200 per vehicle), virtually all FFVs built since 1992 have been produced for the sole purpose of acquiring CAFE credits. About 5 million FFVs have been produced since 1992. There is also no regulatory requirement that FFVs actually use E85, and buyers often are unaware that they own FFVs.

Currently, ethanol has higher value in the light-duty vehicle fuel market as a blending component in E10 than as dedicated E85 fuel. Consequently, the vast majority of the first 16 to 20 billion gallons of ethanol produced per year is projected to be used in E10. When the E10 market is nearly saturated, incremental ethanol production would presumably be consumed as E85, displacing gasoline. The issue is similar for biodiesel. For biodiesel to penetrate the light-duty vehicle fleet beyond the B10 or B5 blending levels, additional biofuel-capable vehicles must be produced and marketed to consumers. Higher consumer demand for biofuels—resulting from evolving market dynamics or government intervention—would encourage expanded production of biofuel-capable vehicles by auto manufacturers.

Market Effects of Government Policy

Federal and State government policy and regulation of biofuels will affect the development of the biofuels industry, both now and in the future. Support for biofuels has resulted in a number of Federal and State policies aimed at reducing their cost, increasing their availability, and ensuring continued market demand during periods of low petroleum prices. The RFS established by EPACT2005 guarantees a market of 7.5 billion gallons per year for ethanol by 2012, providing some long-term stability for the industry. In addition, the blender's tax credits reduce the cost of biofuels, making them more competitive with petroleum fuels. Significant funding is also provided by the Federal Government for research, development, and commercialization of cellulosic ethanol technology.

State support for biofuels varies, but many States have instituted RFSs, reduced fuel taxes, and provided grants and loans for distribution infrastructure. Hawaii, Iowa, Louisiana, Minnesota, Missouri, Montana, and Washington have enacted standards specifying that transportation fuels sold in the State contain a minimum percentage of either ethanol or biodiesel [*160*], and similar legislation has been proposed in California, Colorado, Idaho, Illinois, Indiana, Kansas, New Mexico, Pennsylvania, Virginia, and Wisconsin.

Government support has fueled the rapid growth of the biofuel industry and may have reduced long-term risk for biofuel investments. Changes in laws and regulations can have large impacts on the sector. Preliminary discussions surrounding the 2007 Farm Bill indicate that the final version may contain significant provisions related to the role of energy crops in the agricultural sector and how CRP lands can be used [*161*]. The Federal and State RFS programs may be revised as more experience is gained in their implementation and to accommodate shifts in the political and economic environment. If R&D efforts on cellulosic ethanol significantly reduce the costs of biofuels, tax and regulatory policy may need to be changed to accommodate new market realities.

Finally, Federal and State budgetary issues could affect gasoline taxes and the blender's tax credit. At levels of 16 billion gallons of ethanol and 1 billion gallons of biodiesel, the loss of Federal revenue as a result of the blender's tax credit would be roughly $8 billion for ethanol and $1 billion for biodiesel in nominal terms, as compared with a current total loss of about $2.4 billion. Increasing budgetary impacts may lead to future reconsideration of the subsidy levels.

Market Trends

The projections in the *Annual Energy Outlook 2007* are not statements of what will happen but of what might happen, given the assumptions and methodologies used. The projections are business-as-usual trend estimates, given known technology and technological and demographic trends. *AEO2007* generally assumes that current laws and regulations are maintained throughout the projections. Thus, the projections provide a policy-neutral reference case that can be used to analyze policy initiatives. EIA does not propose, advocate, or speculate on future legislative and regulatory changes. Most laws are assumed to remain as currently enacted; however, the impacts of emerging regulatory changes, when defined, are reflected.

Because energy markets are complex, models are simplified representations of energy production and consumption, regulations, and producer and consumer behavior. Projections are highly dependent on the data, methodologies, model structures, and assumptions used in their development. Behavioral characteristics are indicative of real-world tendencies rather than representations of specific outcomes.

Energy market projections are subject to much uncertainty. Many of the events that shape energy markets are random and cannot be anticipated, including severe weather, political disruptions, strikes, and technological breakthroughs. In addition, future developments in technologies, demographics, and resources cannot be foreseen with certainty. Many key uncertainties in the *AEO2007* projections are addressed through alternative cases.

EIA has endeavored to make these projections as objective, reliable, and useful as possible; however, they should serve as an adjunct to, not a substitute for, a complete and focused analysis of public policy initiatives.

Strong Economic Growth Is Expected To Continue Through 2030

Figure 24. Average annual growth rates of real GDP, labor force, and productivity, 2005-2030 (percent per year)

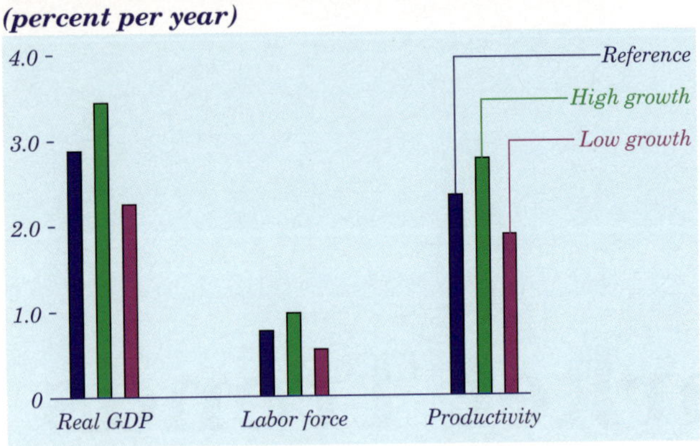

AEO2007 presents three views of economic growth for the projection period from 2005 through 2030. In the reference case, the Nation's economic growth, measured in terms of real GDP, is projected to average 2.9 percent per year (Figure 24). The labor force is projected to grow by 0.8 percent per year on average; labor productivity growth in the nonfarm business sector is projected to average 2.3 percent per year; and investment growth is projected to average 3.8 percent per year. Disposable income grows by 3.1 percent per year in the reference case and disposable income per capita by 2.3 percent per year. Nonfarm employment grows by 1.0 percent per year, while employment in manufacturing shrinks by 0.5 percent per year.

The high and low economic growth cases show the effects of alternative economic growth assumptions on the energy market projections. In the high growth case, real GDP growth is projected to average 3.4 percent per year as a result of higher assumed growth rates for the labor force (1.0 percent per year), nonfarm employment (1.3 percent), and productivity (2.8 percent). With higher productivity gains and employment growth, projected inflation and interest rates are lower than in the reference case. In the low growth case, slower growth in real GDP growth is projected, averaging 2.2 percent per year, as a result of lower assumed growth rates for the labor force (0.5 percent per year), nonfarm employment (0.6 percent per year), and productivity (1.9 percent per year). Consequently, the low growth case projects higher inflation and interest rates and slower growth in industrial output.

Inflation, Interest, and Jobless Rates Fall Below Historical Averages

Figure 25. Average annual inflation, interest, and unemployment rates, 2005-2030 (percent per year)

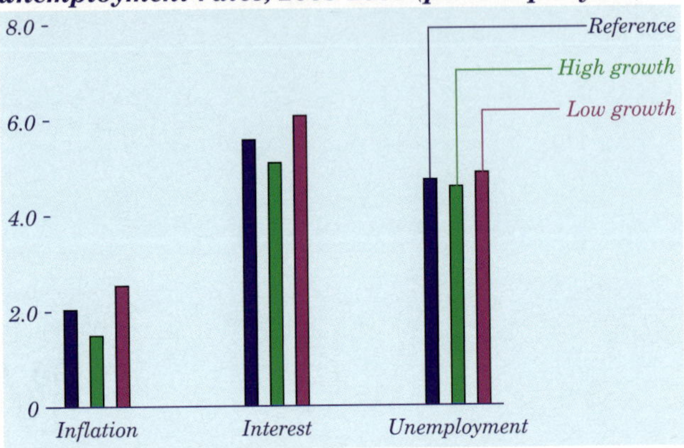

Common indicators for inflation, interest rates and employment are, respectively, the all-urban consumer price index (CPI-U), the interest rate (yield) on 10-year U.S. Treasury notes, and the unemployment rate, which are widely viewed as barometers of conditions in the markets for goods and services, credit, and labor, respectively. In *AEO2007*, the projected average annual inflation rate over the 2005-2030 period, as measured by the all-urban CPI, is 2 percent in the reference case, 1.5 percent in the high economic growth case, and 2.5 percent in the low growth case (Figure 25). Annual yields on the 10-year Treasury note are projected to average 5.6 percent in the reference case, 5.1 percent in the high growth case, and 6.1 percent in the low growth case. The projections for average unemployment rates are 4.7 percent in the reference case, 4.6 percent in the high growth case, and 4.9 percent in the low growth case. Relative to the reference case, the higher inflation, interest, and unemployment rates in the low growth case and the lower rates in the high growth case depend on different assumptions about labor productivity and population growth rates.

Historically, from 1980 to 2005, inflation has averaged 3.5 percent per year, the average yield on 10-year Treasury notes has been 7.7 percent per year, and the unemployment rate has averaged 6.2 percent. In the reference case and also in the high and low economic growth cases for *AEO2007*, projected gains in labor productivity are generally higher than the historical averages of the 1980s, leading to more optimistic projections for inflation, interest, and unemployment rates.

Output Growth for Energy-Intensive Industries Is Expected To Slow

Figure 26. Sectoral composition of industrial output growth rates, 2005-2030 (percent per year)

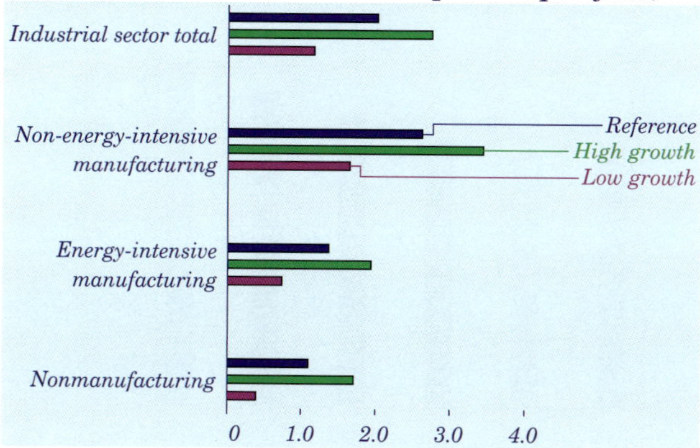

The industrial sector (all non-service industries) has shown slower output growth than the economy as a whole in recent decades, with imports meeting a growing share of demand for industrial goods. That trend is expected to continue in the *AEO2007* reference case. The average annual growth rate for real GDP from 2005 to 2030 is 2.9 percent in the reference case, whereas the industrial sector averages 2.0 percent. Within the industrial sector, manufacturing output is projected to grow more rapidly than nonmanufacturing output (which includes agriculture, mining, and construction). With higher energy prices and more foreign competition expected, the energy-intensive manufacturing sectors are projected to grow by only 1.4 percent per year from 2005 through 2030, compared with a projected 2.6-percent average annual rate of growth for the non-energy-intensive manufacturing sectors (Figure 26). The energy-intensive manufacturing sectors include food, paper, bulk chemicals, petroleum refining, glass, cement, steel, and aluminum.

In the high economic growth case, output from the industrial sector as a whole is projected to grow by an average of 2.8 percent per year, still below the projected average of 3.4 percent for real GDP. In the low economic growth case, with real GDP growth projected to average 2.2 percent per year from 2005 through 2030, industrial output averages 1.2 percent annual growth. In both cases, the highest growth rates are expected for the non-energy-intensive manufacturing segment of the industrial sector, with lower rates projected for the energy-intensive manufacturing and nonmanufacturing segments.

Energy Expenditures Relative to GDP Are Projected To Decline

Figure 27. Energy expenditures in the U.S. economy, 1990-2030 (billion 2005 dollars)

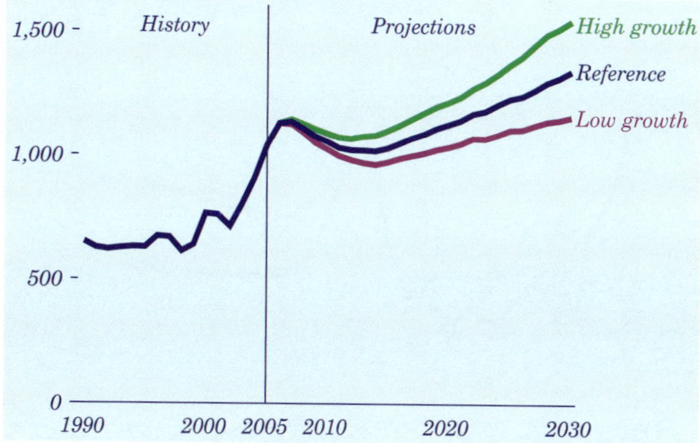

Total expenditures for energy services in the U.S. economy were $1.0 trillion in 2005. In the *AEO2007* projections, energy expenditures in 2030 rise to $1.3 trillion (2005 dollars) in the reference case and $1.5 trillion in the high economic growth case (Figure 27). For the economy as a whole, ratios of energy expenditures to GDP in 2005 were 8.4 percent for all energy, 4.8 percent for petroleum, and 1.6 percent for natural gas. Although recent developments in the world oil market have pushed the expenditure shares upward, in the reference case they are expected to decline from current levels as the energy intensity of the U.S. economy—measured as energy consumption (thousand Btu) per dollar of real GDP—continues to decline and world oil prices return to a relatively lower price path. Total energy expenditures are projected to equal 5.3 percent of GDP in 2030, petroleum expenditures 3.0 percent, and natural gas expenditures less than 1 percent (Figure 28).

Figure 28. Energy expenditures as a share of gross domestic product, 1970-2030 (nominal expenditures as percent of nominal GDP)

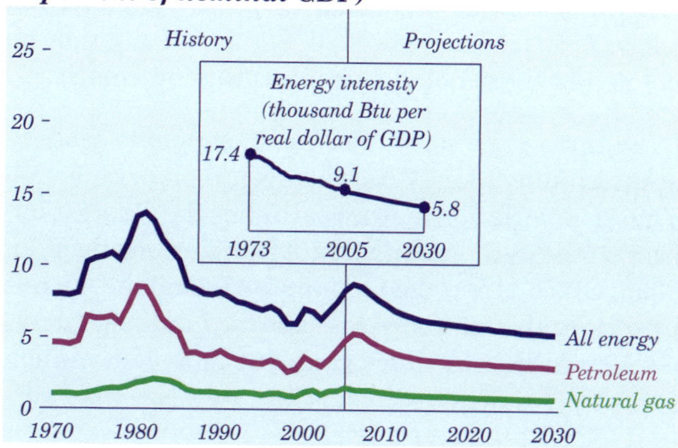

Oil Price Cases Show Uncertainty in Prospects for World Oil Markets

Figure 29. World oil prices, 1980-2030 (2005 dollars per barrel)

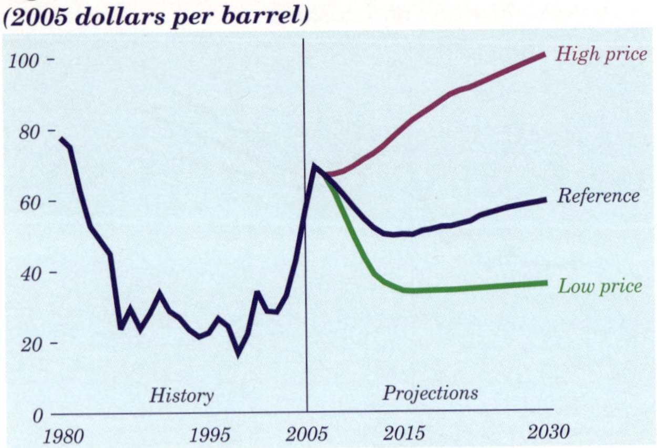

World oil price projections in the *AEO2007*, in terms of the average price of imported low-sulfur, light crude oil to U.S. refiners, are higher for 2006-2014 than those presented in the *AEO2006*. The higher price path reflects lower estimates of oil consumers' sensitivity to higher prices (given that the demand for oil has continued to grow despite the high prices of 2005-2006), an anticipation of lower levels of future investment in production capacity in key resource-rich regions due to political instability, access restrictions, and a reassessment of OPEC producers' ability to influence prices during periods of volatility.

The historical record shows substantial variability in world oil prices, and there is arguably even more uncertainty about future prices in the long term. *AEO2007* considers three price cases to illustrate the uncertainty of prospects for future world oil resources and economics. In the reference case, world oil prices moderate from current levels to about $50 per barrel in 2014, before rising to $59 per barrel in 2030 (2005 dollars). The low and high price cases reflect a wide range of potential world oil price paths, ranging from $36 to $100 per barrel in 2030 (Figure 29), but they do not bound the set of all possible future outcomes.

In all three price cases, non-OPEC suppliers produce at maximum capacity based on world oil price levels. Thus, the variation in price paths has the greatest impact on the need for OPEC supply in the long term. In 2030, OPEC is expected to supply 47.6 million barrels per day in the reference case and 54.7 million barrels per day in the low price case, but only 33.3 million barrels per day in the high price case—less than current OPEC production levels.

Oil Imports in 2030 Approach 18 Million Barrels per Day

Figure 30. U.S. gross petroleum imports by source, 2005-2030 (million barrels per day)

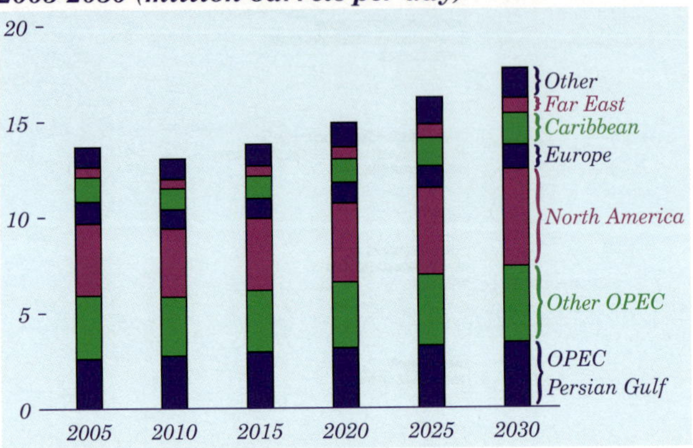

Total U.S. gross petroleum imports increase in the reference case from 13.7 million barrels per day in 2005 to 17.7 million in 2030 (Figure 30), deepening U.S. reliance on imported oil in the long term. In 2030, gross petroleum imports account for 66 percent of total U.S. petroleum supply in the reference case, up from 60 percent in 2005.

U.S. gross petroleum imports in the high world oil price case are 25 percent lower in 2030 than projected in the reference case, at 13.4 million barrels per day. The higher price assumptions lead to increased profitability from domestic production and reduced demand. In the low world oil price case, imports increase to 20.8 million barrels per day in 2030. The projected import shares of total U.S. petroleum supply in 2030 are 54 percent in the high price case and 72 percent in the low price case.

Of the increase in gross imports in the reference case, 37 percent comes from OPEC suppliers. West Coast refiners increase their imports of crude oil from the Far East, to replace a decline in Alaskan oil supplies. Canada and Mexico continue to be important sources of U.S. petroleum supply. Much of the Canadian contribution comes from the development of its enormous oil sands resource base.

Across the three price cases, U.S. gross petroleum imports shift toward heavier crude oil and fewer refined petroleum products. Vigorous growth in demand for lighter, low-sulfur petroleum in developing countries means that U.S. refiners are likely to import smaller volumes of low-sulfur, light crude oil and to increase the technical complexity of their refining operations.

Unconventional Resources Gain Market Share as Prices Rise

Figure 31. Unconventional resources as a share of the world liquids market, 1990-2030 (percent)

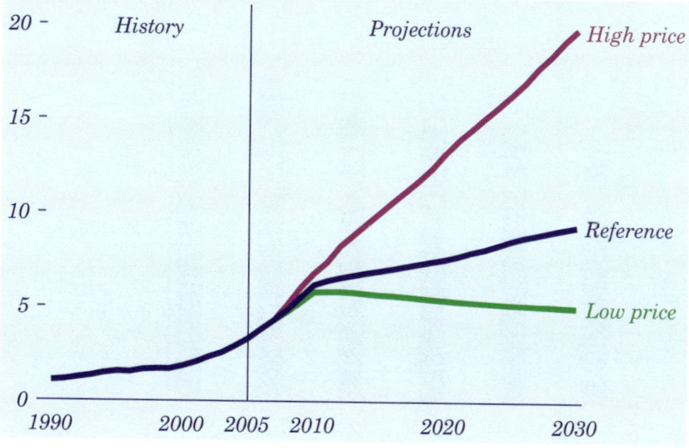

The world's total production of liquid fuels from unconventional resources in 2005 was 2.8 million barrels per day, equal to about 3 percent of total liquids production. Production from unconventional sources included 1.1 million barrels per day from oil sands in Canada, 600,000 barrels per day from very heavy oils in Venezuela, and 260,000 barrels per day of ethanol (172,000 barrels per day oil equivalent) in the United States. In the *AEO2007* reference case, unconventional production is projected to make up 9 percent of total liquids production (10.9 million barrels per day) in 2030.

Unconventional liquids production grows twice as fast in the high price case as in the reference case (Figure 31), because unconventional supplies are more competitive with conventional sources when market prices are higher. In the high price case, unconventional production increases to about 20.1 million barrels per day worldwide in 2030, representing 20 percent of total liquids production around the world. In the low price case, unconventional production totals only 6.4 million barrels per day in 2030, or 5 percent of total production.

More than 80 percent of the world's unconventional resources are controlled by non-OPEC nations. The total volumes of liquids production in non-OPEC countries are fairly constant across the three world oil price cases in *AEO2007*, but non-OPEC unconventional production is significantly higher in the high price case.

World Liquids Supply Is Projected To Remain Diversified in All Cases

Figure 32. World liquids production shares by region, 2005 and 2030 (percent)

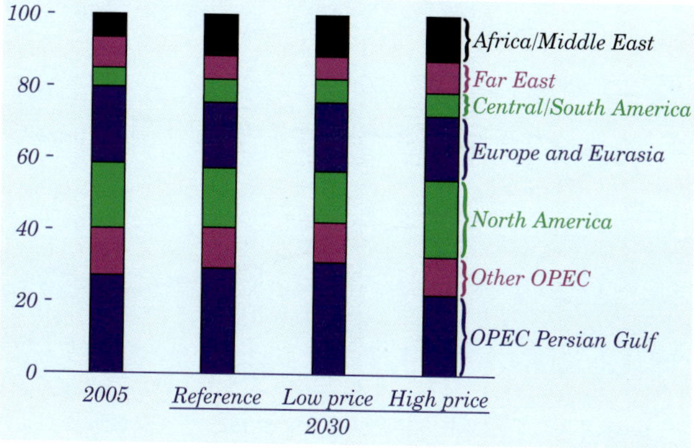

In 2005, OPEC producers in the Persian Gulf accounted for 27 percent of the world's total oil supply, and other OPEC producers accounted for 13 percent. Europe and Eurasia produced 21 percent of the total supply, North America 18 percent, and the rest of the world 20 percent (Figure 32). In the reference case projections, those regional shares remain relatively constant though 2030.

The largest change in regional production share is projected for non-OPEC suppliers in Africa and the Middle East, which increase their share of the world total from 6 percent in 2005 to 11 percent in 2030 in the reference case. OPEC producers in the Persian Gulf are projected to increase their share of the total by 2 percentage points from 2005 to 2030, and the share of OPEC producers in other regions is projected to fall by 2 percentage points.

In the low and high oil price cases, the OPEC Persian Gulf share in 2030 varies from 31 percent to 22 percent, respectively, as compared with 29 percent in the reference case. The changes across the three cases reflect an expectation that OPEC suppliers will vary their production levels in attempts to influence world oil prices. In the projections, OPEC revenues and profits from oil exports vary by less than export volumes across the cases.

Average Energy Use per Person Increases Through 2030

Figure 33. Energy use per capita and per dollar of gross domestic product, 1980-2030 (index, 1980 = 1)

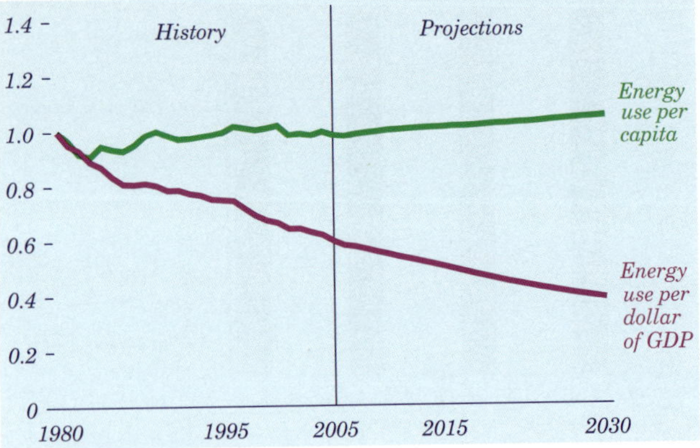

The future path of U.S. energy demand will depend on trends in population, economic growth, energy prices, and technology adoption. *AEO2007* cases developed to illustrate the uncertainties associated with those factors include low and high economic growth cases, low and high price cases, and 2006 and high technology cases (see Appendixes B, C, D, and E).

Population growth is a key determinant of energy demand for housing, services, and travel. Its impact is magnified by changes in energy consumption per capita, which reflect the combined effects of economic growth, energy prices, and other factors. In the reference case, energy consumption per capita grows by 0.3 percent per year from 2005 to 2030, faster than it has in recent history (Figure 33), as a result of projected growth in real disposable income per capita.

Although the Nation's reliance on imported fuel has been growing, the economy is becoming less dependent on energy in general. U.S. energy intensity (energy use per 2000 dollar of GDP) declines by an average of 1.5 percent per year in the low growth case, 1.8 percent in the reference case, and 1.9 percent in the high growth case. Efficiency gains and faster growth in less energy-intensive industries account for most of the projected decline, more than offsetting growth in demand for energy services in buildings, transportation, and electricity generation. The decline is more rapid in the high economic growth case, because the additional growth is concentrated in less energy-intensive industries. In all three growth cases, as energy prices moderate over the longer term, energy intensity declines at a slower rate.

Coal and Liquid Fuels Lead Increases in Primary Energy Use

Figure 34. Primary energy use by fuel, 2005-2030 (quadrillion Btu)

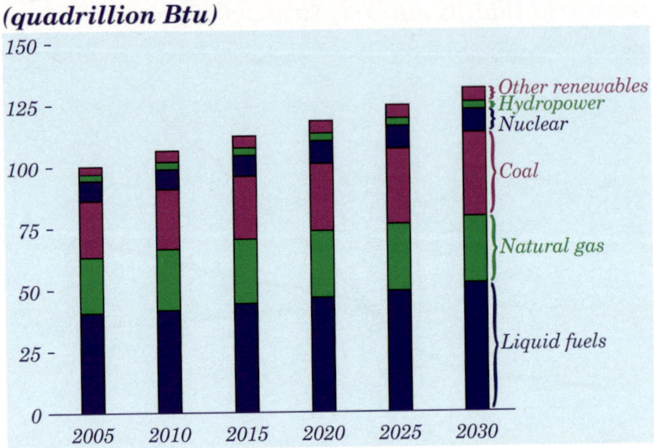

Total primary energy consumption, including energy for electricity generation, grows by 1.1 percent per year from 2005 to 2030 in the reference case (Figure 34). Fossil fuels account for 87 percent of the growth. The increase in coal use occurs mostly in the electric power sector, where strong growth in electricity demand and favorable economics under current environmental policies prompt coal-fired capacity additions. About 61 percent of the projected increase in coal consumption occurs after 2020, when higher natural gas prices make coal the fuel of choice for most new power plants. Over the longer term, growth in natural gas consumption for power generation is restrained by its high price relative to coal, although natural gas use increases in the near term. Industry and buildings account for about 90 percent of the increase in natural gas consumption from 2005 to 2030.

Transportation accounts for 94 percent of the projected increase in liquids consumption, dominated by growth in fuel use for light-duty vehicles. Fuel use by freight trucks, second in energy use among travel modes, grows by 1.8 percent per year on average, the fastest annual rate among the major forms of transport. The remainder of the liquids growth in the *AEO2007* reference case occurs in the industrial sector, primarily in refineries. The projected trend in liquid fuels use in the buildings sectors is relatively flat in the reference case.

AEO2007 projects rapid percentage growth in renewable energy production, partly as a result of State mandates for renewable electricity generation. Additions of new nuclear power plants are also projected, spurred by PTCs available under EPACT2005.

Liquid Fuels and Electricity Lead Growth in Energy Consumption

Figure 35. Delivered energy use by fuel, 1980-2030 (quadrillion Btu)

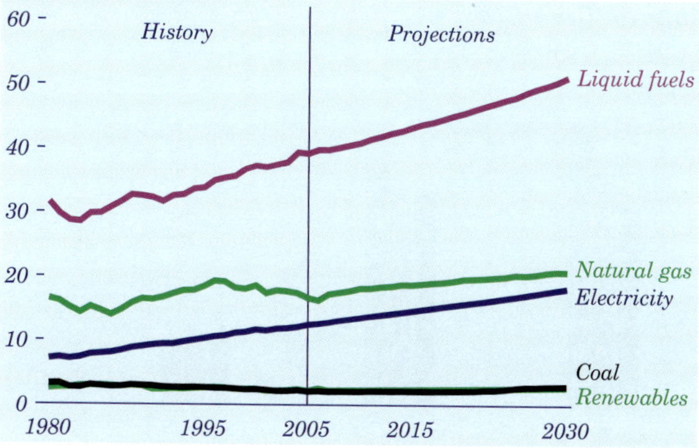

Delivered energy use (excluding losses in electricity generation) grows by 1.1 percent per year from 2005 to 2030 in the reference case. Liquid fuels use, which makes up more than one-half of total delivered energy use, grows by about the same percentage (Figure 35). About 93 percent of the projected increase is in the transportation sector, which depends heavily on liquid fuels. Even in the high price case, liquid fuels use for transportation grows by 0.9 percent per year on average through 2030. Growing population, incomes, and economic output spur travel demand, while fuel efficiency improves only slightly. With varying assumptions on population and economic growth, average annual growth in delivered energy use from 2005 to 2030 ranges from 0.7 percent in the low growth case to 1.5 percent in the high growth case.

Recent trends in electricity use are expected to continue, given strong growth in commercial floorspace, continued penetration of electric appliances, and increases in industrial output. Natural gas use grows more slowly than overall delivered energy demand, in contrast to its more rapid growth during the 1990s. Natural gas consumption in the residential sector is projected to grow by less than 10 percent over the 25-year projection.

End-use demand for energy from marketed renewables, such as wood, grows by 1.1 percent per year. Industrial biomass, mostly a byproduct fuel in the pulp and paper industry, is the largest component of end-use renewable fuel. Renewable energy from solar and geothermal heat pumps more than doubles over the projection, but those sources remain at less than 1 percent of residential delivered energy use.

U.S. Primary Energy Use Climbs to 131 Quadrillion Btu in 2030

Figure 36. Primary energy consumption by sector, 1980-2030 (quadrillion Btu)

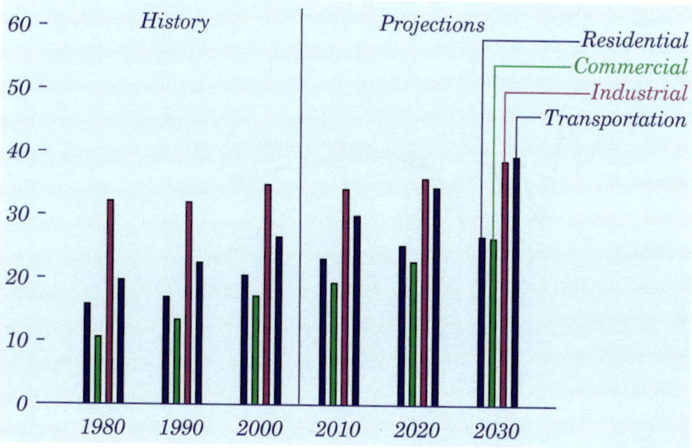

Primary energy use (including electricity generation losses) is projected to increase by 31 percent over the next 25 years in the reference case (Figure 36). The projected growth rate of energy consumption approximately matches the average from 1981 to 2005. Demand for energy in the early 1980s fell in the face of recession, high energy prices, and changing regulations; but beginning in the mid-1980s, declining real energy prices and economic expansion contributed to a marked increase in energy consumption. The long-term upward trend in energy use is projected to continue in the *AEO2007* reference case, but the growth is moderated by rising energy prices.

The most rapid growth in sectoral energy use is in the commercial sector, where services continue to expand more rapidly than the economy as a whole. The growth rate for residential energy use is about half that for the commercial sector, with demographic trends being a dominant factor. Transportation energy use grows by 1.4 percent per year from 2005 to 2030 (about the same as the growth rate from 1980 to 2005), despite relatively high fuel prices. Increases in travel by personal and commercial vehicles are only partially offset by vehicle efficiency gains.

In the reference case, primary energy use grows more slowly in the industrial sector than in the other sectors, with efficiency gains, higher real energy prices, and shifts to less energy-intensive industries moderating the expected growth. In the high economic growth case, however, the projected increase in industrial energy use is almost double that in the reference case.

Residential Energy Use per Capita Varies With Technology Assumptions

Figure 37. Residential delivered energy consumption per capita, 1990-2030 (index, 1990 = 1)

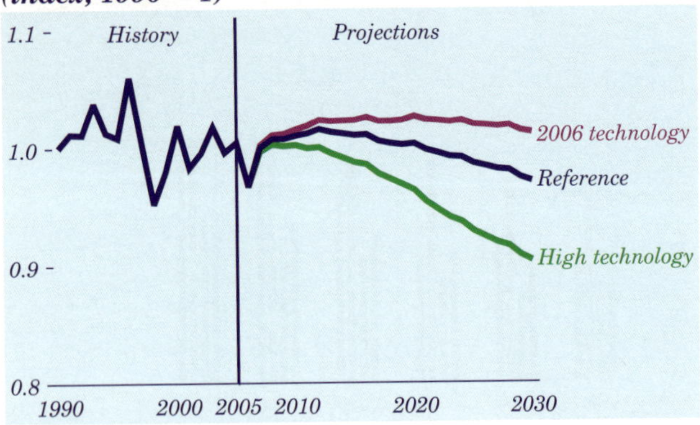

Residential energy use per person has remained fairly constant since 1990 (taking into account year-to-year fluctuations in weather), with increases in energy efficiency offset by consumer preference for larger homes and by new residential uses for energy. As the U.S. population has shifted to the South and West, all-electric homes have become more prevalent and electricity use for air conditioning has increased, leading to a rise in electricity consumption per capita while natural gas use and liquid fuels use per capita have fallen. In the reference case, however, as the population shift to warmer climates continues, slower penetration of new energy-using appliances and increases in efficiency are projected to reduce energy use per capita.

In the *AEO2007* projections, residential energy use per capita changes with assumptions about the rate at which more efficient technologies are adopted. The 2006 technology case assumes no increase in the efficiency of equipment or building shells beyond those available in 2006. The high technology case assumes lower costs, higher efficiencies, and earlier availability of some advanced equipment. In the reference case, residential energy use per capita is projected to fall below the 1990 level after 2020. The 2006 technology case approximates an upper bound on energy use per capita in the future: delivered energy use per capita remains above the 1990 level through 2030, when it is 4 percent higher than projected in the reference case (Figure 37). The high technology case indicates a lower bound for energy use per capita in the cases considered here, falling below the 1990 level after 2013 and reaching a 2030 level that is 7 percent below the reference case projection.

Household Uses for Electricity Continue To Expand

Figure 38. Residential delivered energy consumption by fuel, 2005, 2015, and 2030 (quadrillion Btu)

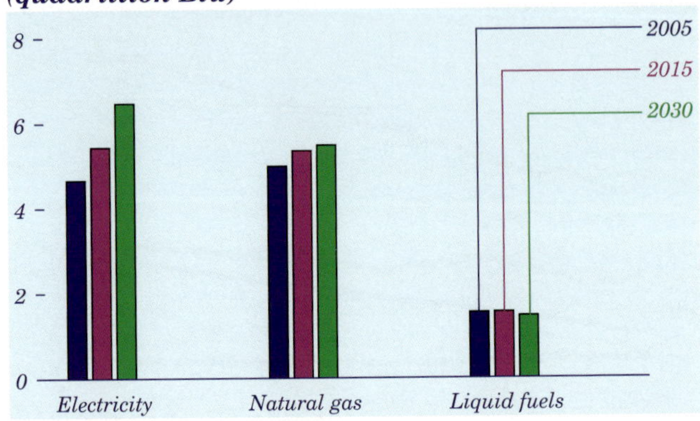

Over the past several decades, residential electricity demand has increased as more uses for electricity have emerged. The reference case projects further increases in residential electricity consumption, averaging 1.3 percent per year from 2005 to 2030 (Figure 38), as more electric devices and larger television sets with digital capability continue to penetrate residential markets. Two alternative cases—the high economic growth case and the high technology case—provide high and low ranges, respectively, for the projections. In the high growth case, population increases lead to more households, which use more electric appliances. In the high technology case, more efficient houses and appliances lead to lower electricity use. The 2030 projections for residential electricity use in the two cases are 0.4 quadrillion Btu higher and 0.6 quadrillion Btu lower, respectively, than the reference case projection of 6.5 quadrillion Btu.

Changes in natural gas and liquid fuels consumption in the residential sector over the past 20 years have been less dramatic. For residential natural gas consumption, the reference case projects annual growth averaging 0.4 percent from 2005 to 2030, and for liquid fuels use a slight decrease is projected. In the high economic growth case, the sector's natural gas use in 2030 is 0.3 quadrillion Btu higher than the reference case level of 5.5 quadrillion Btu; in the high technology case it is 0.3 quadrillion Btu lower. For liquid fuels use, the low and high price cases provide high and low estimates, respectively, both varying in 2030 by 0.1 quadrillion Btu from the reference case level (1.5 quadrillion Btu). With relatively few new homes using oil furnaces, the economic growth cases do not have as much effect on residential liquid fuels use.

Increases in Energy Efficiency Are Projected To Continue

Figure 39. Efficiency indicators for selected residential appliances, 2005 and 2030 (index, 2005 stock efficiency = 1)

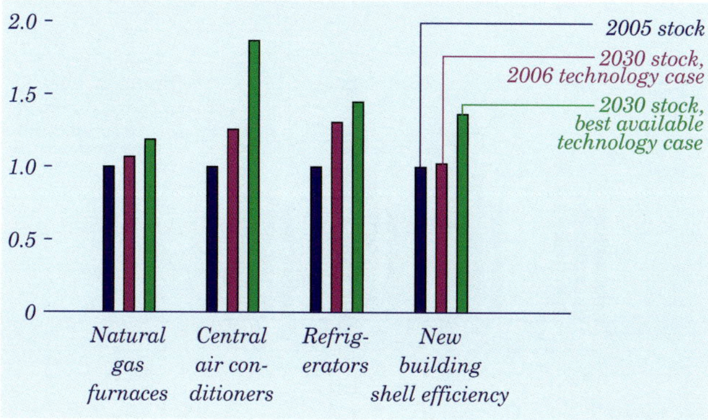

The energy efficiency of new household appliances plays a key role in determining the types and amounts of energy used in residential buildings. As a result of stock turnover and purchases of more efficient equipment, energy use by residential consumers, both per household and per capita, has fallen over time. In the 2006 technology case, which assumes no efficiency improvement for available appliances beyond 2006 levels, normal stock turnover results in higher average energy efficiency for most residential equipment in 2030, as older appliances are replaced with more efficient models from the 2006 stock (Figure 39).

The greatest gains in residential energy efficiency are projected in the best available technology case, which assumes that consumers purchase the most efficient products available at normal replacement intervals regardless of cost, and that new buildings are built to the most energy-efficient specifications available, starting in 2007. In this case, residential delivered energy consumption in 2030 is 27 percent less than projected in the 2006 technology case and 24 percent less than in the reference case. Purchases of more energy-efficient products, such as solid-state lighting and condensing gas furnaces, reduce the amount of energy used without lowering service levels.

Several current Federal programs, including Zero Energy Homes and ENERGY STAR Homes, promote the use of efficient appliances and building envelope components, such as windows and insulation. In the best available technology case, use of the most efficient building envelope components available can reduce heating requirements in an average new home by nearly 30 percent.

Rise in Commercial Energy Use per Capita Is Projected To Continue

Figure 40. Commercial delivered energy consumption per capita, 1980-2030 (index, 1980 = 1)

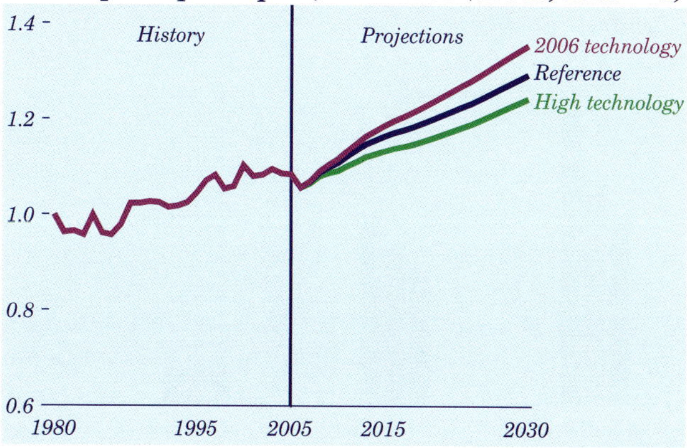

In the commercial sector, delivered energy consumption per capita increased by 8 percent from 1980 to 2005, primarily as a result of rising electricity use as the Nation moved increasingly to a service economy. Commercial energy use per person is projected to increase more rapidly in the reference case, by a total of 19 percent from 2005 to 2030, as the transition to a service economy continues and energy prices moderate from current levels. Depending on assumptions about the availability and adoption of energy-efficient technologies, the size of the projected increase varies from a low of 15 percent in the high technology case to a high of 25 percent in the 2006 technology case (Figure 40).

The reference case assumes future improvements in efficiency for commercial equipment and building shells, as well as increased demand for energy services. While commercial energy use per capita increases by 19 percent from 2005 to 2030 in the reference case, commercial energy intensity (delivered energy consumption per square foot of floorspace) shows little change, increasing by only 1 percent. The 2006 technology case assumes the same laws and regulations as the reference case but with no increase in the energy efficiency of commercial equipment and building shells beyond those available in 2006. The result is a 5-percent increase in commercial delivered energy use in 2030 relative to the reference case. In the high technology case, assuming earlier availability, lower costs, and higher efficiencies for more advanced equipment and building shells, delivered energy consumption in 2030 is 4 percent below the reference case projection.

Electricity Leads Expected Growth in Commercial Energy Use

Figure 41. Commercial delivered energy consumption by fuel, 2005, 2015, and 2030 (quadrillion Btu)

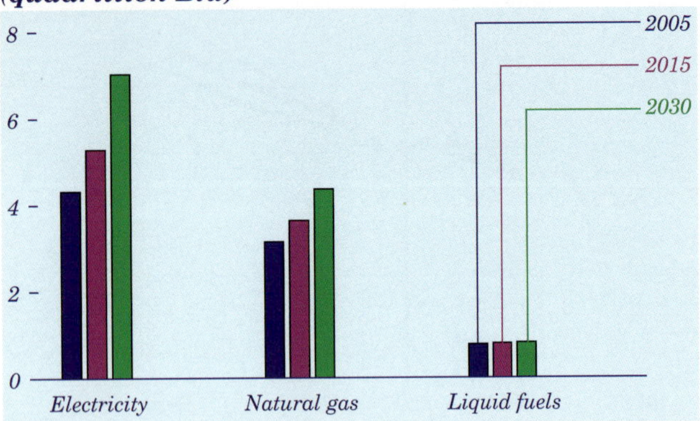

Commercial floorspace growth and, in turn, commercial energy use are driven by trends in economic and population growth. In the *AEO2007* projections, growth in disposable income leads to increased demand for services from hotels, restaurants, stores, theaters, galleries, arenas, and other commercial establishments, which in turn are increasingly dependent on electricity both for basic services and for business services and customer transactions. In addition, the growing share of the population over age 65 increases demand for healthcare and assisted living facilities and for electricity to power medical and monitoring equipment in those facilities. The reference case projects further increases in commercial electricity use, averaging 2.0 percent per year from 2005 to 2030 (Figure 41). The high and low economic growth cases provide high and low ranges for the average annual growth rate of commercial electricity demand from 2005 to 2030, at 2.3 percent and 1.6 percent, respectively.

For commercial natural gas use (primarily for space heating and water heating), the reference case projects average annual growth of 1.3 percent from 2005 to 2030, and for liquid fuels use the projected average annual growth rate is 0.2 percent. The alternative projections for natural gas use in 2030 range from a high of 4.7 quadrillion Btu in the high growth case to 4.0 in the low growth case, compared with 4.4 in the reference case. For liquid fuels use, the low and high oil price cases provide high and low estimates, respectively, which in 2030 vary by 0.2 and 0.1 quadrillion Btu from the reference case projection of 0.8 quadrillion Btu.

Current Technologies Provide Potential Energy Savings

Figure 42. Efficiency indicators for selected commercial energy end uses, 2005 and 2030 (index, 2005 stock efficiency = 1)

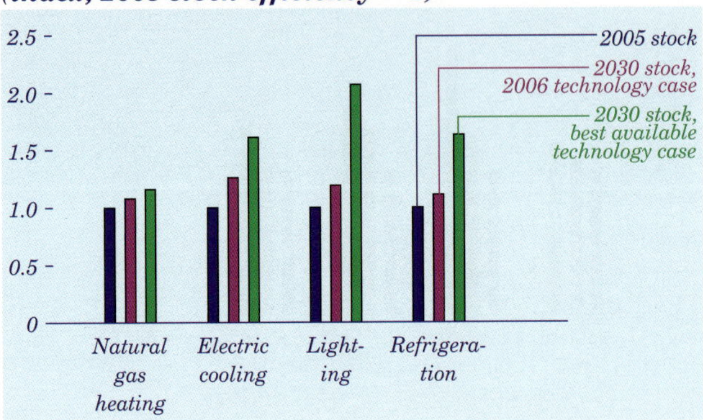

The stock efficiency of energy-using equipment in the commercial sector increases in the *AEO2007* reference case. Adoption of more energy-efficient equipment is expected to moderate the projected growth in demand, in part because of building codes for new construction and minimum efficiency standards, including those in EPACT2005; however, the long service lives of many kinds of energy-using equipment limit the pace of efficiency improvements.

The most rapid increase in overall energy efficiency for the commercial sector is projected in the best technology case, which assumes that only the most efficient technologies are chosen, regardless of cost, and that building shells in 2030 are 50 percent more efficient than projected in the reference case. With the adoption of improved heat exchangers for space heating and cooling equipment, solid-state lighting, and more efficient compressors for commercial refrigeration, commercial delivered energy consumption in 2030 in the best technology case is 13 percent less than projected in the reference case and 17 percent less than in the 2006 technology case.

In the 2006 technology case, which assumes equipment and building shell efficiencies limited to those available in 2006, energy efficiency in the commercial sector still is projected to improve from 2005 to 2030 (Figure 42), because the technologies available in 2006 can provide savings relative to commercial equipment currently in place. When businesses consider equipment purchases, however, the additional capital investment needed to buy the most efficient technologies often carries more weight than do future energy savings.

Advanced Technologies Could Slow Electricity Consumption in Buildings

Figure 43. Buildings sector electricity generation from advanced technologies, 2030 (percent change from reference case)

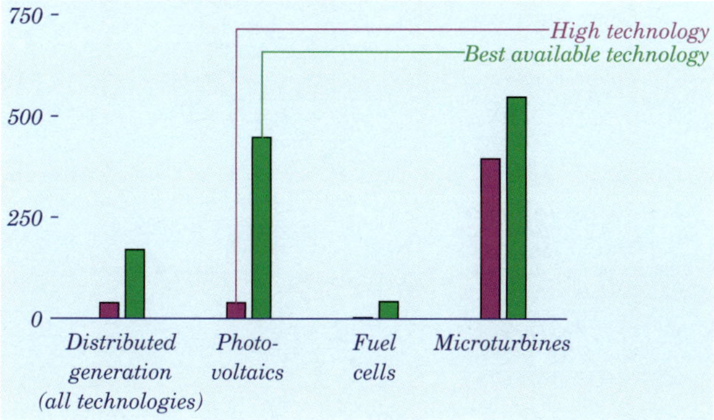

Alternative technology cases for the residential and commercial sectors vary the assumptions about availability and market penetration of distributed generation technologies. In the high technology case, buildings generate 6.2 billion kilowatthours (38 percent) more electricity in 2030 than the 16.1 billion kilowatthours projected in the reference case (Figure 43), most of which offsets residential and commercial electricity purchases. In the best available technology case, electricity generation in buildings in 2030 is 27.4 billion kilowatthours (171 percent) higher than in the reference case, with solar systems responsible for 78 percent of the increase. The optimistic assumptions of the best technology case benefit solar PV systems in particular, because there are no fuel expenses for solar systems. In the 2006 technology case, assuming no technological progress or cost reductions after 2006, electricity generation in buildings in 2030 is 6.1 billion kilowatthours (38 percent) lower than in the reference case.

Some of the heat produced by fossil-fuel-fired generating systems may be used to satisfy heating needs in CHP applications, increasing system efficiency and enhancing the attractiveness of distributed generation technologies for buildings. On the other hand, additional natural gas use for distributed generation systems in the high technology and best technology cases offsets some of the energy cost reductions that result from improvements in end-use equipment and building shells. In addition, if natural gas prices increased substantially, commercial establishments could find electricity purchases more economical than the installation of distributed generation technologies.

Economic Growth Cases Show Range for Projected Industrial Energy Use

Figure 44. Industrial delivered energy consumption, 1980-2030 (quadrillion Btu)

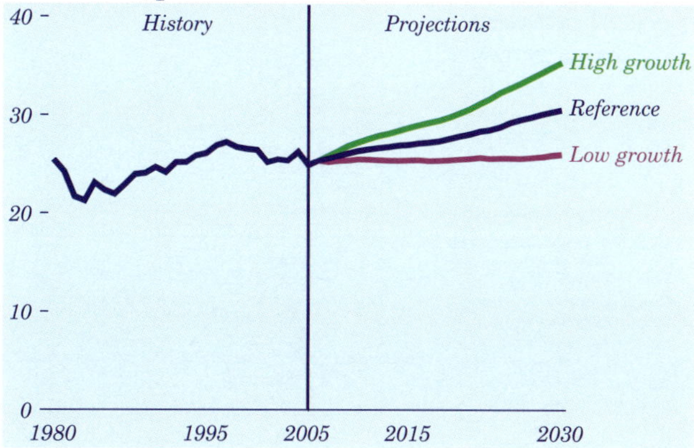

In the *AEO2007* projections, the path of industrial delivered energy consumption varies significantly, depending on the assumptions used in different cases. The projections for industrial sector energy consumption in 2030 range from 26.0 quadrillion Btu in the low economic growth case to 35.3 quadrillion Btu in the high growth case, with the reference case projection approximately midway between the two at 30.5 quadrillion Btu (Figure 44).

In the refining industry, reliance on nonconventional inputs for liquid fuels production is projected to increase rapidly. As a result, energy consumption by refineries in the reference case increases from 3.7 quadrillion Btu in 2005 to 6.3 quadrillion Btu in 2030. More than 60 percent of the increase is the result of increased coal use for CTL production and biomass use for ethanol production.

Non-fuel uses of energy transform normal energy inputs into other, non-energy products. In 2005, the U.S. chemical industry converted petroleum products with an estimated energy value of 3.4 quadrillion Btu into products such as plastics and fertilizers. Such non-fuel use is projected to increase in the reference case, to 3.9 quadrillion Btu in 2030. In addition, petroleum use to make asphalt and road oil, which are necessary components of construction industry activities, is projected to increase from 1.3 quadrillion Btu in 2005 to 1.4 quadrillion Btu in 2030. If energy consumption in the refining sector were excluded, nonfuel uses of petroleum would account for all the projected increase in industrial sector petroleum use in the reference case.

Energy-Intensive Industries Grow Less Rapidly Than Industrial Average

Figure 45. Average output growth in the manufacturing subsectors, 2005-2030 (percent per year)

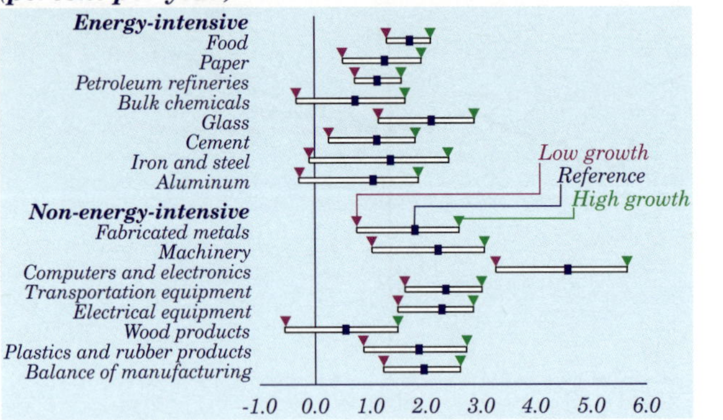

One of the most important determinants of industrial sector energy consumption is growth in value of shipments. In *AEO2007*, average annual growth in value of shipments for the industrial sector from 2005 to 2030 ranges from 1.2 percent per year in the low economic growth case to 2.8 percent per year in the high growth case, with the reference case projection approximately midway between the two at 2.0 percent per year. The range of growth in the individual subsectors is even wider.

By manufacturing subsector, the projected rate of growth in value of shipments in the reference case ranges from a low of 0.6 percent per year (wood products) to a high of 4.6 percent per year (computers). In general, the projected growth rates for the energy-intensive manufacturing subsectors are lower than those for the non-energy-intensive subsectors. Glass is the only energy-intensive subsector with a projected growth rate above 2 percent per year in the reference case.

The projected growth rates for value of shipments in the industrial subsectors in the high and low economic growth cases generally are symmetrical around the reference case (Figure 45). Industries with the most rapid projected growth in the reference case show the widest ranges, so that the pattern of faster value of shipments growth for the non-energy-intensive manufacturing sectors in the reference case is also evident in the high and low economic cases.

Energy Consumption Growth Varies Widely Across Industry Sectors

Figure 46. Average growth of delivered energy consumption in the manufacturing subsectors, 2005-2030 (percent per year)

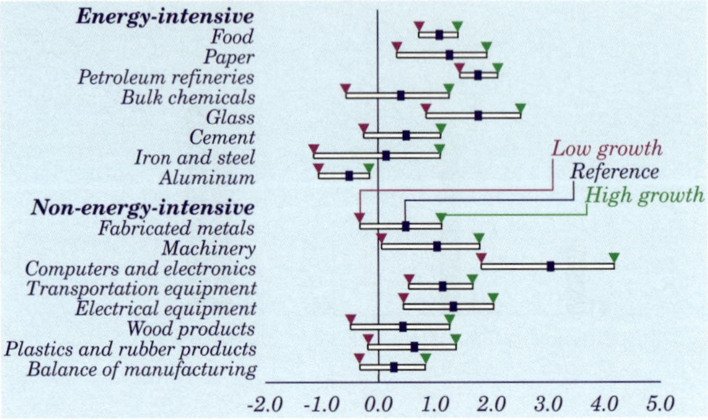

The average annual growth rate for total delivered energy consumption in the industrial sector from 2005 to 2030 ranges from an increase of 0.2 percent per year to an increase of 1.4 percent per year in the alternative cases for *AEO2007*. The widest variation is across the economic growth cases. Again, the range of the projections for individual subsectors is wider.

In the reference case, energy consumption growth rates for the manufacturing subsectors range from an increase of 3.0 percent per year (computers and electronics) to a decrease of 0.5 percent per year (aluminum). Delivered energy consumption growth in some of the energy-intensive industries (aluminum and steel) is held down by expected changes in production technology over the projection period. In general, the subsectors with the highest projected growth rates in energy consumption are those with the highest projected growth rates in value of shipments (computers and glass). The petroleum refining sector is an exception. As more refineries shift to alternative feedstocks for liquids production (biomass, coal, heavier crude oil) they use more energy per unit of output than is used for traditional petroleum-based refining.

The projected rates of growth in energy consumption in the alternative economic growth cases are generally symmetric around the reference case (Figure 46); however, the rate of growth is moderated by the level of investment in each case.

Energy Intensity in the Industrial Sector Continues To Decline

Figure 47. Industrial delivered energy intensity, 1980-2030 (thousand Btu per 2000 dollar value of shipments)

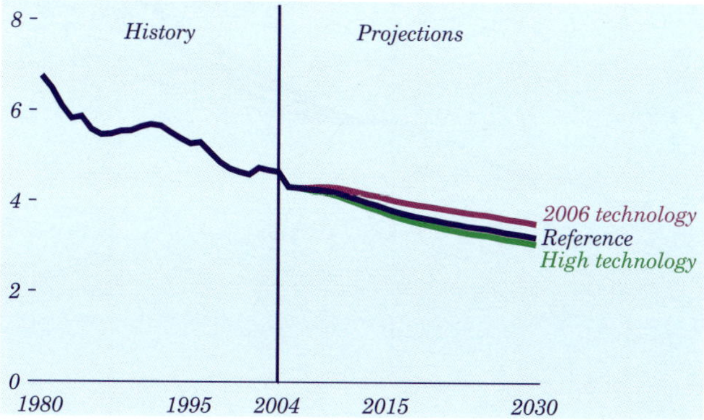

From 1980 to 2004 [*162*], energy consumption in the industrial sector was virtually unchanged, growing by a total of 3.1 percent, while value of shipments increased by 50 percent. Thus, industrial delivered energy use per dollar of industrial value of shipments declined by an average of 1.6 percent per year from 1980 to 2004 (Figure 47). Since 1990, however, the rate of decline in the sector's energy intensity has slowed: a 29-percent increase in industrial output from 1990 to 2004 resulted in a 6.5-percent increase in energy use and a 1.4-percent average annual decline in energy intensity.

Factors contributing to the decline in industrial energy intensity include a greater focus on energy efficiency after the energy price shocks of the 1970s and 1980s and a reduction in the share of industrial activity accounted for by the most energy-intensive industries. The energy-intensive industries' share of industrial output fell from 24 percent in 1980 to 21 percent in 2004.

The industrial value of shipments is projected to grow by 65 percent overall from 2005 to 2030, and the share attributed to the energy-intensive industries is projected to fall from 20 percent in 2005 to 17 percent in 2030. Consequently, even if no specific industry showed a reduction in energy intensity, the aggregate energy intensity of the industrial sector as a whole would decline [*163*].

Expected Declines in Energy Intensity Vary by Industry and Technology

Figure 48. Average change in energy intensity in the manufacturing subsectors, 2005-2030 (percent per year)

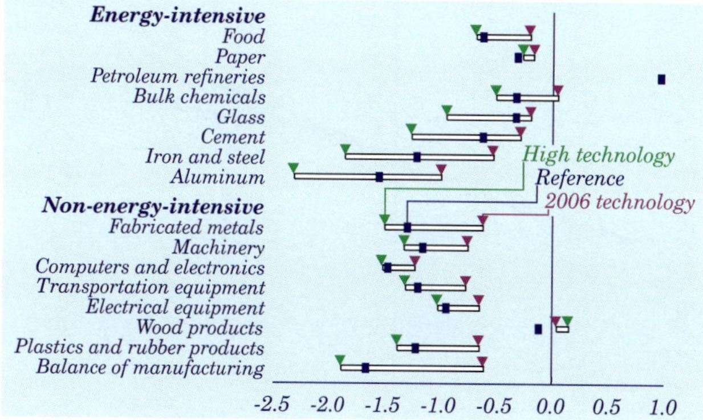

Energy intensity in the industrial subsectors can change for a variety of reasons. For example, no new primary smelting capacity is expected to be constructed in the U.S. aluminum industry, and secondary smelting, a less energy-intensive process of melting scrap, is expected to become the subsector's dominant technology. As a result, the reference case projection for energy intensity in the aluminum industry in 2030 is nearly one-third less than the 2005 level. In the petroleum refining industry, projected increases in coal use for CTL production result in increasing energy intensity, at an average rate of 1.0 percent per year from 2005 to 2030 [*164*].

A range of potential energy intensity and energy consumption outcomes for the industrial sector were developed for *AEO2007*. Energy intensity in the refining industry does not change in the 2006 technology and high technology cases [*165*]. Excluding refineries, projected average annual decreases in aggregate industrial energy intensity range from 1.0 percent per year in the 2006 technology case to 1.7 percent per year in the high technology case (Figure 48). In the high technology case, industrial delivered energy consumption in 2030 (excluding refining) is 1.4 quadrillion Btu less than in the reference case for the same level of output; in the 2006 technology case, it is 2.9 quadrillion Btu higher than in the reference case. Although the energy efficiency of new equipment is assumed to remain at 2006 levels in the 2006 technology case, average efficiency improves as old equipment is retired.

Transportation Energy Use Is Expected To Increase

Figure 49. Delivered energy consumption for transportation, 1980-2030 (quadrillion Btu)

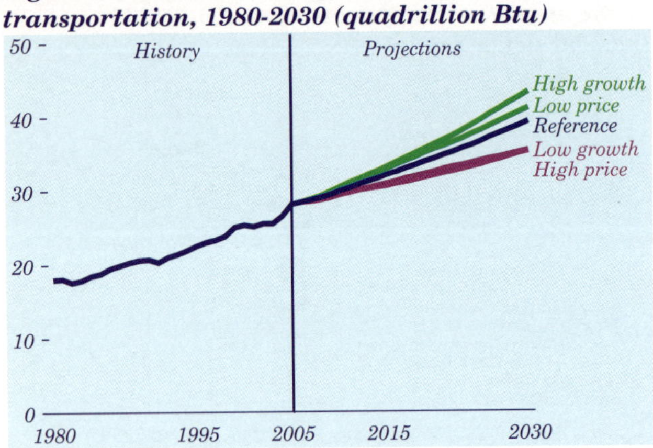

Total delivered energy consumption in the transportation sector is projected to grow at an average annual rate of 1.4 percent in the *AEO2007* reference case, from 28.1 quadrillion Btu in 2005 to 39.3 quadrillion Btu in 2030 (Figure 49). The reference case projection is consistent with recent historical trends.

Energy demand for transportation is influenced by a variety of factors, including economic growth, population growth, fuel prices, and vehicle fuel efficiency (for example, economic growth drives energy demand for heavy vehicle travel, and fuel prices and economic growth drive energy demand for light-duty vehicle travel). *AEO2007* provides several cases to examine the impacts of those factors on delivered energy demand. In 2030, the sector's delivered energy demand is about 10 percent higher in the high economic growth case and 10 percent lower in the low economic growth case than projected in the reference case, and it is about 10 percent lower in the high world oil price case and 5 percent higher in the low oil price case than in the reference case.

By transportation mode, the most rapid increase in the share of total delivered energy demand for transportation is projected for heavy vehicle travel, which includes medium and large freight trucks and buses. Energy demand for heavy vehicles accounted for 18 percent of the sector's total delivered energy demand in 2005, and in 2030 it accounts for 20 percent of the total in the reference case. Energy demand for air travel accounts for 10 percent of the total in 2030, the same as in 2005, because infrastructure constraints limit the potential growth of air travel in the United States.

Higher Prices Slow Increase in Demand for Light-Duty Vehicle Fuels

Figure 50. Delivered energy consumption in light-duty vehicles, 1980-2030 (quadrillion Btu)

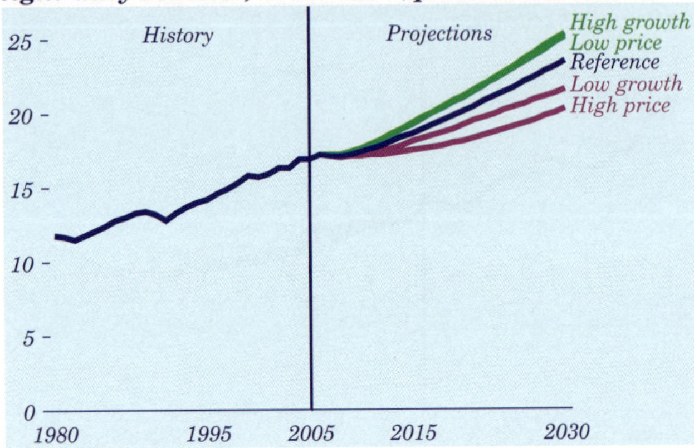

Delivered energy consumption for light-duty vehicle travel is projected to grow at an average annual rate of 1.3 percent in the reference case, from 16.9 quadrillion Btu in 2005 to 23.5 quadrillion Btu in 2030 (Figure 50). In 1980, energy use for light-duty vehicle travel totaled 11.8 quadrillion Btu.

The two factors that have the greatest impact on energy demand for light-duty vehicles in *AEO2007* are fuel price and, to a lesser extent, disposable income. The high economic growth case and high world oil price case provide higher and lower ranges, respectively, for the projections. The high growth case projects 25.3 quadrillion Btu, and the high price case projects 20.3 quadrillion Btu, for light-duty vehicle energy use in 2030.

The projections in the low world oil price case are nearly the same as those in the high economic growth case. As compared with the reference case, increased travel demand in the high growth case results in an 8-percent increase in energy use for light-duty vehicles in 2030; in the low price case, the combination of lower vehicle fuel economy and higher travel demand leads to a 7-percent increase.

Energy demand for light-duty vehicle travel in 2030 is lower in both the high price case and the low growth case than projected in the reference case, by 14 percent and 8 percent, respectively. Lower travel demand is the chief reason for the decrease in both cases. In addition, the high price case projects a 9-percent increase in light-duty vehicle fuel economy in 2030 relative to the reference case projection.

New Technologies Promise Improved Fuel Economy for Light-Duty Vehicles

Figure 51. Average fuel economy of new light-duty vehicles, 1980-2030 (miles per gallon)

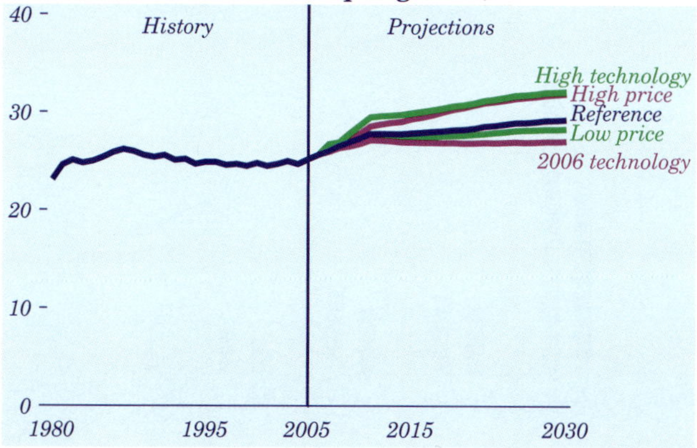

In 2005, U.S. sales of light trucks account for more than one-half of all new light-duty vehicles sold (as compared with only one-quarter of all new light-duty vehicle sales in 1980) [*166*]. Consequently, despite fuel economy improvements for cars and light trucks over the past years, the average fuel economy of new light-duty vehicles has declined from a 1987 peak of 26.2 miles per gallon to 25.2 miles per gallon in 2005 (Figure 51).

In March 2006, NHTSA finalized a new fuel economy standard for light trucks, based on vehicle footprint and product mix offered by the manufacturer (see "Legislation and Regulations"). The new CAFE standard, coupled with technological advances, is expected to have a positive impact on the fuel economy of new light-duty vehicles. In the reference case, average fuel economy for new light-duty vehicles is projected to increase to 29.2 miles per gallon in 2030. Additional improvement is projected in the high technology and high price cases, as a result of consumer demand for more fuel-efficient cars and improved economics that make producing them more profitable.

In the 2006 technology and low oil price cases, the projections for light-duty vehicle fuel economy in 2030 are lower than those in the reference case, but they still are higher than the 2005 CAFE standard for cars and the 2011 CAFE standard for light trucks. In the low price case, fuel economy for new light-duty vehicles in 2030 is 3.3 percent lower than projected in the reference case—due to consumer preference for more powerful vehicles over fuel economy—and in the 2006 technology case it is 7 percent lower than in the reference case.

Unconventional Vehicle Technologies Exceed 27 Percent of Sales in 2030

Figure 52. Sales of unconventional light-duty vehicles by fuel type, 2005, 2015, and 2030 (thousand vehicles sold)

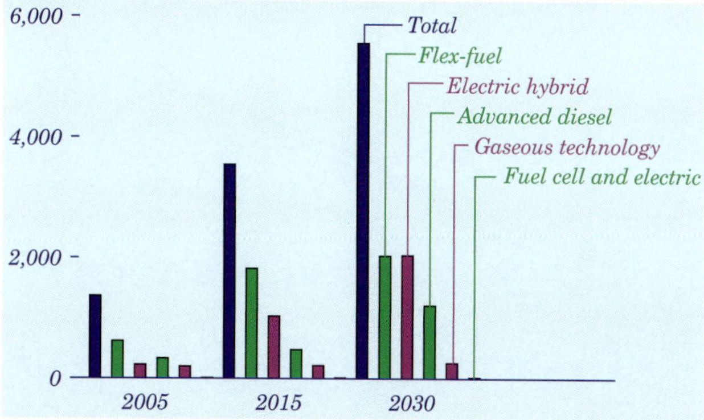

Concerns about oil supply, fuel prices, and emissions continue to drive the development and market penetration of unconventional vehicles (vehicles that can use alternative fuels or employ electric motors and advanced electricity storage, advanced engine controls, or other new technologies). Without new legislation or regulation, sales of unconventional vehicles total 5.5 million units in 2030 in the reference case (Figure 52), making up more than 27 percent of total new light-duty vehicle sales. In the high oil price case, unconventional vehicle sales total 8.1 million units, or more than 40 percent of new light-duty vehicle sales, as compared with 28 percent of sales in the low economic growth case.

Hybrid vehicles are becoming more popular, and in the reference case they are projected to top 2 million vehicles sold in 2030, as manufacturers continue to introduce new product lines. Light-duty diesel engines with advanced direct injection, which can significantly reduce exhaust emissions, are projected to capture 6 percent of the new light-duty vehicle market in 2030. The availability of ULSD and biodiesel fuels, along with advances in emission control technologies that reduce criteria pollutants, increase the projected sales of unconventional diesel vehicles.

Currently, manufacturers selling FFVs receive fuel economy credits that count toward their compliance with CAFE regulations. Continued commitment to the technology and increased product offerings are expected to increase sales of FFVs to 2 million units in 2030 in the reference case, from the 2005 level of 612,400 units.

Continued Growth in Electricity Use Is Expected in All Sectors

Figure 53. Annual electricity sales by sector, 1980-2030 (billion kilowatthours)

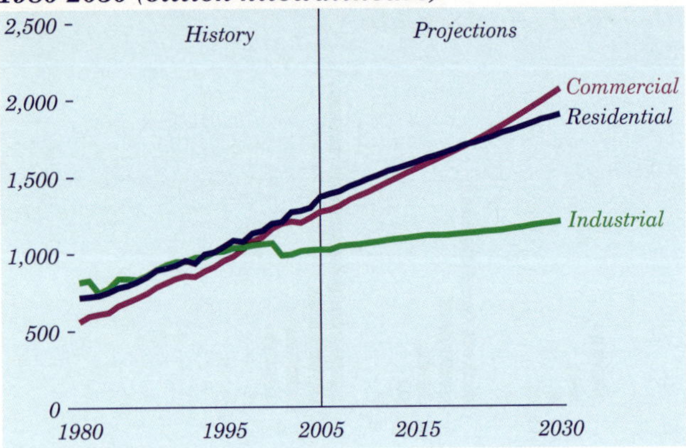

Total electricity sales increase by 41 percent in the *AEO2007* reference case, from 3,660 billion kilowatthours in 2005 to 5,168 billion kilowatthours in 2030. The largest increase is in the commercial sector (Figure 53), as service industries continue to drive growth. Electricity sales, which are strongly affected by the rate of economic growth, are projected to grow by 54 percent in the high growth case, to 5,654 billion kilowatthours in 2030, but by only 28 percent in the low growth case, to 4,682 billion kilowatthours in 2030.

By end-use sector, electricity demand in the reference case is projected to grow by 39 percent from 2005 to 2030 in the residential sector, by 63 percent in the commercial sector, and by 17 percent in the industrial sector. Growth in population and disposable income is expected to lead to increased demand for products, services, and floorspace, with a corresponding increase in demand for electricity for space heating and cooling and to power the appliances and equipment used by buildings and businesses. Population shifts to warmer regions will also increase the need for cooling.

The growth in demand for electricity is expected to be potentially offset by efficiency gains in both the residential and commercial sectors, and higher energy prices are expected to encourage investment in energy-efficient equipment. In both sectors, continuing efficiency gains are expected for electric heat pumps, air conditioners, refrigerators, lighting, cooking appliances, and computer screens. In the industrial sector, increases in electricity sales are offset by rapid growth in on-site generation.

Coal-Fired Power Plants Provide Largest Share of Electricity Supply

Figure 54. Electricity generation by fuel, 2005 and 2030 (billion kilowatthours)

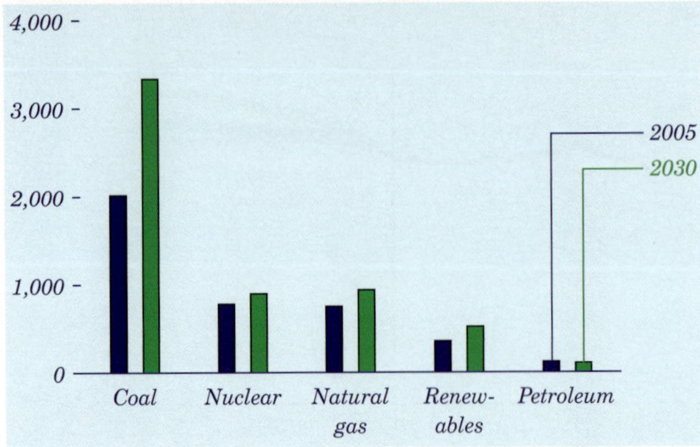

Coal-fired power plants (including utilities, independent power producers, and end-use CHP) continue to supply most of the Nation's electricity through 2030 (Figure 54). In 2005, coal-fired plants accounted for 50 percent of generation and natural-gas-fired plants for 19 percent. Most capacity additions over the next 10 years are natural-gas-fired plants, increasing the natural gas share to 22 percent and lowering the coal share to 49 percent in 2015. As natural gas becomes more expensive, however, more coal-fired plants are built. In 2030, the generation shares for coal and natural gas are 57 percent and 16 percent, respectively.

Nuclear and renewable generation increase as new plants are built, stimulated by Federal tax incentives and rising fossil fuel prices. Nuclear generation also increases modestly with improvements in plant performance and expansion of existing facilities, but the nuclear share of total generation falls from 19 percent in 2005 to 15 percent in 2030. The generation share from renewable capacity (about 9 percent of total electricity supply in 2005) remains roughly constant at about 9 percent.

Relative fuel costs, particularly for natural gas and coal, affect both the utilization of existing capacity and technology choices for new plants. Natural-gas-fired plants are projected to provide 27 percent of total electricity supply in 2030 in the low price case but only 11 percent in the high price case, while the projected share of total generation from coal-fired plants is 45 percent in the low price case but increases to 61 percent in the high price case. Changes in environmental policies would also affect the *AEO2007* projections for capacity additions.

Early Capacity Additions Use Natural Gas, Coal Plants Are Added Later

Figure 55. Electricity generation capacity additions by fuel type, including combined heat and power, 2006-2030 (gigawatts)

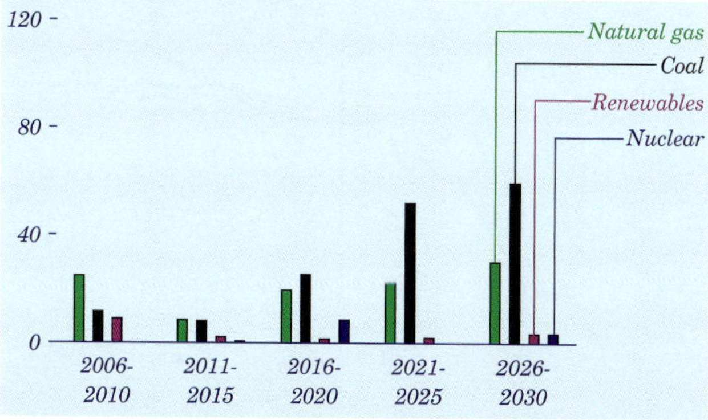

In the reference case, 292 gigawatts of new generating capacity (including end-use CHP) is required by 2030 to meet growth in electricity demand and to replace inefficient, older generating plants that are retired. Capacity decisions depend on the costs and operating efficiencies of different options, fuel prices, demand growth, and the availability of Federal tax credits for investments in some technologies.

Coal-fired capacity, which typically is expensive to build but has relatively low operating costs, accounts for about 54 percent of the total capacity additions from 2006 to 2030 (Figure 55). Natural-gas-fired plants, which generally are the least expensive capacity to build but have comparatively high fuel costs, represent 36 percent of the projected additions. Renewable and nuclear plants, which have high investment costs and low operating costs, account for 6 percent and 4 percent of total additions, respectively. Of the 12 gigawatts of new nuclear capacity expected by 2030, 3 gigawatts is added after the EPACT2005 PTC expires in 2020.

Different fuel price paths or growth rates for electricity demand can affect the quantity and mix of capacity additions. In the low and high price cases, variations in fuel prices have little impact on total capacity additions but do affect the mix of capacity types. Because fuel costs are a larger share of total expenditures for new natural-gas-fired capacity, higher fuel prices lead to more coal-fired additions. In the economic growth cases, capacity additions range from 191 gigawatts in the low growth case to 398 gigawatts in the high growth case, but with similar shares for the different generating technologies in both cases.

Least Expensive Technology Options Are Likely Choices for New Capacity

Figure 56. Levelized electricity costs for new plants, 2015 and 2030 (2005 mills per kilowatthour)

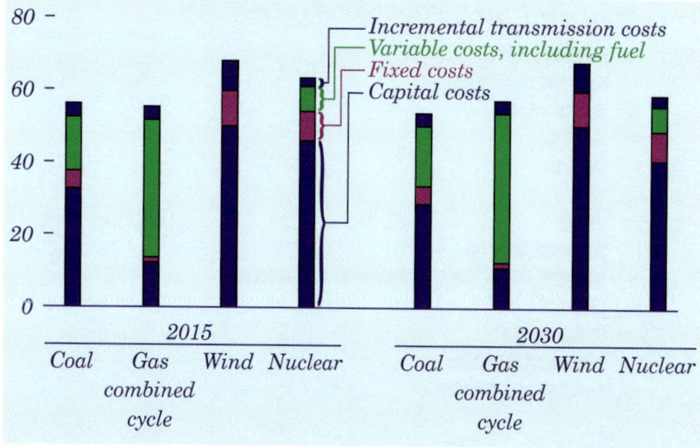

Technology choices for new generating capacity are made to minimize cost while meeting local and Federal emissions constraints. The choice of technology for capacity additions is based on the least expensive option available (Figure 56) [*167*]. The *AEO2007* reference case assumes a capital recovery period of 20 years. In addition, the cost of capital is based on competitive market rates, to account for the risks of siting new units.

Capital costs decline over time (Table 16), at rates that depend on the current stage of development for each technology. For the newest technologies, capital costs are initially adjusted upward to reflect the optimism inherent in early estimates of project costs. As project developers gain experience, the costs are assumed to decline. The decline continues at a progressively slower rate as more units are built. The efficiency of new plants is also assumed to improve through 2015, with heat rates for advanced combined cycle and coal gasification units declining from 6,572 and 8,309 Btu per kilowatthour, respectively, in 2005 to 6,333 and 7,200 Btu per kilowatthour in 2015.

Table 16. Costs of producing electricity from new plants, 2015 and 2030

Costs	2015 Advanced coal	2015 Advanced combined cycle	2030 Advanced coal	2030 Advanced combined cycle
	2005 mills per kilowatthour			
Capital	32.64	12.16	28.71	11.12
Fixed	4.89	1.44	4.89	1.44
Variable	14.82	37.97	16.49	41.17
Incremental transmission	3.72	3.67	3.64	3.49
Total	**56.07**	**55.24**	**53.73**	**57.22**

Largest Capacity Additions Expected in the Southeast and the West

Figure 57. Electricity generation capacity additions, including combined heat and power, by region and fuel, 2006-2030 (gigawatts)

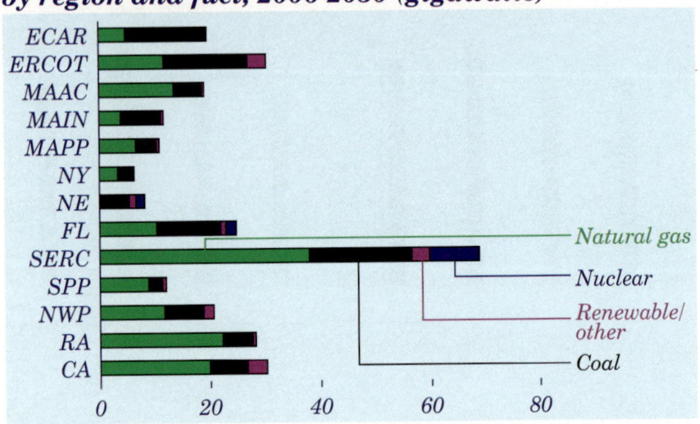

Most areas of the United States currently have excess generation capacity, but all electricity demand regions (see Appendix F for definitions) are expected to need additional, currently unplanned, capacity by 2030. The largest amounts of new capacity are expected in the Southeast (FL and SERC) and the West (NWP, RA, and CA). In the Southeast, electricity demand represents a relatively large share of total U.S. electricity sales, and its need for new capacity is greater than in other regions (Figure 57).

With natural gas prices rising in the reference case, coal-fired plants make up most of the capacity additions through 2030, given the assumption that current environmental policies are maintained indefinitely. The largest concentrations of new coal-fired plants are in the Southeast and the West. In the Southeast, new coal-fired plants are built in view of the size of the electricity market and the corresponding need for additional capacity. In the West, where the capacity requirement is much smaller, the choice to build mostly coal-fired plants is based on the region's lower-than-average coal prices and higher-than-average natural gas prices.

Nationwide, some new natural-gas-fired plants are built to maintain a diverse capacity mix or to serve as reserve capacity. Most are located in the Midwest (MAPP, MAIN, and ECAR) and Southeast (FL and SERC). The Midwest has a surplus of coal-fired generating capacity and does not need to add many new coal-fired plants. In the Southeast, natural-gas-fired plants are needed along with coal-fired plants to maintain diversity in the capacity mix.

EPACT2005 Tax Credits Are Expected To Stimulate New Nuclear Builds

Figure 58. Electricity generation from nuclear power, 1973-2030 (billion kilowatthours)

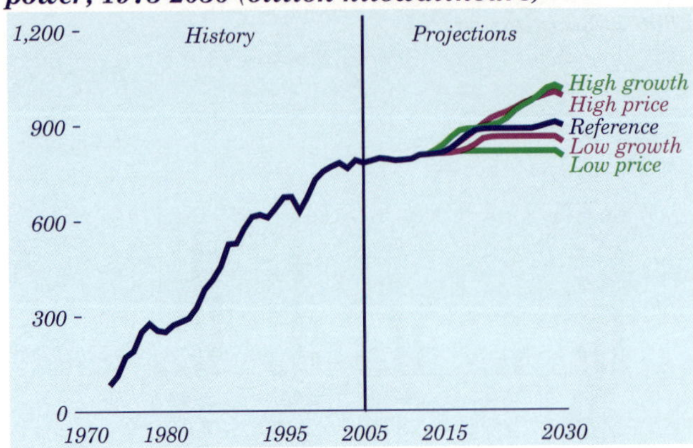

In the *AEO2007* reference case, nuclear capacity increases from 100.0 gigawatts in 2005 to 112.6 gigawatts in 2030. The change includes 2.7 gigawatts of capacity expansion at existing plants, 12.5 gigawatts of capacity at new plants, and 2.6 gigawatts of retirements of older units. EPACT2005 provides an 8-year PTC of 1.8 cents per kilowatthour for up to 6 gigawatts of new nuclear capacity built before 2021; however, the credit can be shared for additional capacity at a lower credit value. The reference case assumes that 9.0 gigawatts will be built by 2020 and will receive tax credits worth 1.2 cents per kilowatthour. The increase in capacity at existing units assumes that all uprates approved, pending, or expected by the NRC will be carried out.

Most existing nuclear units are expected to continue operating through 2030, based on the assumption that they will apply for and receive license renewals. Four units, totaling 2.6 gigawatts, are projected to be retired in 2030, when the date of their original licenses plus a 20-year renewal is reached.

Projected nuclear capacity additions vary, depending on overall demand for electricity and the prices of other fuels. Across the five main *AEO2007* cases, nuclear generation grows from 780 billion kilowatthours in 2005 to between 799 and 1,010 billion kilowatthours in 2030 (Figure 58). In the low price case, the delivered price of natural gas in 2030 is 10 percent lower than in the reference case, and new nuclear plants are not economical. In the high price and high growth cases, respectively, 24 and 27 gigawatts of new nuclear capacity are projected, because more capacity is needed and the cost of alternatives is higher.

When Lower Costs Are Assumed, New Nuclear Plants Are More Competitive

Figure 59. Levelized electricity costs for new plants by fuel type, 2015 and 2030 (2005 cents per kilowatthour)

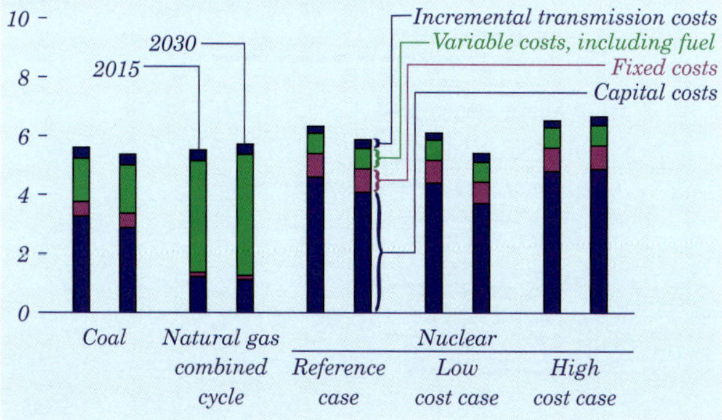

The reference case assumptions for the cost and performance characteristics of new technologies are based on cost estimates by government and industry analysts, allowing for uncertainties about new designs. Because no new nuclear plants have been ordered in this country since 1977, there is no reliable estimate of what they might cost. To test the significance of uncertainty in the assumptions, alternative cases vary key parameters. The low nuclear cost case assumes capital and operating costs 10 percent below those in the reference case in 2030, reflecting a 25-percent reduction in overnight capital costs from 2006 to 2030. The high nuclear cost case assumes no change in capital costs for advanced nuclear technologies from their 2006 levels.

Nuclear generating costs in the low nuclear cost case are more competitive with the generating costs for new coal- and natural-gas-fired units toward the end of the projection period (Figure 59). (The figure shows average generating costs, assuming generation at the maximum capacity factor for each technology; the costs and relative competitiveness of the technologies could vary by region.) In the reference case, Federal tax credits result in 9.0 gigawatts of new nuclear capacity by 2020, leading to lower costs in the future and an additional 3.5 gigawatts after the tax credits expire. In the low nuclear cost case, 28.5 gigawatts of new nuclear capacity is added between 2005 and 2030. The additional nuclear capacity displaces primarily new coal-fired capacity. In the high nuclear cost case, where capital costs are higher than expected, only 6 gigawatts of nuclear capacity is projected to be built, all due to the Federal tax credits.

Biomass and Wind Lead Projected Growth in Renewable Generation

Figure 60. Nonhydroelectric renewable electricity generation by energy source, 2005-2030 (billion kilowatthours)

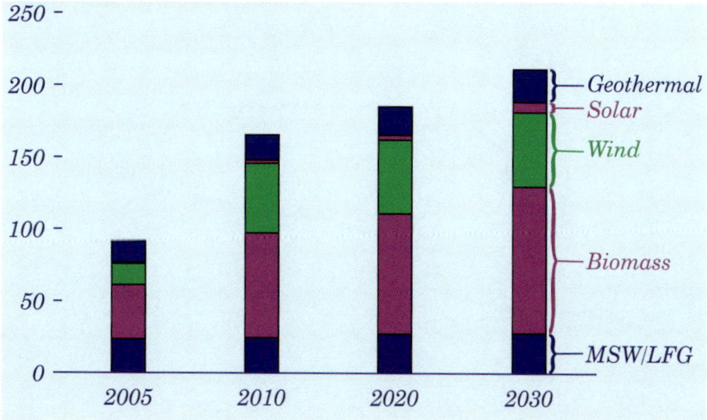

There is considerable uncertainty about the growth potential of wind power, which depends on a variety of factors, including fossil fuel costs, State renewable energy programs, technology improvements, access to transmission grids, public concerns about environmental and other impacts, and the future of the Federal PTC, which was set to expire at the end of 2007 but has been extended to 2008. In the *AEO2007* reference case, generation from wind power increases from 0.4 percent of total generation in 2005 to 0.9 percent in 2030 (Figure 60). Generation from geothermal facilities, while increasing, is not projected to gain market share and remains at its 2005 level of 0.4 percent of total generation in 2030, because opportunities for the development of new sites are limited. Most of the suitable sites, restricted mainly to Nevada and California, involve relatively high up-front costs and performance risks; and although geothermal power plants are eligible for the Federal PTC, the long construction lead times required make it unlikely that significant new capacity could be built in time to benefit from the current credit.

Among the other alternative fuel technologies, generation from municipal solid waste (MSW) and LFG stays at 0.5 percent of total generation. Solar technologies in general remain too costly for grid-connected applications, but demonstration programs and State policies support some growth in central-station solar PV, and small-scale customer-sited PV applications grow rapidly [*168*]. Grid-connected solar generation increases to 0.1 percent of total generation in 2030.

Technology Advances, Tax Provisions Increase Renewable Generation

Figure 61. Grid-connected electricity generation from renewable energy sources, 1990-2030 (billion kilowatthours)

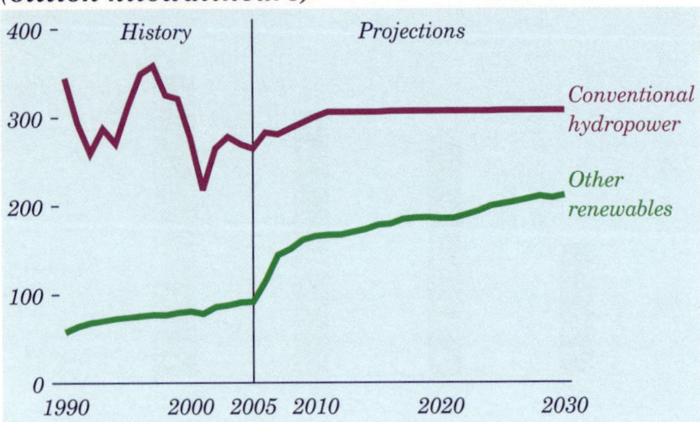

Despite technology improvements, rising fossil fuel costs, and public support, the contribution of renewable fuels to U.S. electricity supply remains relatively small in the *AEO2007* reference case at 9.0 percent of total generation in 2030—about the same as their share in 2005 (Figure 61). Although conventional hydropower remains the largest source of renewable generation through 2030, environmental concerns and the scarcity of untapped large-scale sites limit its growth, and its share of total generation falls from 6.6 percent in 2005 to 5.3 percent in 2030. Electricity generation from nonhydroelectric alternative fuels increases, however, bolstered by technology advances and State and Federal supports. The share of nonhydropower renewable generation increases by 60 percent, from 2.3 percent of total generation in 2005 to 3.6 percent in 2030.

Biomass is the largest source of renewable electricity generation among the nonhydropower renewable fuels. Co-firing with coal is relatively inexpensive when low-cost biomass resources are available. As low-cost feedstocks begin to be exhausted, however, more costly biomass resources are used, and new dedicated biomass facilities, such as IGCC plants, are built. Electricity generation from biomass increases from 1.0 percent of total generation in 2005 to 1.8 percent in 2030, with approximately 47 percent of the increase coming from biomass co-firing, 29 percent from dedicated power plants, and 25 percent from new on-site CHP capacity.

Renewables Are Expected To Become More Competitive Over Time

Figure 62. Levelized and avoided costs for new renewable plants in the Northwest, 2030 (2005 mills per kilowatthour)

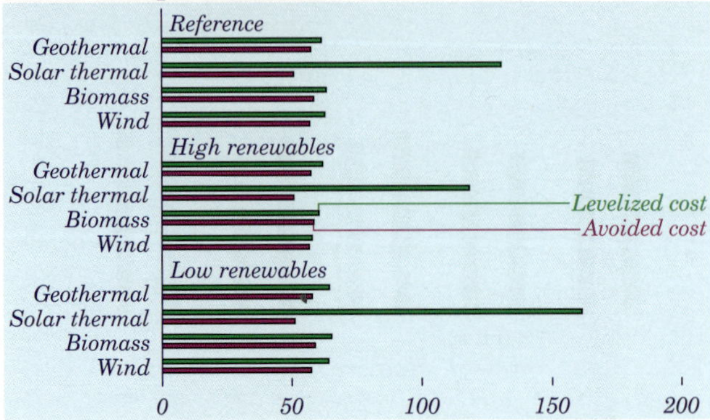

The competitiveness of both conventional and renewable generation resources is based on the most cost-effective mix of capacity that satisfies the demand for electricity across all hours and seasons. Baseload technologies tend to have low operating costs and set the market price for power only during the hours of least demand. Dispatchable geothermal and biomass resources compete directly with new coal and nuclear plants, which to a large extent determine the avoided cost [*169*] for baseload energy. In some regions and years, new geothermal or biomass plants may be competitive with new coal-fired plants, but their development is limited by the availability of geothermal resources or competitive biomass fuels.

Wind and solar are intermittent technologies that can be used only when resources are available. With relatively low operating costs and limited resource availability, their avoided costs are determined largely by the operating costs of the most expensive units in operation when their resources are available. Solar generators tend to operate during peak load periods, when natural-gas-fired combustion turbines and combined-cycle units with higher fuel costs determine avoided costs. The levelized cost of solar thermal generation is significantly higher than its avoided cost through 2030 (Figure 62). The availability of wind resources varies among regions, but wind plants tend to displace intermediate load generation. Thus, the avoided costs of wind power are determined largely by the low-to-moderate operating costs of combined-cycle and coal-fired plants, which set power prices during intermediate load hours. In some regions and years, levelized costs for wind power are approximately equal to its avoided costs.

State Portfolio Standards Increase Generation from Renewable Fuels

Figure 63. Renewable electricity generation, 2005-2030 (billion kilowatthours)

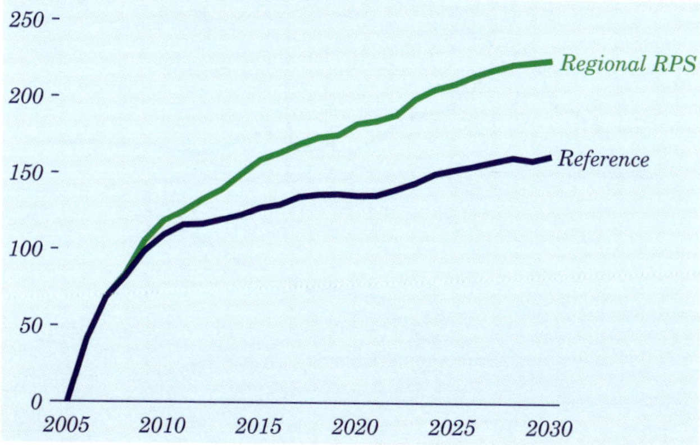

In 2005, 23 States and the District of Columbia had RPS or similar programs in effect. An alternative case was prepared for *AEO2007* to examine the potential impacts of full compliance with those programs. Because NEMS does not provide projections at the State level, the *AEO2007* regional RPS case assumed that all States would reach their goals within each program's legislative framework, and the results were aggregated at the regional level. In some States, however, compliance could be limited by authorized funding levels for the programs. For example, California is not expected to meet its renewable energy targets because of restraints on the funding of its RPS program.

In the regional RPS case, State renewable energy programs are projected to result in a national total of 61 billion kilowatthours of additional nonhydropower renewable generation in 2030 relative to the reference case, a 29-percent increase (Figure 63). Most of the additional generation is projected to come from biomass resources, with smaller increases for wind, municipal waste, and geothermal generation, which together account for 8 percent of the projected increase.

Nearly 5 gigawatts of additional new dedicated biomass capacity is projected for the mid-Atlantic region in the RPS case, as a result of the implementation of aggressive standards and the limited availability of other renewable resources. Florida, New York, and New England each would add 500 megawatts or more biomass capacity, whereas States in the West would add little new capacity beyond that projected in the reference case.

Fossil-Fired Capacity Additions Vary With Cost and Performance

Figure 64. Cumulative new generating capacity by technology type, 2006-2030 (gigawatts)

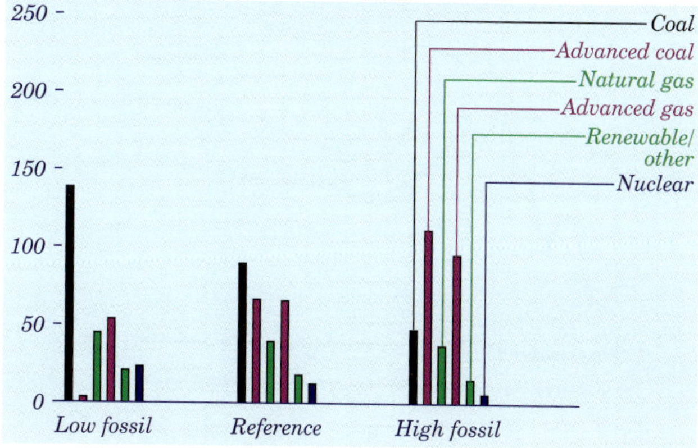

The cost and performance of various generating technologies in the reference case are determined in consultation with industry and government specialists. To test the significance of uncertainty in the assumptions, alternative cases vary key parameters. In the high fossil technology case, capital costs, heat rates, and operating costs for advanced fossil-fired generating technologies in 2030 are assumed to be 10 percent lower than in the reference case. The low fossil technology case assumes no change from the 2006 capital costs and heat rates for advanced technologies.

With different cost and performance assumptions, the mix of generating technologies changes (Figure 64). In all cases, assuming continuation of current environmental policies, coal technologies account for at least 50 percent of new capacity additions; in the high fossil technology case, 70 percent of coal-fired additions use advanced technologies, compared with only 2 percent in the low fossil case. Natural-gas-fired capacity makes up 35 to 42 percent of new additions in all cases. Advanced technologies represent 72 percent of those additions in the high fossil case and 55 percent in the low fossil case. The improved economics of advanced fossil technologies in the high fossil case result in fewer nuclear and renewable builds and more retirements of older steam units. Electricity prices are 2 percent lower in 2030 in the high fossil case than in the reference case. Because fossil-fired capacity is more costly in the low fossil case, more nuclear capacity (11 gigawatts) and slightly more renewable capacity are added; however, the higher costs of operating less efficient fossil-fired capacity in the low fossil technology case cause projected electricity prices in 2030 to be 2 percent higher than in the reference case.

Electricity Prices

Fuel Costs Drop from Recent Highs, Then Increase Gradually

Figure 65. Fuel prices to electricity generators, 1995-2030 (2005 dollars per million Btu)

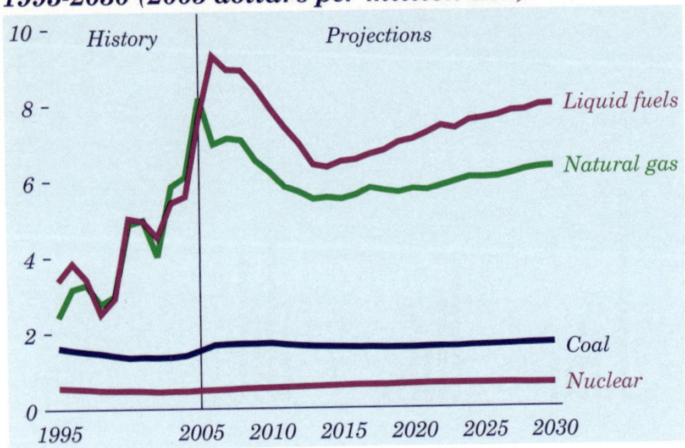

Electricity production costs are a function of fuel, operation and maintenance, and capital costs. In the reference case, fuel costs account for about two-thirds of production costs for new natural-gas-fired plants, less than one-third for new coal-fired units, and about one-tenth for new nuclear power plants in 2030. Generation from natural-gas-fired power plants increased in the early 2000s, but rising natural gas prices have increased their generation costs. After a 34-percent jump from 2004, natural gas prices were $8.18 per million Btu (2005 dollars) in 2005.

In the reference case, the price of natural gas delivered to the electric power sector drops to $5.50 per million Btu in 2013, then rises to $6.33 per million Btu in 2030 (Figure 65). Coal prices to the electric power sector remain relatively low, peaking at $1.71 per million Btu in 2010, falling to $1.69 per million Btu in 2018, and remaining at that level through 2030. Accordingly, the natural gas share of generation (including utilities, independent power producers, and end-use CHP) peaks at 22 percent in 2016, then drops to 16 percent in 2030 as prices rise, while the coal share increases from 50 percent in 2016 to 57 percent in 2030. Nuclear fuel costs rise steadily, to $0.62 per million Btu in 2030.

In the low and high price cases, coal prices to the power sector in 2030 are $1.51 and $1.80 per million Btu, respectively, and natural gas prices are $5.71 and $7.79 per million Btu. As a result, the respective coal and natural gas shares of total generation in 2030 are projected to be 45 percent and 27 percent in the low price case, as compared with 61 percent and 11 percent in the high price case.

Electricity Prices Moderate in the Near Term, Then Rise Gradually

Figure 66. Average U.S. retail electricity prices, 1970-2030 (2005 cents per kilowatthour)

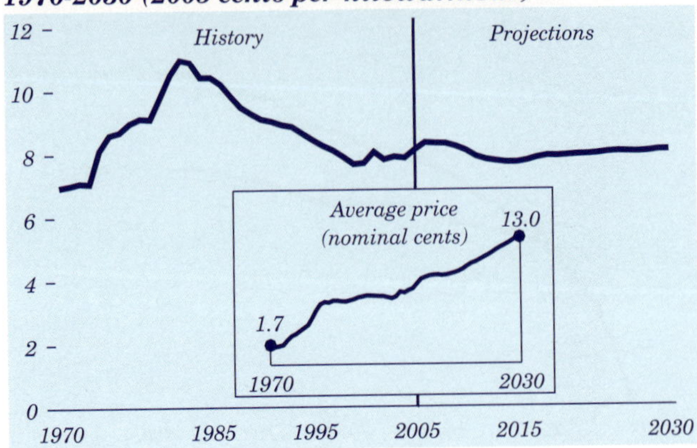

In the reference case, retail electricity prices peak at 8.3 cents per kilowatthour (2005 dollars) in 2006, then fall to 7.7 cents per kilowatthour in 2015 as new sources of natural gas and coal are brought on line. After 2013, fossil fuel prices rise slowly but steadily, and retail electricity prices also rise gradually after 2015, to 8.1 cents per kilowatthour in 2030 (Figure 66). Customers in States with competitive retail markets for electricity are expected to see the effects of changes in natural gas prices in their electricity bills more rapidly than those in regulated States, because competitive prices are determined by the marginal cost of energy rather than the average of all plant costs, and natural-gas-fired plants, with their higher operating costs, often set hourly marginal prices.

Electricity distribution costs are projected to decline by 8 percent from 2005 to 2030, as technology improvements and a growing customer base lower the cost of the distribution infrastructure. Transmission costs, on the other hand, increase by 29 percent, because additional investment is needed to meet consumers' growing demand for electricity and to facilitate competition in wholesale energy markets.

Economic expansion increases electricity consumption by businesses, factories, and residents as they buy and use more electrical equipment. Thus, over the long term, the rate of economic growth has a greater effect on the range of electricity prices than do oil and natural gas prices, because power suppliers can substitute coal, nuclear, and renewable fuels for expensive natural gas. In the low and high economic growth cases, electricity prices are 7.8 and 8.4 cents per kilowatthour, respectively, in 2030.

Projected Natural Gas Use for Electricity Generation Peaks in 2020

Figure 67. Natural gas consumption by sector, 1990-2030 (trillion cubic feet)

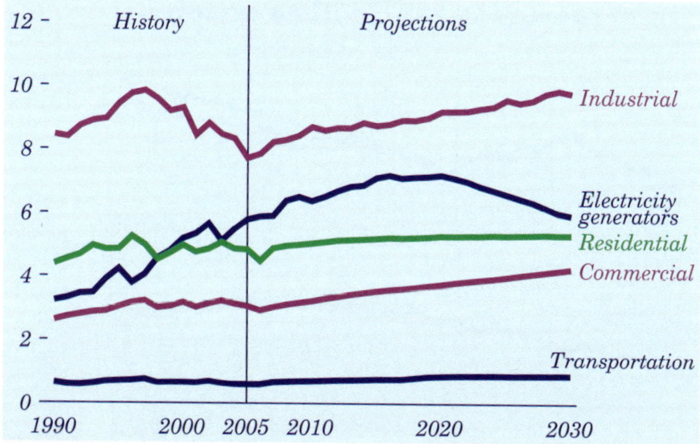

Total natural gas consumption in the United States is projected to increase from 22.0 trillion cubic feet in 2005 to 26.1 trillion cubic feet in 2030 in the *AEO2007* reference case. Much of the growth is expected before 2020, with demand for natural gas in the electric power sector growing from 5.8 trillion cubic feet in 2005 to a peak of 7.2 trillion cubic feet in 2020 (Figure 67). Natural gas use in the electric power sector declines after 2020, to 5.9 trillion cubic feet in 2030, as new coal-fired generating capacity displaces natural-gas-fired generation. Much of the projected decline in natural gas consumption for electricity generation results from higher delivered prices for natural gas in the reference case projection after 2020.

Continued growth in residential, commercial, and industrial consumption of natural gas is roughly offset by the projected decline in natural gas demand for electricity generation. As a result, overall natural gas consumption is almost flat between 2020 and 2030 in the *AEO2007* reference case, and the natural gas share of total projected energy consumption drops from 23 percent in 2005 to 20 percent in 2030.

Natural Gas Consumption Varies with Fuel Prices and Economic Growth

Figure 68. Total natural gas consumption, 1990-2030 (trillion cubic feet)

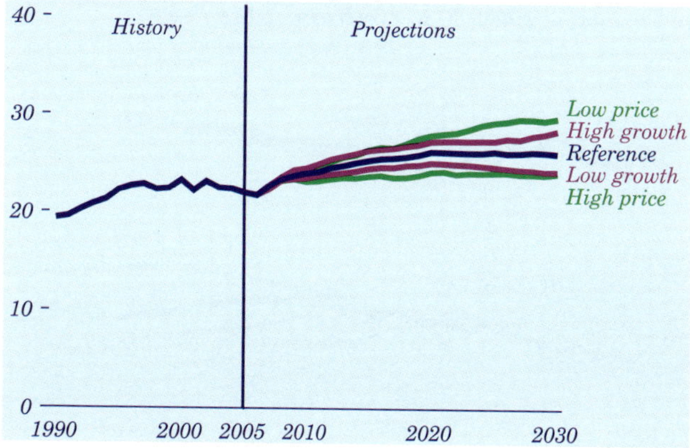

In the *AEO2007* projections, domestic natural gas consumption is influenced by the level of natural gas prices and the rate of economic growth. Higher (or lower) natural gas prices reduce (or increase) consumption, while higher (or lower) rates of economic growth increase (or reduce) gas consumption. The greatest variation occurs in the high and low price cases, where natural gas consumption in 2030 ranges from 29.7 trillion cubic feet in the low price case to 24.1 trillion cubic feet in the high price case (Figure 68). The high and low economic growth cases project natural gas consumption in 2030 at 28.4 trillion cubic feet and 24.2 trillion cubic feet, respectively.

The effects of economic growth on natural gas consumption are not as large as the effects of prices, because only a part of the incremental change in disposable personal income in the high and low economic growth cases is directed toward energy purchases. For example, when higher GDP growth is assumed, energy purchases make up a smaller proportions of GDP and of personal expenditures.

In contrast, the price of natural gas directly affects the level of natural gas consumption. High prices provide a direct economic incentive for users to reduce their natural gas consumption, and low prices encourage more consumption. The strength of the relationship between natural gas prices and consumption depends on the short- and long-term capabilities for fuel conservation and substitution in each consuming sector.

Natural Gas Demand

Natural Gas Use in the Electric Power Sector Is Sensitive to Prices

Figure 69. Natural gas consumption in the electric power and other end-use sectors in alternative price cases, 1990-2030 (trillion cubic feet)

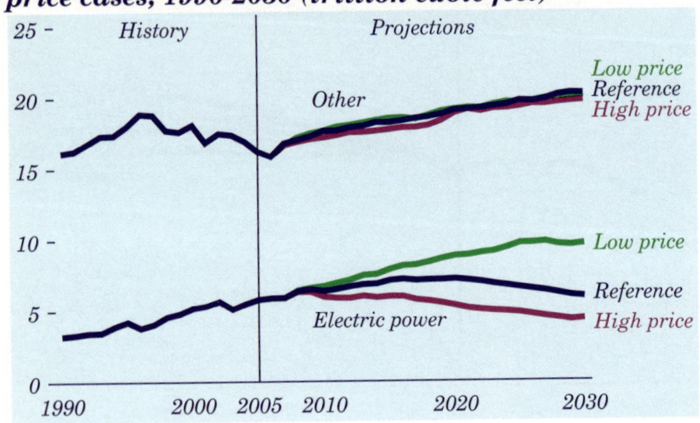

In the *AEO2007* projections, the largest variation in sectoral demand for natural gas in response to high and low price assumptions occurs in the electric power sector (Figure 69). Natural gas consumption by electricity producers in 2030, projected at 5.9 trillion cubic feet in the reference case, increases to 9.6 trillion cubic feet in the low price case but falls to 4.4 trillion cubic feet in the high price case.

Much of the variation in projected natural gas demand in the electric power sector between the low and high price cases is the result of different projections for the amount of natural-gas-fired generating capacity built—and consequently the amount of electricity generated from natural gas—from 2006 to 2030. In the high price case, a cumulative 70 gigawatts of new natural-gas-fired generating capacity is added between 2006 and 2030. In the low price case, cumulative natural-gas-fired capacity additions total 192 gigawatts over the same period. The projected totals for electricity generation from natural gas in 2030 are 649 billion kilowatthours in the high price case and 1,548 billion kilowatthours in the low price case.

In the residential, commercial, industrial, and transportation sectors, fuel price assumptions have a considerably smaller effect on natural gas consumption, because fuel substitution options are limited and the stocks of equipment that use natural gas have relatively slow turnover rates.

Natural Gas Use in Other Sectors Is Sensitive to Economic Growth

Figure 70. Natural gas consumption in the electric power and other end-use sectors in alternative growth cases, 1990-2030 (trillion cubic feet)

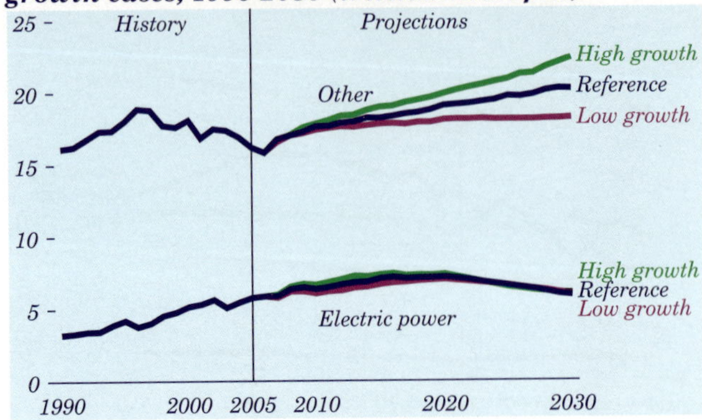

The largest variation in natural gas consumption in the residential, commercial, industrial, and transportation end-use sectors results from different assumptions about economic growth rates. In the high economic growth case, natural gas consumption in the other end-use sectors is projected to total 22.4 trillion cubic feet in 2030. In the low growth case, the projected total in 2030 is 18.2 trillion cubic feet (Figure 70). Most of the difference between the projections in the two cases is attributable to the industrial sector, where projected natural gas consumption in 2030 varies from 7.4 trillion cubic feet in the low growth case to 10.1 trillion cubic feet in the high growth case.

Natural gas consumption in the electric power sector is sensitive to natural gas prices because other fuels, such as coal, can be substituted directly for natural gas in generating electricity. In the high and low economic growth cases, however, natural gas consumption in the electric power sector shows little variation from the reference case projection. In the three cases (reference, high growth, and low growth), natural gas use for electricity generation in 2030 remains roughly constant, at about 6 trillion cubic feet. In the high economic growth case, when natural gas consumption in the electric power sector begins to rise, natural gas prices increase significantly, and in response coal and nuclear power are substituted for natural gas.

Projected Natural Gas Prices Remain Above Historical Levels

Figure 71. Lower 48 wellhead and Henry Hub spot market prices for natural gas, 1990-2030 (2005 dollars per thousand cubic feet)

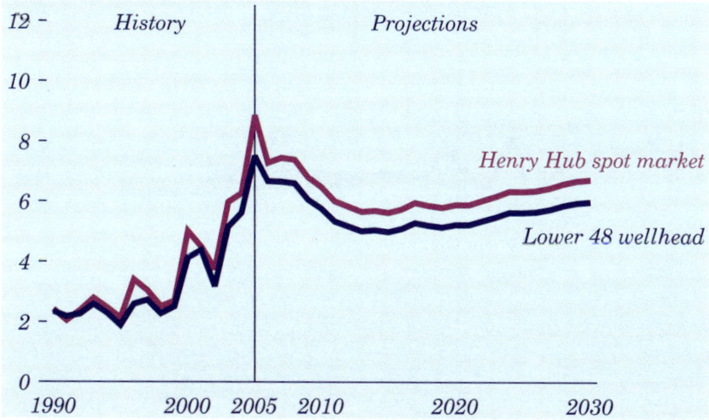

In the *AEO2007* reference case, lower 48 wellhead prices for natural gas are projected to decline from current levels to an average of $5.01 per thousand cubic feet (2005 dollars) in 2013, then rise to $5.98 per thousand cubic feet in 2030. Henry Hub spot market prices are projected to decline to $5.49 per million Btu ($5.33 per thousand cubic feet) in 2013 and then rise to $6.52 per million Btu ($6.33 per thousand cubic feet) in 2030 (Figure 71).

Current high natural gas prices are expected to stimulate the construction of new LNG terminal capacity, resulting in a significant increase in LNG import capacity. Projected natural gas prices in the reference case also are expected to stimulate the construction of an Alaska natural gas pipeline (projected to begin operation in 2018), as well as increased unconventional natural gas production.

On the demand side, current natural gas prices are sufficiently high to reduce growth in consumption. The combination of increased natural gas supply and slower growth in demand leads to a decline in natural gas prices through 2013. After 2013, wellhead natural gas prices increase largely as a result of rising costs, as technically recoverable U.S. natural gas resources decline from the current level (Table 17).

Table 17. Technically recoverable U.S. natural gas resources as of January 1, 2005 (trillion cubic feet)

Proved	*Unproved*	*Total*
192.5	1,148.5	1,341.0

Prices Vary With Resource Size and Technology Progress Assumptions

Figure 72. Lower 48 wellhead natural gas prices, 1990-2030 (2005 dollars per thousand cubic feet)

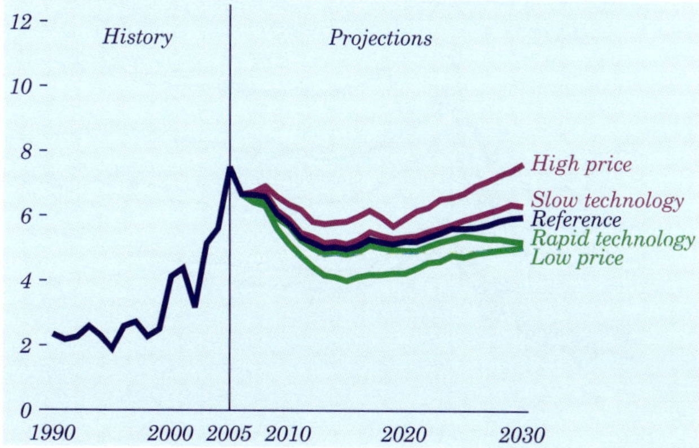

The high and low price cases assume that the unproven U.S. natural gas resource base is 15 percent lower and higher, respectively, than the estimate used in the reference case (Table 17). As a result, E&P costs—and wellhead prices—are higher in the high price case and lower in the low price case than projected in the reference case (Figure 72). In the low price case, wellhead natural gas prices increase to $5.06 per thousand cubic feet in 2030 (2005 dollars), as compared with $5.98 per thousand cubic feet in 2030 in the reference case. In the high price case, wellhead prices rise to $7.63 per thousand cubic feet in 2030.

Technological progress affects the future production of natural gas by reducing production costs and expanding the economically recoverable resource base. In the *AEO2007* reference case, the rate of improvement in natural gas production technology is based on the historical rate. The slow oil and natural gas technology case assumes an improvement rate 50 percent lower than in the reference case. As a result, future capital and operating costs are higher, causing the projected average wellhead price of natural gas to increase to $6.32 per thousand cubic feet in 2030. The rapid technology case assumes a rate of technology improvement 50 percent higher than in the reference case, reducing natural gas development and production costs. In the rapid technology case, wellhead natural gas prices are projected to average $5.21 per thousand cubic feet in 2030.

Delivered Natural Gas Prices Follow Trends in Wellhead Prices

Figure 73. Natural gas prices by end-use sector, 1990-2030 (2005 dollars per thousand cubic feet)

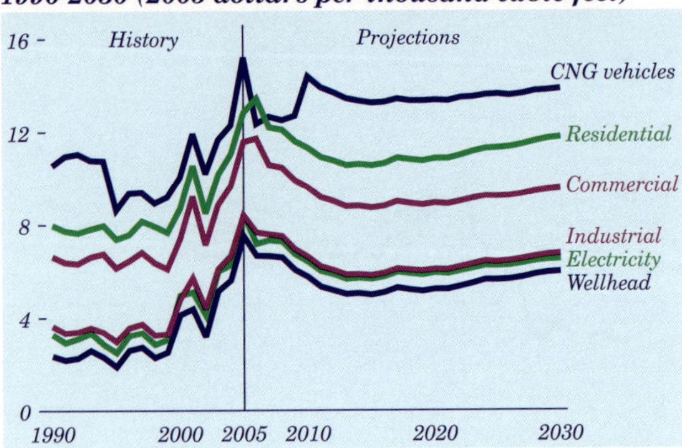

Trends in delivered natural gas prices largely reflect changes in projected wellhead prices. In the *AEO2007* reference case, prices for natural gas delivered to the end-use sectors decline through 2015 as wellhead gas prices decline, then increase along with wellhead prices over the rest of the projection period (Figure 73).

On average, projected end-use transmission and distribution margins remain relatively constant, because the cost of adding new facilities largely offsets the reduced depreciation expenses of existing facilities. Transmission and distribution margins in the end-use sectors reflect both the volumes of natural gas delivered and the infrastructure arrangements of the different sectors. The industrial and electricity generation sectors have the lowest end-use prices, because they receive most of their natural gas directly from interstate pipelines, avoiding local distribution charges. In addition, summer-peaking electricity generators reduce transmission costs by using interruptible transportation services during the summer, when there is spare pipeline capacity. As power generators take a larger share of the natural gas market, however, they are expected to rely more on higher cost firm transportation service.

The reference case assumes that sufficient transmission and distribution capacity will be built to accommodate the projected growth in natural gas consumption. If public opposition were to prevent infrastructure expansion, however, delivered prices could be higher than projected in the reference case.

Transmission and Distribution Costs Are Reduced With Higher Volumes

Figure 74. Average natural gas transmission and distribution margins, 1990-2030 (2005 dollars per thousand cubic feet)

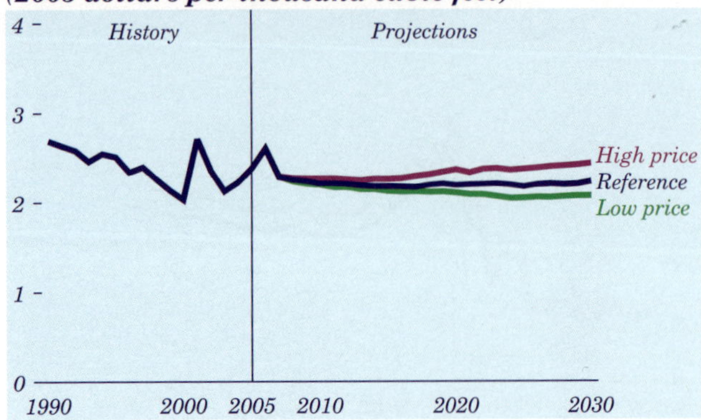

The transmission and distribution margin for natural gas delivered to end users is the difference between the average delivered price and the average source price, which is the quantity-weighted average of the lower 48 wellhead price and the average import price. It reflects both the capital and operating costs for pipelines and the volume of natural gas transported. Although operating costs vary with the level of pipeline utilization, capital costs are fixed for the most part. Variations in pipeline throughput result in higher or lower transmission and distribution costs per thousand cubic feet of natural gas transported. Thus, because the high and low price case projections show the greatest variation in total natural gas consumption, the greatest variation in transmission and distribution margins is also seen in those cases.

In the high price case, total natural gas consumption in 2030 is projected to be only 24.1 trillion cubic feet. As a result, the average transmission and distribution margin for delivered natural gas is projected to increase from \$2.38 per thousand cubic feet in 2005 to \$2.44 per thousand cubic feet in 2030 (2005 dollars). In the low price case, total natural gas consumption in 2030 grows to 29.7 trillion cubic feet, and the average transmission and distribution margin in 2030 drops to \$2.07 per thousand cubic feet. In the reference case, with projected natural gas consumption of 26.1 trillion cubic feet in 2030, the projected average transmission and distribution margin in 2030 is \$2.24 per thousand cubic feet (Figure 74).

Unconventional Production Is a Growing Source of U.S. Gas Supply

Figure 75. Natural gas production by source, 1990-2030 (trillion cubic feet)

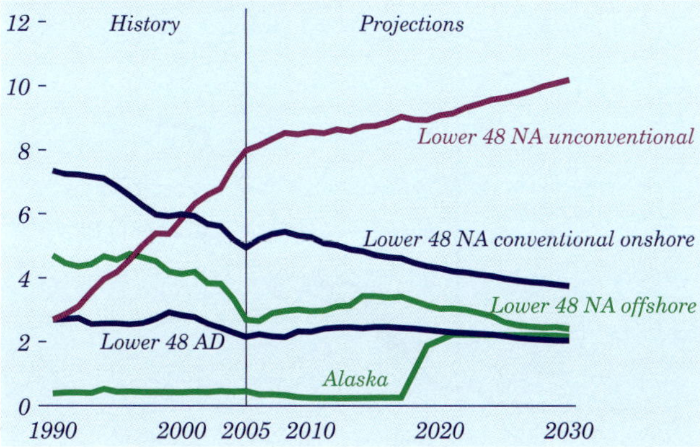

A large proportion of the onshore lower 48 conventional natural gas resource base has been discovered. Discoveries of new conventional natural gas reservoirs are expected to be smaller and deeper, and thus more expensive and riskier to develop and produce. Accordingly, total lower 48 onshore conventional natural gas production declines in the *AEO2007* reference case from 6.4 trillion cubic feet in 2005 to 4.9 trillion cubic feet in 2030 (Figure 75).

Incremental production of lower 48 onshore natural gas comes primarily from unconventional resources, including coalbed methane, tight sandstones, and gas shales. Lower 48 unconventional production increases in the reference case from 8.0 trillion cubic feet in 2005 to 10.2 trillion cubic feet in 2030, when it accounts for 50 percent of projected domestic U.S. natural gas production.

The Alaska natural gas pipeline is expected to begin transporting natural gas to the lower 48 States in 2018. In 2030, Alaska's natural gas production totals 2.2 trillion cubic feet in the reference case.

Considerable natural gas resources remain in the offshore Gulf of Mexico, especially in the deep waters. Deepwater natural gas production in the Gulf of Mexico increases in the reference case from 1.4 trillion cubic feet in 2005 to a peak volume of 3.1 trillion cubic feet in 2015, then declines to 2.1 trillion cubic feet in 2030. Production in the shallow waters declines throughout the projection period, from 2.0 trillion cubic feet in 2005 to 1.1 trillion cubic feet in 2030.

Natural Gas Supply Projections Reflect Rates of Technology Progress

Figure 76. Total U.S. natural gas production, 1990-2030 (trillion cubic feet)

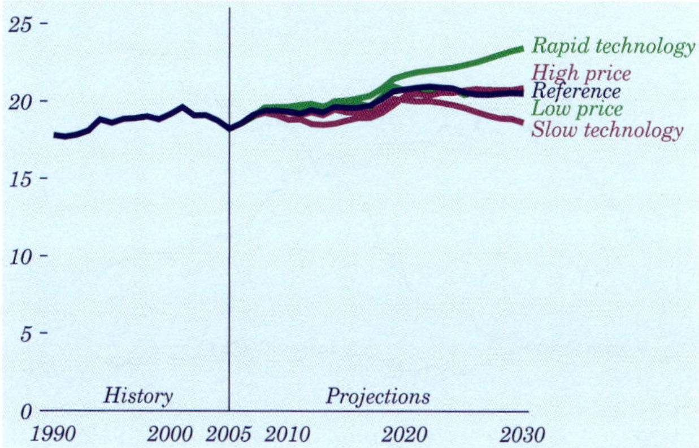

Exploration for and production of natural gas becomes more profitable when prices increase and when exploration and development costs decline. The rapid and slow technology cases show the effects of different assumed rates of technology improvement in the oil and natural gas industries. The high and low price cases show the effects of different assumptions for oil prices and unproved oil and natural gas resources.

Technological progress generally reduces the cost of natural gas production, leading to lower wellhead prices, more end-use consumption, and more production. More rapid progress increases domestic natural gas production and slower progress lowers production in the technology cases. U.S. natural gas production in 2030 is 14.2 percent higher in the rapid technology case and 9.1 percent lower in the slow technology case than in the reference case (Figure 76).

The high and low price cases show smaller effects on total production than do the technology cases. Domestic natural gas production is determined by balancing total U.S. natural gas supply and demand. Higher world oil prices—in combination with a smaller world natural gas resource base—lead to higher costs for developing domestic resources, higher wellhead natural gas prices, and lower levels of U.S. consumption and imports of natural gas. Lower world oil prices—and a larger oil and natural gas resource base—lead to lower resource development costs, lower prices, and higher levels of consumption and imports. The net effect in each case is a small variation in U.S. natural gas production, as changes in production costs, consumption, and imports counter the impacts of higher or lower natural gas prices.

Net Imports of Natural Gas Grow in the Projections

Figure 77. Net U.S. imports of natural gas by source, 1990-2030 (trillion cubic feet)

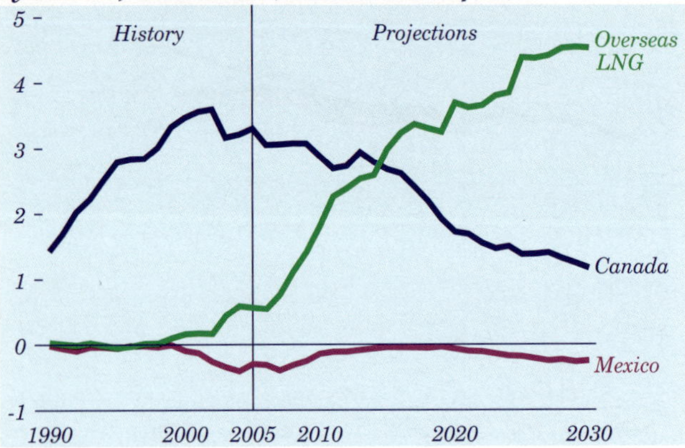

With U.S. natural gas production remaining relatively constant, imports of natural gas are projected to rise to meet an increasing share of domestic consumption. Most of the expected growth in U.S. natural gas imports is in the form of LNG. The total capacity of U.S. LNG receiving terminals increases from 1.4 trillion cubic feet in 2005 to 6.5 trillion cubic feet in 2030 in the reference case, and net LNG imports grow from 0.6 trillion cubic feet in 2005 to 4.5 trillion cubic feet in 2030 (Figure 77). Nevertheless, the U.S. LNG market is expected to be tight until 2012, because of supply constraints at a number of liquefaction facilities, delays in the completion of new liquefaction projects, and rapid growth in global LNG demand.

A projected decline in Canada's non-Arctic conventional natural gas production is only partly offset by an increase in its Arctic and unconventional production. Although a MacKenzie Delta natural gas pipeline is expected to begin transporting natural gas in 2012 in the *AEO2007* reference case, its impact is offset by an expected decline in conventional natural gas resources in Alberta and increases in Canada's domestic consumption. Accordingly, net imports of natural gas from Canada are projected to fall in the reference case from 3.3 trillion cubic feet in 2005 to 1.2 trillion cubic feet in 2030.

Net exports of U.S. natural gas to Mexico are projected to decline from nearly 400 billion cubic feet in 2007 to 35 billion in 2019. After 2019 they are expected to increase steadily to nearly 250 billion cubic feet in 2030.

LNG Imports Are the Source of Supply Most Affected in the Price Cases

Figure 78. Net U.S. imports of liquefied natural gas, 1990-2030 (trillion cubic feet)

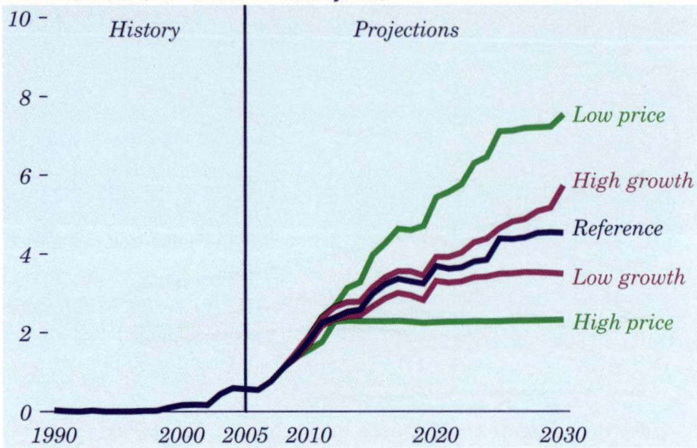

Changes in LNG imports account for most of the variation in net U.S. natural gas imports across the alternative price and economic growth cases. Unlike the situation in Canada and the United States, in much of the rest of the world the natural gas resource base has not been significantly exploited. Thus, there is ample potential for growth in LNG supply.

The *AEO2007* reference case projects net U.S. imports of LNG totaling 4.5 trillion cubic feet in 2030. The alternative projections of net LNG imports in 2030 are 7.5 trillion cubic feet in the low price case, 2.3 trillion cubic feet in the high price case, 5.7 trillion cubic feet in the high economic growth case, and 3.5 trillion cubic feet in the low economic growth case (Figure 78).

Higher oil prices are expected to reduce world petroleum consumption and increase natural gas consumption. In addition, some LNG contract prices are tied directly to crude oil prices, which could exert upward pressure on LNG prices. Higher oil prices are also projected to spur greater GTL production around the world, further increasing the pressure on natural gas prices. Collectively, these trends are expected to increase natural gas and LNG prices in both U.S. and international energy markets. Higher LNG prices, in turn, are projected to slow the rate of expansion of U.S. LNG terminal capacity and lower the capacity utilization rates at existing LNG terminals.

U.S. Crude Oil Production Is Expected To Grow Over the Next Decade

Figure 79. Domestic crude oil production by source, 1990-2030 (million barrels per day)

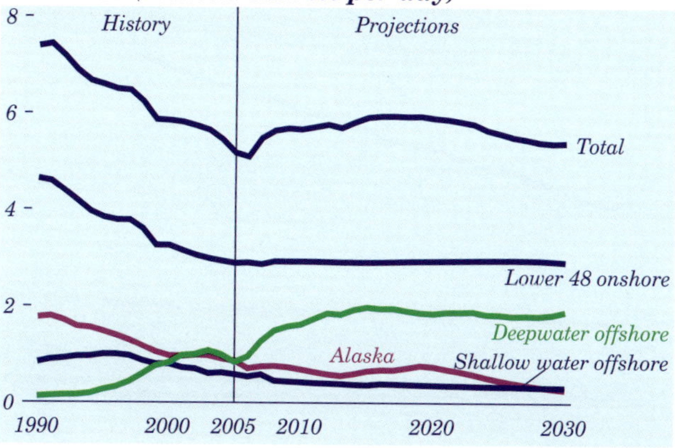

A large portion of the total U.S. resource base of onshore conventional oil has been produced. New oil reservoir discoveries are likely to be smaller, more remote (e.g., Alaska), and increasingly costly to exploit. However, higher oil prices, increased production with enhanced oil recovery techniques, and recent resource discoveries in the Bakken shale formation in Montana allow lower 48 onshore production to remain relatively constant at about 2.9 million barrels per day over the projection period in the *AEO2007* reference case (Figure 79).

Because drilling currently is prohibited in the Arctic National Wildlife Refuge (ANWR), the reference case does not project any production from ANWR. Alaska's projected oil production declines from 860,000 barrels per day in 2005 to 270,000 barrels per day in 2030.

Considerable oil resources remain offshore, especially in the deep waters of the Gulf of Mexico. Deepwater oil production in the Gulf of Mexico is projected to increase from 840,000 barrels per day in 2005 to a peak of 2.0 million barrels per day in 2015 and then fluctuate between 1.8 and 1.9 million barrels per day over the last 15 years of the projection. Production from the shallow waters of the Gulf is projected to continue declining, from 470,000 barrels per day in 2005 to 290,000 barrels per day in 2030. As a result, total domestic offshore oil production increases in the reference case from 1.4 million barrels per day in 2005 to a peak of 2.3 million barrels per day in 2015, then declines to 2.2 million barrels per day in 2030.

More Rapid Technology Advances Could Raise U.S. Oil Production

Figure 80. Total U.S. crude oil production, 1990-2030 (million barrels per day)

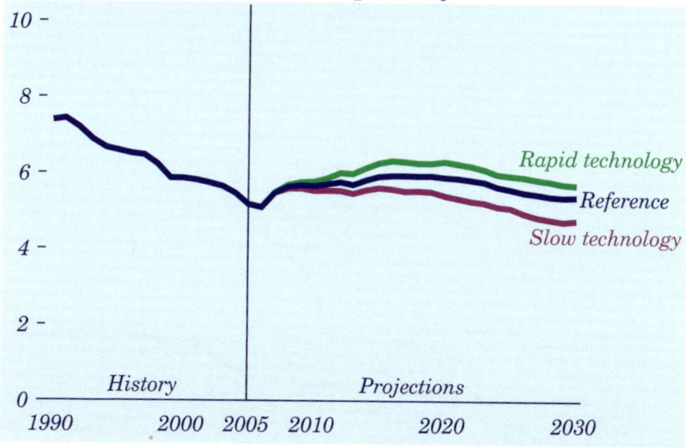

The rapid and slow oil and gas technology cases assume rates of technological progress in the petroleum industry that are 50 percent higher and 50 percent lower, respectively, than the historical rate. The rate of technological progress determines the projected cost of developing and producing the remaining domestic oil resource base. Higher (or lower) rates of technological progress result in lower (or higher) oil development and production costs, which in turn allow more (or less) oil production.

With domestic oil consumption determined largely by oil prices and economic growth rates, oil consumption does not change significantly in the technology cases. Domestic crude oil production in 2030, which is 5.4 million barrels per day in the reference case, increases to 5.7 million barrels per day in the rapid technology case and drops to 4.8 million barrels per day in the slow technology case (Figure 80). The projected changes in domestic oil production result in different projections for petroleum imports. In 2030, projected net crude oil and petroleum product imports range from 16.0 million barrels per day in the rapid technology case to 17.0 million barrels per day in the slow technology case, as compared with 16.4 million barrels per day in the reference case.

Cumulative U.S. crude oil production from 2006 through 2030 is projected to be 2.6 billion barrels (4.9 percent) higher in the rapid technology case and 3.3 billion barrels (6.4 percent) lower in the slow technology case than the reference case projection of 51.8 billion barrels.

Unconventional Liquids Production Increases With Higher Oil Prices

Figure 81. Total U.S. unconventional oil production, 2005-2030 (thousand barrels per day)

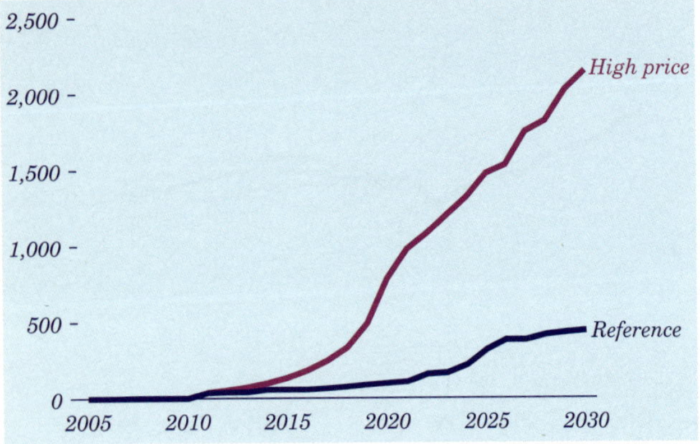

The future of unconventional oil and liquids production (such as oil shale, CTL, and GTL) will depend on oil prices. For example, CTL production is projected in both the reference and high price cases; GTL and oil shale production are projected only in the high price case; and no unconventional oil production of any kind is projected in the low price case.

In the reference case, CTL production is projected to start at about 40,000 barrels per day in 2011 and increase to about 440,000 barrels per day in 2030. In the high price case, CTL, oil shale, and GTL production are projected to be economically feasible, and total domestic production of unconventional oil is projected to reach 2.1 million barrels per day in 2030 (Figure 81). Of that total, CTL is projected to account for 1.6 million barrels per day and oil shale 405,000 barrels per day. Because natural gas prices are relatively high throughout the projections, GTL production reaches only about 100,000 barrels per day in 2030 in the high price case.

The costs of unconventional oil production are uncertain. As an example, current CTL technology produces significant amounts of CO_2, and if Federal restrictions on CO_2 emissions were enacted in the future, CTL production costs could rise substantially.

Transportation Uses Lead Growth in Liquid Fuels Consumption

Figure 82. Liquid fuels consumption by sector, 1990-2030 (million barrels per day)

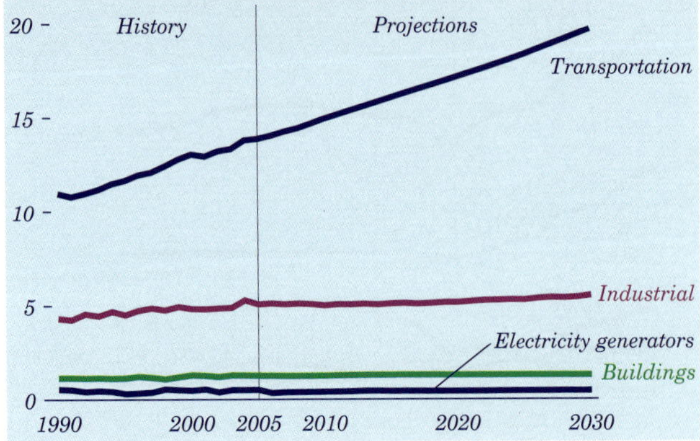

U.S. consumption of liquid fuels—including fuels from petroleum-based sources and, increasingly, those derived from such nonpetroleum primary fuels as coal, biomass, and natural gas—is projected to total 26.9 million barrels per day in 2030, an increase of 6.2 million barrels per day over the 2005 total. Most of the increase is in the transportation sector, which is projected to account for 73 percent of total liquid fuels consumption in 2030, up from 67 percent in 2005 (Figure 82).

Liquid fuels use for transportation increases by 5.8 million barrels per day from 2005 to 2030 in the *AEO2007* reference case, by 7.8 million barrels per day in the high economic growth case, and by 3.8 million barrels per day in the high price case. Gasoline, ULSD, and jet fuel are the main transportation fuels. The reference case includes the effects of technology improvements that are expected to increase the efficiency of motor vehicles and aircraft, but the projected growth in demand for each mode outpaces those improvements as the demand for transportation services grows in proportion to increases in population and GDP.

Consumption of liquid fuels from nonpetroleum sources increases substantially over the projection period. Ethanol, which made up 3 percent of the motor gasoline pool in 2005, increases to approximately 8 percent of the total motor gasoline pool in 2030. Total production of liquid fuels from CTL plants, which are expected to commence operation in 2011, increases in the reference case to 440,000 barrels per day—equivalent to 7 percent of the total pool of distillate fuel—in 2030.

Imports of Liquid Fuels Increase With Rising U.S. Demand

Figure 83. Net import share of U.S. liquid fuels consumption, 1990-2030 (percent)

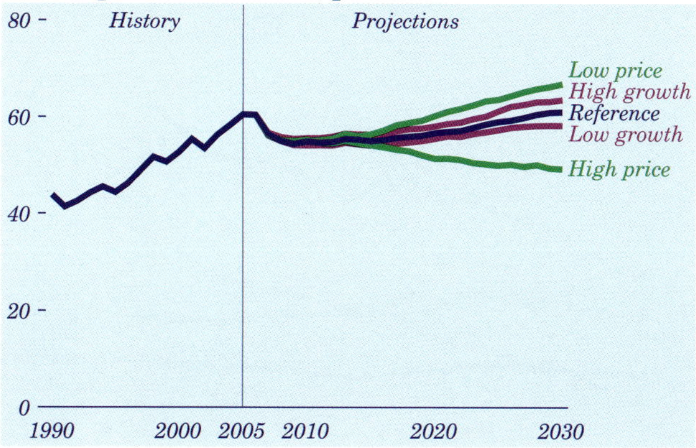

In 2005, net imports of liquid fuels, primarily petroleum, accounted for 60 percent of domestic consumption. The United States is expected to continue its dependence on liquid fuel imports in the *AEO2007* reference case. The import share of domestic consumption declines slightly to 55 percent in 2015 before climbing to 61 percent in 2030 (Figure 83). Dependence on imports is tied to total consumption. In the high price case, net imports as a share of domestic consumption of liquid fuels fall to 49 percent in 2030. In the low price case, dependence on petroleum imports increases to 67 percent in 2030 as U.S. demand for lower priced fuels increases more rapidly than domestic production.

In the reference case, demand for refined products continues to increase more rapidly than refining capacity. Historically, the availability of product imports has been limited by a lack of foreign refineries capable of meeting the stringent U.S. standards for liquids products. More recently, however, liquids demand has grown rapidly in some countries of Eastern Europe and Asia, and those nations are moving to adopt the same quality standards as the developed world. As a result, refineries throughout the world are becoming more sophisticated, and in the future more of them will be able to provide products suitable for the U.S. market, which they may do if it is profitable.

U.S. Motor Gasoline Prices Rise and Fall With Changes in World Oil Price

Figure 84. Average U.S. delivered prices for motor gasoline, 1990-2030 (2005 dollars per gallon)

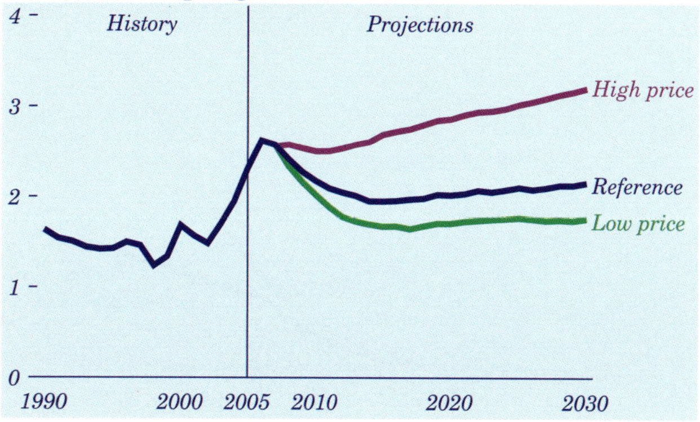

The retail prices of petroleum products largely follow changes in crude oil prices. In the reference case, the world oil price path reaches a low of about $50 per barrel in 2014, then increases slowly to about $59 in 2030 (2005 dollars). The reference case projections for average U.S. average motor gasoline prices follow the same trend, rising from $1.95 per gallon in 2014 to $2.15 in 2030.

In the high price case, with the price of imported crude oil projected to rise to more than $100 per barrel in 2030, the average price of U.S. motor gasoline follows the higher price path of world oil prices, increasing from $2.61 per gallon in 2014 to a high of $3.20 per gallon in 2030. In the low price case, gasoline prices decline to a low of $1.64 per gallon in 2017, increase slowly through the early 2020s, and level off at about $1.76 per gallon through 2030 (Figure 84).

Because changes from the reference case assumptions for economic growth rates have less pronounced effects on projected motor gasoline prices than do changes in oil price assumptions, the projected average prices for U.S. motor gasoline in the high and low economic growth cases are close to those in the reference case. In the high growth case, the average gasoline price falls to a low of $2.00 per gallon in 2016, then rises to $2.21 per gallon in 2030. In the low growth case, the average price reaches a low of $1.92 per gallon in 2014, then rises to $2.08 per gallon in 2030.

Lower Costs, Greater Demand Could Spur Cellulose Ethanol Production

Figure 85. Cellulose ethanol production, 2005-2030 (billion gallons per year)

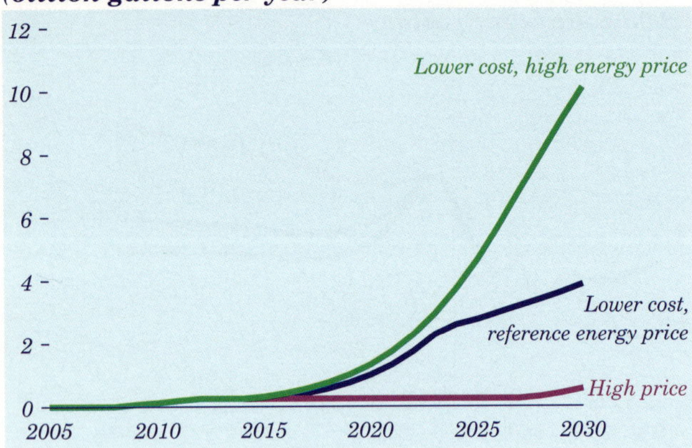

For *AEO2007*, two alternative ethanol cases examine the potential impact on ethanol demand of lower costs for cellulosic ethanol production, in combination with policies that increase sales of FFVs [*170*]. The reference case projects that 10.5 percent of new light-duty vehicles will be capable of burning E85 in 2016. The lower cost ethanol case using reference energy prices assumes that capital and operating costs for cellulose ethanol plants in 2018 are 20 percent lower than projected in the reference case, that at least 80 percent of new light-duty vehicles in 2016 can run on E85, and that energy prices will be the same as projected in the reference case. The lower cost ethanol case using high energy prices is based on the same assumptions for cellulose ethanol plant costs and FFV sales but with energy prices from the high price case.

E85 is projected to be competitive with gasoline in both alternative ethanol cases, and projected demand for ethanol fuels increases accordingly. In the lower cost ethanol case with reference prices, E85 demand in 2030 is projected to be 1.9 billion gallons, or 1.7 billion gallon higher than in the reference case. In the lower cost ethanol case with high energy prices, E85 demand in 2030 is projected to be 27.9 billion gallons, or 24.7 billion gallons higher than in the high price case. Increased demand for E85 and reduced production costs in the alternative ethanol cases result in increased production of cellulosic ethanol, which exceeds the mandated level in 2015 in both cases, growing to 3.9 billion gallons per year in 2030 in the lower cost ethanol case with reference prices and to 10.1 billion gallons per year in 2030 in the lower cost ethanol case with high energy prices (Figure 85).

Western Coal Production Continues To Increase Through 2030

Figure 86. Coal production by region, 1970-2030 (quadrillion Btu)

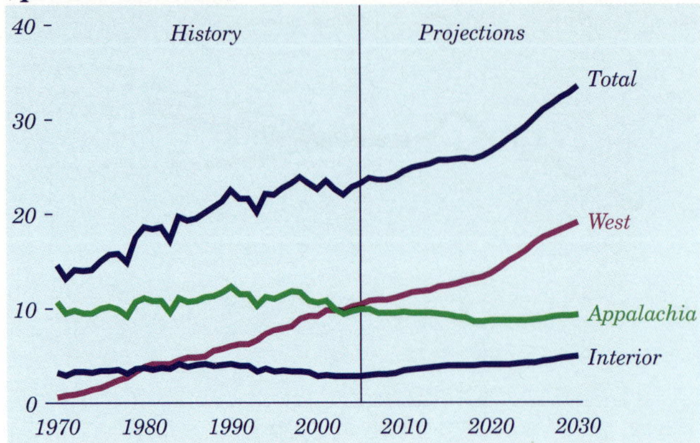

In the *AEO2007* reference case, increasing coal use for electricity generation at existing plants and construction of a few new coal-fired plants lead to annual production increases that average 1.1 percent per year from 2005 to 2015, when total production is 25.7 quadrillion Btu. The growth in coal production is even stronger from 2015 to 2030, averaging 1.8 percent per year, as substantial amounts of new coal-fired generating capacity are added and several CTL plants are brought on line.

Western coal production, which has grown steadily since 1970, continues to increase through 2030 (Figure 86). Much of the projected growth is in output from the Powder River Basin, where producers are well positioned to increase production from the vast remaining surface-minable reserves. Constraints on rail capacity limited growth in coal production from the Basin during 2005 and 2006, but recent and planned maintenance and investment in the rail infrastructure serving the region should allow for substantial growth in future production.

Appalachian coal production declines slightly in the reference case. Although producers in Central Appalachia are well situated to supply coal to new generating capacity in the Southeast, the Appalachian basin has been mined extensively, and production costs have been increasing more rapidly than in other regions. The eastern portion of the Interior coal basin (Illinois, Indiana, and western Kentucky), with extensive reserves of mid- and high-sulfur bituminous coals, benefits from the new coal-fired generating capacity in the Southeast.

Eastern Power Plants Are Expected To Use More Western Coal

Figure 87. Distribution of coal to domestic markets by supply and demand regions, including imports, 2005 and 2030 (quadrillion Btu)

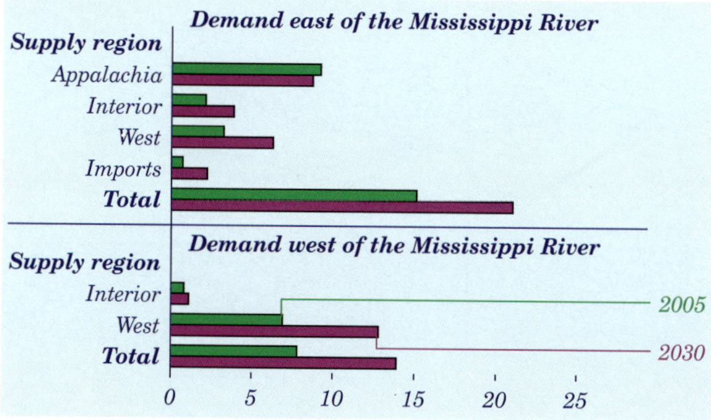

In the reference case, coal use is expected to grow substantially throughout the United States. For States east of the Mississippi River, coal demand is projected to increase by 5.9 quadrillion Btu, or 39 percent, from 2005 to 2030. Much of that increase is expected to be met by western coal—particularly in those States that are relatively close to the Powder River Basin supply region. Coal supply from Appalachian producers to markets east of the Mississippi River remains close to current levels, but increases in shipments from mines in the Eastern Interior region and in coal imports contribute to the overall decline in Appalachia's share of the market east of the Mississippi, from 61 percent in 2005 to 42 percent in 2030.

West of the Mississippi River, coal demand is projected to increase by 6.1 quadrillion Btu, or 79 percent, from 2005 to 2030, with western coal producers as the primary source of supply (Figure 87). Most of the remainder is expected to be supplied from lignite mines in the Gulf Coast area, primarily in Texas.

East of the Mississippi River, an increase in utilization rates for existing coal-fired power plants—from 71 percent in 2005 to 82 percent in 2030—accounts for approximately 30 percent of the projected increase in coal demand for the electric power sector. In contrast, west of the Mississippi, existing coal-fired plants already are operating at an average utilization rate of 80 percent. Therefore, increased utilization accounts for only a small amount of the projected increase in the region's coal demand over the projection period.

Long-Term Production Outlook Varies Considerably Across Cases

Figure 88. U.S. coal production, 2005, 2015, and 2030 (quadrillion Btu)

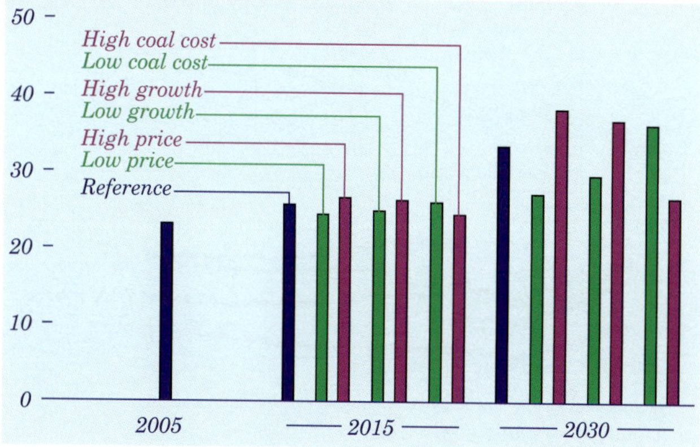

In all the *AEO2007* cases, U.S. coal production is projected to increase from 2005 to 2030; however, different assumptions about economic growth and the costs of producing fossil fuels lead to different results. The reference case projects a 44-percent increase from 2005 to 2030, whereas the alternative cases show increases ranging from as little as 15 percent to as much as 65 percent (Figure 88). Because the level of uncertainty is higher in the longer term, the projected increases in coal production from 2005 to 2015 show significantly less variation, ranging from 6 percent to 15 percent.

Regional coal production trends generally follow the national trend. For example, production of subbituminous coal in Wyoming's Powder River Basin is projected to increase by 73 percent from 2005 to 2030 in the reference case, as compared with 45 percent in the low price case and 95 percent in the high price case. The projected regional shares of total coal production in 2030 (from the Appalachian, Interior, and Western supply regions) do not vary by much among the reference, high and low price, and high and low economic growth cases.

In the high coal cost case, higher mining and transportation costs for coal from the Powder River Basin hold the projected increase in the region's annual coal production from 2005 to 2030 to a relatively small 0.2 quadrillion Btu, or 2 percent. As a result, the Wyoming Powder River Basin share of total U.S. coal production in 2030 is 26 percent in the high coal cost case, as compared with 33 percent to 36 percent in the other cases.

Coal Prices

Minemouth Coal Prices in the Western and Interior Regions Increase Slowly

Figure 89. Average minemouth price of coal by region, 1990-2030 (2005 dollars per million Btu)

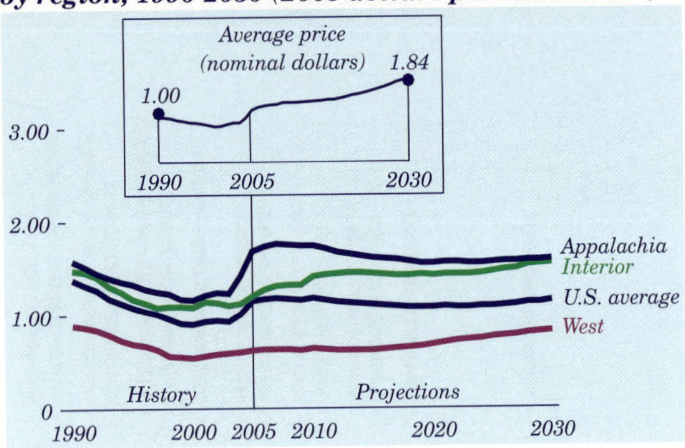

From 1990 to 1999, the average minemouth price of coal declined by 4.5 percent per year, from $1.38 per million Btu (2005 dollars) to $0.91 per million Btu (Figure 89). Increases in U.S. coal mining productivity of 6.3 percent per year helped to reduce mining costs and contributed to the price decline. Since 1999, U.S. coal mining productivity has declined by 0.6 percent per year, and the average minemouth coal price has increased by 3.9 percent per year, to $1.15 per million Btu in 2005.

In the reference case, the average minemouth coal price drops slightly from 2010 to 2019, as mine capacity utilization declines and production shifts away from higher cost Central Appalachian mines. After 2019, rising natural gas prices and the need for additional generating capacity result in the construction of 119 gigawatts of new coal-fired generating plants. The substantial investment in new mining capacity required to meet increasing demand during the period, combined with low productivity growth and rising utilization of mining capacity, leads to an increase in the average minemouth price, from $1.08 per million Btu in 2019 to $1.15 per million Btu in 2030. In the projection, the increasing share of lower rank coals (subbituminous and lignite) in the U.S. production mix tempers the price increase.

Strong growth in production in the Interior and Western supply regions, combined with limited improvement in coal mining productivity, results in minemouth price increases of 1.0 and 1.1 percent per year, respectively, for the two regions from 2005 through 2030. Average minemouth prices in Appalachia decline by 0.2 percent per year over the same period.

Higher Mining and Transportation Costs Raise Delivered Coal Prices

Figure 90. Average delivered coal prices, 1990-2030 (2005 dollars per million Btu)

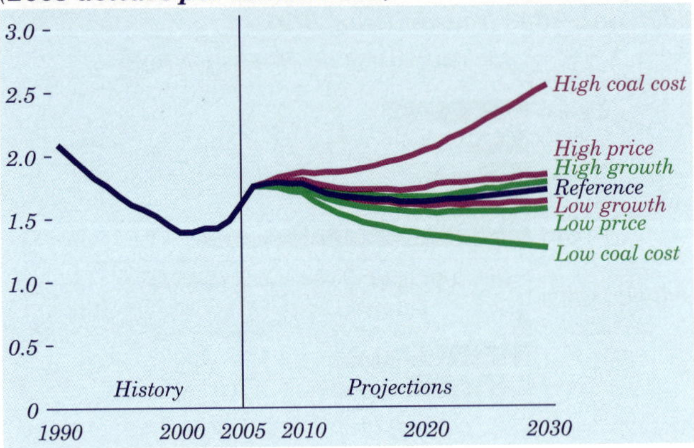

Alternative assumptions for coal mining and transportation costs affect coal prices and demand. Two alternative coal cost cases developed for *AEO2007* examine the impacts on U.S. coal markets of alternative assumptions about mining productivity, labor costs, and mine equipment costs on the production side, and about railroad productivity and rail equipment costs on the transportation side.

In the high coal cost case, the average delivered coal price in 2005 dollars is $2.54 per million Btu in 2030—49 percent higher than in the reference case (Figure 90). As a result, U.S. coal consumption is 6.4 quadrillion Btu (18 percent) lower than in the reference case in 2030, reflecting both a switch from coal to natural gas, nuclear, and renewables in the electricity sector and reduced CTL production. In the low coal cost case, the average delivered coal price in 2030 is $1.25 per million Btu—27 percent lower than in the reference case—and total coal consumption is 2.3 quadrillion Btu (9 percent) higher than in the reference case.

Because the high and low economic growth and high and low price cases use the reference case assumptions for coal mining and rail transportation productivity and equipment costs, they show smaller variations in average delivered coal prices than do the two coal cost cases. Different coal price projections in the high and low economic growth cases and high and low price cases result mainly from higher and lower projected levels of demand for coal. In the price cases, higher and lower fuel costs for both coal producers and railroads contribute to the variations in projected coal prices.

CTL Production Increases Coal Use Outside the Electric Power Sector

Figure 91. Coal consumption in the industrial and buildings sectors and at coal-to-liquids plants, 2005, 2015, and 2030 (quadrillion Btu)

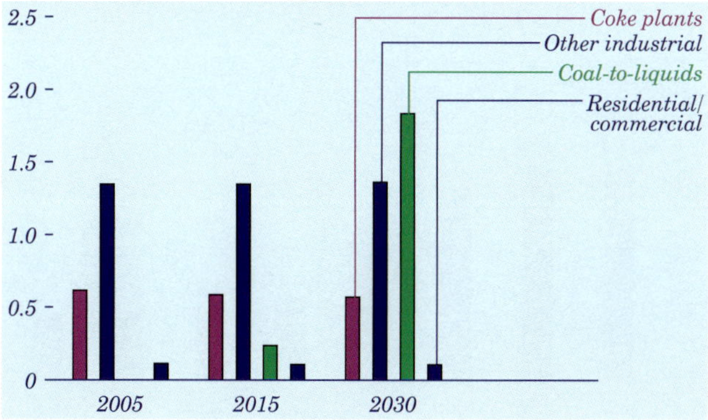

Although the electric power sector accounts for the bulk of U.S. coal consumption, 2.1 quadrillion Btu of coal currently is consumed in the industrial and buildings (residential and commercial) sectors (Figure 91). In the industrial sector, steam coal is used to manufacture or produce cement, paper, chemicals, food, primary metals, and synthetic fuels; as a boiler fuel to produce process steam and electricity; as a direct source of heat; and as a feedstock. Coal consumption in the other industrial sector (excluding CTL production) increases slightly in the *AEO2007* reference case.

Coal is also used to produce coke, which in turn is used as a source of energy and as a raw material input at blast furnaces to produce steel. A continuing shift from coke-based production at integrated steel mills to electric arc furnaces, combined with a relatively flat outlook for U.S. steel production, leads to a slight decline in consumption of coal at coke plants.

Outside the electric power sector, most of the increase in coal demand in the reference case is for production of coal-based synthetic liquids. High world oil prices spur investment in the CTL industry, leading to the construction of new plants in the West and Midwest that produce a total of 440,000 barrels of liquids per day in 2030. In *AEO2007*, CTL technology is represented as an IGCC coal plant equipped with a Fischer-Tropsch reactor to convert the synthesis gas to liquids. Of the total amount of energy from coal consumed at each plant, 49 percent is retained in the liquid product, and the remainder is used to produce electricity—with 40 percent used at the plant and 60 percent available for sale to the grid.

Rising Energy Consumption Increases Carbon Dioxide Emissions

Figure 92. Carbon dioxide emissions by sector and fuel, 2005 and 2030 (million metric tons)

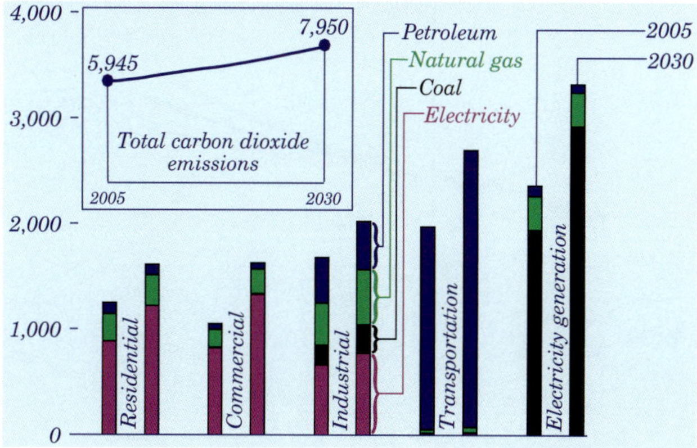

CO_2 emissions from the combustion of fossil fuels are proportional to fuel consumption and the carbon content of the fuel. Among commonly used fossil fuel types, coal has the highest carbon content and natural gas the lowest, with petroleum in between. In the *AEO2007* reference case, the shares of these fuels change slightly from 2005 to 2030, with more coal and less natural gas. The combined share of carbon-neutral renewable and nuclear energy is stable from 2005 to 2030 at 14 percent. As a result, CO_2 emissions increase by an average of 1.2 percent per year over the period, slightly higher than the average annual increase in total energy use (Figure 92). At the same time, the economy becomes less carbon intensive: the percentage increase in CO_2 emissions is almost one-third the increase in GDP, and emissions per capita increase by only 9 percent over the 25-year period.

The factors that influence growth in CO_2 emissions are the same as those that drive increases in fossil energy demand. Among the most significant are population and economic growth; increased penetration of computers, electronics, appliances, and office equipment; increases in commercial floorspace; increases in highway, rail, and air travel; and continued reliance on coal for electric power generation. The increases in demand for energy services are partially offset by efficiency improvements and shifts toward less energy-intensive industries. New CO_2 mitigation programs, more rapid improvements in technology, or more rapid adoption of voluntary CO_2 emissions reduction programs could result in lower CO_2 emissions levels than projected here.

Emissions Projections Change With Economic Growth Assumptions

Figure 93. Carbon dioxide emissions, 1990-2030 (million metric tons)

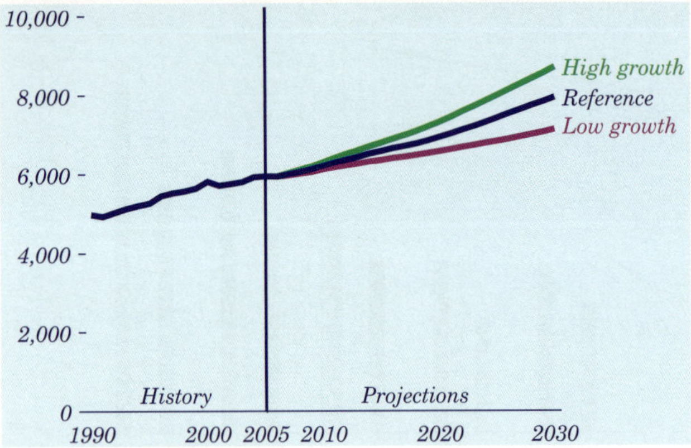

Higher growth in population, labor force, and productivity is assumed in the high economic growth case than in the *AEO2007* reference case, leading to higher industrial output, higher disposable income, lower inflation, and lower interest rates. The low economic growth case assumes the reverse. In the high and low growth cases, GDP projections vary by about 15 percent and population projections by about 8 percent from the reference case projections for 2030.

Alternative projections for industrial output, commercial floorspace, housing, and transportation in the economic growth cases influence the demand for energy and result in variations in CO_2 emissions (Figure 93). Emissions in 2030 are 10 percent lower in the low economic growth case and 10 percent higher in the high growth case than in the reference case. The strength of the relationship between economic growth and emissions varies by end-use sector. It is strongest for the industrial sector and, to a lesser extent, the transportation sector, where economic activity strongly influences energy use and emissions, and where fuel choices are limited. It is weaker in the commercial and residential sectors, where population and building characteristics have large influences and vary less across the three cases.

In the electricity sector, changes in electricity sales across the cases affect the amount of new, more efficient generating capacity required, reducing somewhat the sensitivity of energy use to GDP. However, the choice of coal for most new baseload capacity increases CO_2 intensity in the high growth case while decreasing it in the low growth case, offsetting the effects of changes in efficiency across the cases.

Clean Air Interstate Rule Reduces Sulfur Dioxide Emissions

Figure 94. Sulfur dioxide emissions from electricity generation, 1995-2030 (million short tons)

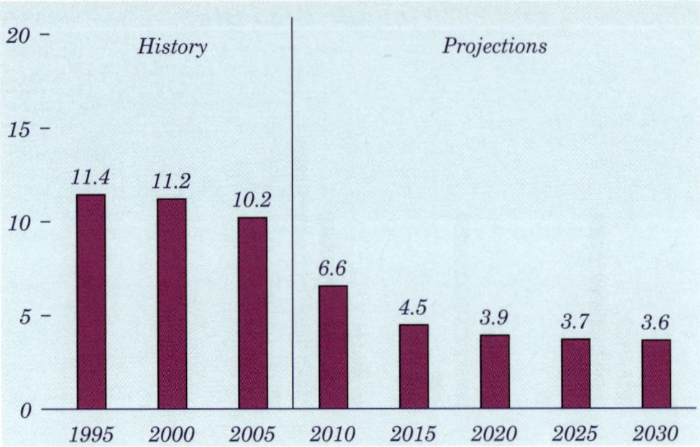

In March 2005, EPA promulgated the CAIR to limit formation of fine particles and ozone in nonattainment areas [*171*]. States can achieve mandated emissions reductions in two ways: by requiring power plants to participate in the EPA's national cap and trade program or by requiring them to meet State-specific emissions milestones through measures chosen by the State.

The reference case projects a drop in national SO_2 emissions from electricity generation, from 10.2 million short tons in 2005 to 3.6 million in 2030 (Figure 94). The reduction results from both use of lower sulfur coal and projected additions of flue gas desulfurization equipment on 143 gigawatts of capacity. SO_2 allowance prices are projected to rise to $900 per ton in 2015, remain between $900 and $1,100 per ton until 2025, and then fall to $800 per ton in 2030.

SO_2 emissions projections are not greatly affected by economic growth assumptions. In the *AEO2007* high growth case, with more coal-fired power plants added, the new plants are equipped for SO_2 capture before beginning operation, which is less costly than retrofitting existing plants. Therefore, allowance prices do not differ by much from those in the reference case. Fuel price assumptions have a greater effect on SO_2 allowance prices. More CTL plants are constructed in the high price case, and they are expected to have more efficient SO_2 capture equipment than advanced pulverized coal plants. Thus, in the last few years of the projections, SO_2 allowance prices are nearly 50 percent lower in the high price case than in the reference case, while the inflexible CAIR cap keeps emissions at nearly the same level in all cases.

Nitrogen Oxide Emissions Also Fall As CAIR Takes Effect

Figure 95. Nitrogen oxide emissions from electricity generation, 1995-2030 (million short tons)

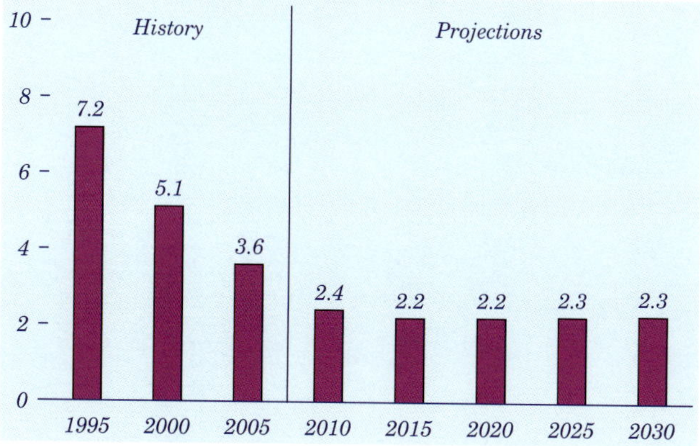

CAIR also mandates NO_x emission reductions in 28 States and the District of Columbia [*172*]. The required reductions are intended to reduce the formation of ground-level ozone, for which NO_x emissions are a major precursor. As with the CAIR-mandated SO_2 reductions, each State can determine a preferred method for reducing NO_x emissions. Options include joining the EPA's cap and trade program and enforcing individual State regulations. Each State will be subject to two NO_x limits: a 5-month summer season limit and an annual limit.

In the reference case, national NO_x emissions from the electric power sector are projected to fall from 3.6 million short tons in 2005 to 2.3 million short tons in 2030 (Figure 95). Because the CAIR caps are inflexible, different assumptions in the high and low growth and high and low fuel price cases do not affect the projections for aggregate NO_x emissions.

Between 2009 and 2030, after mandatory compliance begins, NO_x allowance prices are projected to range from $2,400 to $3,300 per ton emitted in the reference case, tending to rise as the emission caps tighten. By 2030, selective catalytic reduction equipment is projected to be added to an additional 116 gigawatts of coal-fired generating capacity. In the high price case, with more CTL capacity built, allowances are projected to be less costly, because CTL plants emit less NO_x than the coal-fired power plants they would displace. In the high economic growth case, with more coal-fired capacity in operation, allowance prices are projected to be slightly higher than in the reference case, even with the requirement for NO_x emission controls on all new plants.

Clean Air Mercury Rule Reduces Mercury Emissions

Figure 96. Mercury emissions from electricity generation, 1995-2030 (short tons)

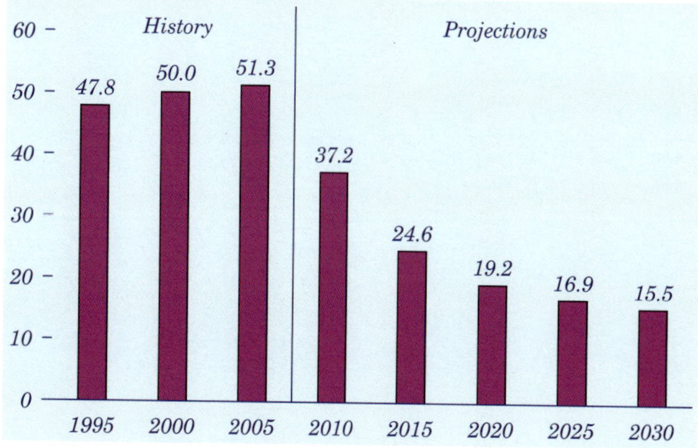

EPA's CAMR, also promulgated in 2005, imposes a national cap on emissions of mercury, to be implemented in two phases [*173*]. As with CAIR, States can enact their own programs or participate in the EPA cap and trade system. Although no States have made final decisions, more stringent regulations have been proposed by several States in the East where many power plants use coal with higher mercury content.

AEO2007 assumes that all States will participate in the cap and trade program and meet the CAMR restrictions, with no mandates for further reductions. In the reference case, national mercury emissions are projected to be reduced by 70 percent, from 51.3 short tons in 2005 to 15.5 short tons in 2030 (Figure 96). Nationally, power producers are projected to retrofit 133 gigawatts of coal-fired capacity with activated carbon injection technology. (Mercury controls also are expected to help the States to meet CAIR targets, because the retrofits reduce SO_2 and NO_x emissions as well.) The 2030 projection is slightly higher than the final EPA cap of 15 short tons, however, because allowances banked from earlier years could be used by some power plants. Allowance prices are expected to climb to a high of $68,000 per pound in 2030.

Overall trends in mercury allowance prices are not greatly affected by economic growth or fuel price assumptions. The *AEO2007* high growth case projects more coal-fired generation than the reference case, causing allowance prices to rise more rapidly than in the reference case. In the high price case, more efficient CTL facilities are built, leading to a 6-percent decrease in total annual mercury emissions in 2030 relative to the reference case projection.

Comparison with Other Projections

Comparison with Other Projections

Only Global Insights, Inc. (GII) produces a comprehensive energy projection with a time horizon similar to that of *AEO2007*. Other organizations, however, address one or more aspects of the energy markets. The most recent projection from GII, as well as others that concentrate on economic growth, international oil prices, energy consumption, electricity, natural gas, petroleum, and coal, are compared here with the *AEO2007* projections.

Economic Growth

In the *AEO2007* reference case, the projected growth in real GDP, based on 2000 chain-weighted dollars, is 2.9 percent per year from 2005 to 2030. The *AEO2007* projections for economic growth are based on the August short-term projection of GII, extended by EIA through 2030 and modified to reflect EIA's view on energy prices, demand, and production.

Projections of the average annual GDP growth rate for the United States from 2005 through 2010 range from 2.9 percent to 3.2 percent (Table 18). The *AEO2007* reference case projects annual growth of 3.0 percent over the period, matching the projection made by the Social Security Administration (SSA) and GII, but it is slightly lower than the 3.2-percent real GDP growth projected by the Office of Management and Budget (OMB), the CBO, and Energy Ventures Analysis, Inc. (EVA). The consensus Blue Chip projection is for 3.0-percent average annual growth from 2005 to 2010. Three other organizations—Interindustry Forecasting at the University of Maryland (INFORUM), the Bureau of Labor Statistics (BLS), and the International Energy Agency (IEA)—project somewhat lower annual growth of 2.9 percent over the same period. The IEA projection of 2.9-percent average annual growth covers the period from 2004 through 2015.

Over the period from 2010 to 2015, the uncertainty in the projected rate of GDP growth is greater, with projections ranging from 2.2 to 3.0 percent per year (excluding the *AEO2007* alternative cases); however, all but one projection falls in the range of 2.7 to 3.0 percent—SSA with projected average growth of 2.2 percent per year. The *AEO2007* reference case projection of 2.8 percent average annual economic growth from 2010 to 2015 is in the middle of the range, excluding the SSA projection. The Blue Chip consensus projection is 3.0 percent, and both BLS and the IEA project 2.9 percent, from 2010 to 2015. Projections slightly below the *AEO2007* reference case, at 2.7 percent, include GII, CBO, INFORUM, and EVA.

There are few public or private projections of GDP growth rates for the United States that extend to 2030. The *AEO2007* reference case projection reflects a slowing of the GDP growth rate after 2020, consistent with an expected slowing of population growth.

Table 18. Projections of annual average economic growth, 2005-2030

Projection	Average annual percentage growth 2005-2010	2010-2015	2015-2020	2020-2030
AEO2006	3.2	2.9	3.1	2.8
AEO2007				
Reference	3.0	2.8	3.0	2.8
Low growth	2.3	2.2	2.6	2.1
High growth	3.7	3.4	3.4	3.4
GII	3.0	2.7	2.9	2.8
OMB	3.2	NA	NA	NA
CBO	3.2	2.7	NA	NA
Blue Chip	3.0	3.0	NA	NA
INFORUM	2.9	2.7	2.7	NA
SSA	3.0	2.2	2.1	1.9
BLS	2.9	2.9	NA	NA
EVA	3.2	2.7	2.3	NA
IEA	2.9	2.9	1.9	1.9

NA = not available.

World Oil Prices

Comparisons of the *AEO2007* projections with other oil price projections are shown in Table 19. The world oil prices in the *AEO2007* reference case generally are higher than other world oil price projections available for comparison. Three of the six publicly available long-term projections—Deutsche Bank AG (DB), Strategic Energy and Economic Research, Inc. (SEER), and EVA—anticipate that world oil prices will decline faster than in the *AEO2007* reference case in the near term, with their projections for 2010 falling below that in the *AEO2007* low price case. All

Table 19. Projections of world oil prices, 2010-2030 (2005 dollars per barrel)

Projection	2010	2015	2020	2025	2030
AEO2006 (reference case)	48.72	49.24	52.24	55.72	58.69
AEO2007					
Reference	57.47	49.87	52.04	56.37	59.12
Low price	49.21	33.99	34.10	34.89	35.68
High price	69.21	79.57	89.12	94.40	100.14
GII	57.11	46.54	45.06	43.21	40.25
IEA (reference)	51.50	47.80	50.20	52.60	55.00
EEA	56.94	49.80	47.42	45.16	NA
DB	39.66	40.11	39.73	39.95	40.16
SEER	44.21	45.27	45.87	46.23	46.60
EVA	42.28	42.35	45.76	49.45	NA

the projections—except for the price projection from EVA, which was not available for comparison in last year's outlook—have raised their price expectations for 2010 and in the longer term relative to last year's releases.

The world oil price measures are, by and large, comparable across projections. For *AEO2007*, EIA reports the price of imported low-sulfur, light crude oil, approximately the same as the WTI prices that are widely cited as a proxy for world oil prices in the trade press. The only series that does not report projections in WTI terms is the IEA's *World Energy Outlook 2006*, where prices are expressed as the IEA crude oil import price.

Recent variability in crude oil prices demonstrates the uncertainty inherent in the projections. The *AEO2007* reference case and DB define the range of projected prices among the comparative series throughout the projection period. The range among the projections is $18 per barrel in 2010 (from a low price of $39.66 per barrel to a high of $57.47 per barrel), declining to $10 per barrel in 2015 and then widening to $19 per barrel in 2030 (from a low of $40.16 per barrel to a high of $59.12 per barrel).

Excluding the *AEO2007* high and low price cases, there are four distinct views proffered by the comparative series beginning in 2010: (1) prices moderate by 2015 before beginning a steady increase; (2) prices do not moderate over the mid-term but increase toward the end of the projection; (3) prices decline throughout the projection; and (4) prices remain relatively flat throughout. In the *AEO2007* reference case, prices decline from about $57 per barrel in 2010 to $50 per barrel in 2015 and rise steadily to $59 per barrel in 2030 (all prices expressed in real 2005 dollars). IEA projects a similar trend. In the EVA projection, prices remain flat until after 2015, then begin to rise. Although GII and Energy and Environmental Analysis, Inc. (EEA) anticipate a (rather sharper) decline in prices over the 2010 to 2015 period compared to the *AEO2007* reference case, both expect the decline to continue, albeit slowly, through the end of their respective projection periods. Finally, DB and SEER expect oil prices to remain relatively flat or increase slightly from 2010 to 2030.

Total Energy Consumption

The *AEO2007* reference case projects higher growth in end-use sector consumption of petroleum, natural gas, and coal than occurred from 1980 to 2005 but lower growth in electricity consumption (Table 20). Much of the projected growth in petroleum consumption is driven by increased demand in the transportation sector, with continued growth in personal travel and freight transport projected to result from demographic trends and economic expansion. Natural gas consumption is expected to increase in the residential, commercial, and industrial sectors, despite relatively high prices. Natural gas is cleaner than other fuels, does not require on-site storage, and has tended to be priced competitively with oil for heating. Coal consumption as a boiler fuel in the commercial and industrial sectors is expected to decline slightly, with potential use in new boilers limited by environmental restrictions; however, the projections for industrial coal include its use in CTL plants, a technology that is expected to become competitive at the level of oil prices assumed in the *AEO2007* reference case.

While strong growth in electricity use is projected to continue in the *AEO2007* reference case, the pace slows from historical rates. Some rapidly growing applications, such as air conditioning and computers, slow as penetration approaches saturation levels. Electrical efficiency also continues to improve, due in large part to efficiency standards, and the impacts tend to accumulate with the gradual turnover of appliance stocks.

The *AEO2007* reference case generally includes greater growth in primary energy consumption through 2030 than is shown in the outlook from GII. GII projects little growth in end-use natural gas consumption, whereas the *AEO2007* reference case projects continued growth in the industrial and buildings sectors. Some of the difference can be attributed to the higher natural gas price assumptions in the GII projection. End-use natural gas prices in

Table 20. Projections of average annual growth rates for energy consumption, 2005-2030 (percent)

Energy use	History 1980-2005	Projections: AEO2007	Projections: GII
*Petroleum**	0.9	1.0	0.9
*Natural gas**	0.0	0.9	0.1
*Coal**	-1.7	1.4	-0.2
Electricity	2.2	1.4	1.3
Delivered energy	0.7	1.1	0.8
Electricity losses	1.9	1.0	0.7
Primary energy	1.0	1.1	0.7

**Excludes consumption by electricity generators in the electric power sector; includes consumption for end-use combined heat and power generation.*

the *AEO2007* reference case decline rapidly from 2006 to 2013 before resuming a slow upward trend. In contrast, GII projects a more moderate decline in natural gas prices from 2005 to 2015, with little further change by 2025. GII projects an industrial natural gas price of $7.91 per thousand cubic feet in 2025, compared with $6.40 per thousand cubic feet in the *AEO2007* reference case (2005 dollars). GII's projected growth rates for petroleum and electricity consumption are similar to those in the *AEO2007* reference case. Differences between the *AEO2007* reference case and the GII projections for end-use coal consumption result from a projected increase in coal use for CTL in the *AEO2007* reference case.

Electricity

The *AEO2007* projections of retail electricity prices are based on average costs for electricity. The projections include supply regions that still are regulated, regions that are competitive and where marginal rather than average prices are assumed, and regions with a mix of regulated and competitive markets where average and marginal prices are weighted by the amount of load that serves regulated and competitive markets. As of 2005, 4 of the 13 electricity market regions had fully competitive retail markets in operation, 7 regions had mixed competitive and regulated retail markets, and 2 regions had fully regulated markets. The *AEO2007* cases assume that no additional retail markets will be restructured and that partial restructuring (in wholesale markets) will lead to increased competition in the electric power industry. Competition is assumed to lower operating and maintenance costs and to cause the retirement of uneconomical generating units. The *AEO2007* electricity projections assume continuation of current laws and regulations. Other projections may reflect explicit assessments of the nature and likelihood of policy developments over the next 25 years.

Comparisons of the *AEO2007* projections and those from other organizations are shown in Table 21. The projections for electricity sales in 2015 range from a low of 4,133 billion kilowatthours in the *AEO2007* low economic growth case to a high of 4,433 billion kilowatthours in the EVA projection. EVA projects higher sales in the commercial and residential sectors, with somewhat less growth in industrial sales, than are projected by the *AEO2007* reference case, GII, and EEA. The projections for total electricity sales in 2030 range from 4,682 billion kilowatthours (*AEO2007* low economic growth case) to 5,654 billion kilowatthours (*AEO2007* high economic growth case). The annual rate of demand growth ranges from 1.0 percent (*AEO2007* low economic growth case) to 1.8 percent (*AEO2007* high economic growth case). GII projects lower growth in the commercial sector and higher growth in the industrial and, to a lesser extent, residential sectors in 2030 than is projected in the *AEO2007* reference case.

The *AEO2007* reference case shows a decline in real electricity prices early in the projection period and then rising prices at the end of the period because of increases in the cost of fuels used for generation and increases in capital expenditures for construction of new capacity. The rising fossil fuel prices and increased capital outlays in the *AEO2007* reference case lead to an increase in average electricity prices, from 7.7 cents per kilowatthour in 2015 to 8.1 cents per kilowatthour in 2030. GII projects increases in prices initially and then a slight decline at the end of the period.

Projections of total electricity generation in 2015 are similar for the *AEO2007* reference case, EVA, and EEA. In contrast, the projection by GII is lower than the others because of lower projected growth in electricity sales. The GII projection of total electricity generation in 2015 is similar to that in the *AEO2007* low economic growth case. Although GII projects a lower level of total electricity generation in 2030 than is projected in the *AEO2007* reference and high economic growth cases, its projection for renewable generation in 2030 is considerably higher than the *AEO2007* reference case projection.

The need for new generating capacity is driven by growth in electricity sales and the need to replace existing units that are no longer economical to operate. Consistent with its projection of higher growth in electricity sales, EVA projects greater growth in requirements for new fossil-fuel-fired generating plants as well as nuclear plants in 2015 compared with the *AEO2007* reference case and GII. Except for nuclear plants, the EVA projections for generating capacity are similar to EEA's projections for 2015. As noted above, the GII projections for renewable capacity in 2030 are higher than those in the *AEO2007* reference and high and low economic growth cases. The projections for nuclear capacity additions from 2005 to 2030 as a result of the incentives in EPACT2005 range from 28 gigawatts in the *AEO2007* high economic growth case to 6 gigawatts in the *AEO2007* low economic growth case. The *AEO2007* cases assume

Table 21. Comparison of electricity projections, 2015 and 2030 (billion kilowatthours, except where noted)

Projection	2005	AEO2007			Other projections		
		Reference	Low economic growth	High economic growth	GII	EVA	EEA
		2015					
Average end-use price (2005 cents per kilowatthour)	**8.1**	**7.7**	**7.5**	**7.9**	**8.6**	**NA**	**NA**
Residential	9.4	8.9	8.7	9.1	10.0	9.7	NA
Commercial	8.6	8.0	7.7	8.2	9.2	8.7	NA
Industrial	5.7	5.6	5.4	5.9	5.8	5.6	NA
Total generation plus imports	**4,063**	**4,729**	**4,597**	**4,865**	**4,610**	**4,720**	**4,766**
Coal	2,015	2,295	2,235	2,353	2,244	2,336	NA
Oil	122	103	100	105	92	NA	NA
Natural gas [a]	756	1,023	959	1,068	934	982	NA
Nuclear	780	812	809	837	829	860	NA
Hydroelectric/other [b]	365	487	485	493	492	523	NA
Net imports	25	8	8	10	20	20	NA
Electricity sales	**3,660**	**4,251**	**4,133**	**4,370**	**4,186**	**4,433**	**4,302**
Residential	1,365	1,591	1,560	1,622	1,592	1,665	1,597
Commercial/other [c]	1,274	1,557	1,532	1,583	1,485	1,709	1,507
Industrial	1,021	1,103	1,041	1,165	1,110	1,059	1,198
Capability, including CHP (gigawatts) [d]	**988**	**997**	**981**	**1,018**	**1,011**	**1,045**	**1,035**
Coal	315	329	322	336	333	351	346
Oil and natural gas	448	430	422	440	429	453	457
Nuclear	100	102	102	106	104	108	102
Hydroelectric/other	125	136	135	136	146	132	130
		2030					
Average end-use price (2005 cents per kilowatthour)	**8.1**	**8.1**	**7.8**	**8.4**	**8.5**	**NA**	**NA**
Residential	9.4	9.1	8.8	9.6	9.9	NA	NA
Commercial	8.6	8.3	7.9	8.7	9.1	NA	NA
Industrial	5.7	5.9	5.6	6.3	5.6	NA	NA
Total generation plus imports	**4,065**	**5,810**	**5,255**	**6,375**	**5,586**	**NA**	**NA**
Coal	2,015	3,330	2,871	3,672	2,999	NA	NA
Oil	122	107	104	112	76	NA	NA
Natural gas [a]	756	942	924	1,010	952	NA	NA
Nuclear	780	896	845	1,010	826	NA	NA
Hydroelectric/other [b]	365	522	499	555	719	NA	NA
Net imports	25	13	12	15	15	NA	NA
Electricity sales	**3,660**	**5,168**	**4,682**	**5,654**	**5,071**	**NA**	**NA**
Residential	1,365	1,896	1,773	2,016	1,921	NA	NA
Commercial/other [c]	1,274	2,073	1,907	2,234	1,872	NA	NA
Industrial	1,021	1,199	1,003	1,403	1,278	NA	NA
Capability, including CHP (gigawatts) [d]	**988**	**1,220**	**1,112**	**1,331**	**1,157**	**NA**	**NA**
Coal	315	465	403	511	443	NA	NA
Oil and natural gas	448	500	464	544	404	NA	NA
Nuclear	100	113	106	127	109	NA	NA
Hydroelectric/other	125	142	138	149	202	NA	NA

[a]Includes supplemental gaseous fuels. For EVA, represents total oil and natural gas. [b]"Other" includes conventional hydroelectric, pumped storage, geothermal, wood, wood waste, municipal waste, other biomass, solar and wind power, batteries, chemicals, hydrogen, pitch, purchased steam, sulfur, petroleum coke, and miscellaneous technologies. [c]"Other" includes sales of electricity to government, railways, and street lighting authorities. [d]EIA capacity is net summer capability, including combined heat and power plants. GII capacity is nameplate, excluding cogeneration plants.

CHP = combined heat and power. NA = not available.

Sources: ***2005 and AEO2007:*** AEO2007 National Energy Modeling System, runs AEO2007.D112106A (reference case), LM2007.D112106A (low economic growth case), and HM2007.D112106A (high economic growth case). **GII:** Global Insight, Inc., *2006 U.S. Energy Outlook* (November 2006). **EVA:** Energy Ventures Analysis, Inc., *FUELCAST: Long-Term Outlook* (August 2006). **EEA:** Energy and Environmental Analysis, Inc., *EEA's Compass Service Base Case* (October 2006).

that 2.6 gigawatts of nuclear capacity will be retired by 2030 because their operating licenses will have expired.

Environmental regulations have an important influence on the technology choices made for electricity generation. EVA assumes that legislation similar to the Clear Skies Act (including new restrictions on SO_2, NO_x, and mercury emissions) will be in effect by 2010. EVA also includes a tax of \$6 per ton on CO_2 emissions beginning in 2013. The combination of stronger environmental restrictions, a tax on CO_2 emissions, and aggregate State-level RPS program requirements leads to greater growth in nonhydroelectric renewable generation in the EVA projection than in the other projections in 2015. The *AEO2007* cases reflect EPA's recently enacted CAIR and CAMR regulations. Because *AEO2007* generally includes only current laws and regulations, it does not assume any policies to address CO_2 emissions. As noted above, restrictions on CO_2 emissions could change the mix of technologies used to generate electricity.

Natural Gas

In the *AEO2007* reference case, natural gas consumption is projected to grow steadily through 2020 and then level off as higher projected natural gas prices cause natural gas to lose market share to coal for electricity generation. With the exception of GII, this is a major difference between the *AEO2007* reference and high price cases and the other projections (Table 22), which show natural gas consumption generally increasing throughout the projection period, both overall and for electricity generation. The lowest projected overall growth is in the GII projection, with 2030 consumption that is 2.4 trillion cubic feet less than in the *AEO2007* reference case. The DB, SEER, and Altos projections expect natural gas consumption in 2030 to exceed the *AEO2007* reference case projection by 1.1, 4.1, and 4.8 trillion cubic feet, respectively; the two latter projections even exceed the *AEO2007* low price case projection. Although GII projects less total natural gas consumption than does the *AEO2007* reference case, the GII projection for consumption by electricity generators exceeds that in the *AEO2007* reference case, further highlighting a fundamental difference between the *AEO2007* reference case and the other projections.

Natural gas consumption by electricity generators grows from 2005 to 2015 in all the projections. With the exception of the *AEO2007* reference and high price cases, the projected growth continues through 2025. DB is the only projection with less growth in natural gas consumption by electricity generators than the *AEO2007* reference case from 2005 to 2015. Natural gas consumption in the DB projection in 2015 is 6 percent below the *AEO2007* reference case value, and the other projections are between 2 percent (GII) and 32 percent (Altos) above the *AEO2007* reference case. In 2025, natural gas consumption by electricity generators in all the other projections exceeds that in the *AEO2007* reference case by 6 percent (DB) to 69 percent (Altos). In 2030, consumption in the other projections is 20 percent (DB) to 109 percent (Altos) higher than in the *AEO2007* reference case. Only the GII and DB projections for natural gas consumption by electricity generators are consistently lower than those in the *AEO2007* low price case.

All the projections show steady growth in natural gas consumption in the combined residential and commercial sectors, with the exception of GII, which expects a slight decline in consumption from 2025 to 2030. The *AEO2007* reference case shows higher industrial natural gas consumption than all the other projections over the entire 2005-2030 period. With the exception of GII and EEA, all the other organizations project growth in industrial natural gas consumption from 2005 to 2015 and through the end of the projection period. Growth in residential, commercial, and industrial natural gas consumption in the *AEO2007* reference case is offset, however, by the decline in natural gas consumption by electricity generators.

Domestic natural gas production is projected to decline in the GII, EVA, and Altos projections over the next decade; in all the other projections it increases over the same period. GII and EVA expect the decline to be reversed in 2025, with production slightly exceeding 2005 production levels. DB and Altos are more pessimistic, projecting that natural gas production will have declined by about 10 percent in 2025 relative to 2005 levels. Altos expects domestic natural gas production in 2030 to be 21 percent below 2005 levels. The *AEO2007* high price case shows domestic natural gas production of 20.9 trillion cubic feet in 2030, one of the more optimistic projections. It is exceeded only by the SEER projection of 21.2 trillion cubic feet in 2030.

With the exception of the *AEO2007* high price case, net imports increase significantly from 2005 to 2030 in all the projections, with increases ranging from approximately 50 percent in the *AEO2007* reference case and GII projections to 255 percent in the Altos projection. The increase is expected to come from LNG. With the exception of the DB projection and the *AEO2007* high price case, all the projections show higher LNG imports than the *AEO2007* reference case in 2015. Net LNG imports in 2015 in the Altos projection, at 6.8 trillion cubic feet, are significantly higher than those in the other projections; and Altos remains the most optimistic projection in 2030, at 12.0 trillion cubic feet of net LNG imports. Net LNG imports are 4.5 trillion cubic feet in 2030 in the *AEO2007* reference case, by far the lowest level of imports of any of the projections, with DB and Altos projecting more than double that level. LNG imports in the *AEO2007* high price case are even lower, at 2.3 trillion cubic feet in 2030. The *AEO2007* reference case also projects the lowest percentage of consumption accounted for by LNG imports. LNG imports account for slightly under 17 percent of total natural gas consumption in 2025 in the *AEO2007* reference case—about the same as in the EEA projection—whereas the other organizations expect LNG imports to account for between 21 and 40 percent of consumption.

Table 22. Comparison of natural gas projections, 2015, 2025, and 2030 (trillion cubic feet, except where noted)

Projection	2005	AEO2007			Other projections					
		Reference	Low price	High price	GII[a]	EVA	EEA[b]	DB	SEER	Altos
						2015				
Dry gas production [c]	18.23	19.60	19.82	18.77	17.45	17.93	20.42	19.53	18.87	18.19
Net imports	3.57	5.62	6.46	4.81	5.73	7.95	5.40	4.90	6.50	8.06
Pipeline	3.01	2.63	2.48	2.52	NA	3.73	2.26	2.60	2.80	1.22
LNG	0.57	2.99	3.98	2.29	NA	4.22	3.13	2.29	3.70	6.84
Consumption	21.98	25.32	26.40	23.71	23.38	25.67	26.00	24.20	25.37	25.87 [d]
Residential	4.84	5.19	5.29	5.11	4.99	5.14	5.49	5.30	5.12	5.41
Commercial	3.05	3.53	3.65	3.42	3.05	3.11	3.40	3.42	3.18	3.54
Industrial [e]	6.64	7.67	7.71	7.44	6.48	6.95	6.39	7.19	6.99	7.53 [f]
Electricity generators [g]	5.78	7.11	7.89	5.99	7.26	8.50	8.53	6.71	8.10	9.39
Other [h]	1.66	1.83	1.86	1.75	1.61	1.97	2.18	1.59	1.99	NA
Lower 48 wellhead price (2005 dollars per thousand cubic feet) [i]	7.51	4.99	4.01	5.83	6.10	5.55	6.51	6.07	5.12	5.60
End-use prices (2005 dollars per thousand cubic feet)										
Residential	12.80	10.55	9.48	11.48	11.28	NA	10.95	NA	10.59	NA
Commercial	11.54	8.73	7.68	9.64	10.05	NA	9.98	NA	8.83	NA
Industrial [j]	8.41	5.82	4.80	6.70	7.87	NA	7.95	NA	6.45	NA
Electricity generators	8.42	5.66	4.74	6.40	6.68	NA	7.54	NA	6.11	NA
						2025				
Dry gas production [c]	18.23	20.59	20.44	20.73	18.26	18.82	22.61	16.67	20.91	16.41
Net imports	3.57	5.58	8.70	3.32	5.23	9.93	6.28	9.54	7.80	12.59
Pipeline	3.01	1.20	1.58	1.03	NA	2.11	1.24	1.25	1.80	1.10
LNG	0.57	4.38	7.11	2.29	NA	7.82	5.04	8.29	6.00	11.49
Consumption	21.98	26.30	29.27	24.13	23.69	28.53	29.03	26.18	28.71	29.01 [d]
Residential	4.84	5.29	5.38	5.21	4.97	5.18	5.86	5.85	5.64	6.03
Commercial	3.05	3.98	4.06	3.82	3.07	3.42	3.55	3.87	3.45	4.14
Industrial [e]	6.64	8.42	8.09	7.94	6.53	7.95	7.10	7.94	7.66	7.69 [f]
Electricity generators [g]	5.78	6.59	9.71	4.82	7.45	9.74	10.25	6.98	9.78	11.15
Other [h]	1.66	2.02	2.03	1.99	1.67	2.24	2.27	1.54	2.18	NA
Lower 48 wellhead price (2005 dollars per thousand cubic feet) [i]	7.51	5.62	4.75	6.70	6.21	6.06	6.83	5.71	5.61	6.96
End-use prices (2005 dollars per thousand cubic feet)										
Residential	12.80	11.30	10.32	12.43	11.21	NA	10.95	NA	11.19	NA
Commercial	11.54	9.23	8.29	10.34	10.02	NA	10.08	NA	9.51	NA
Industrial [j]	8.41	6.40	5.52	7.51	7.91	NA	8.22	NA	7.12	NA
Electricity generators	8.42	6.22	5.56	7.18	6.78	NA	7.85	NA	6.78	NA

NA = not available. See notes and sources at end of table.

Comparison with Other Projections

For the most part, all the projections expect natural gas wellhead prices to decline significantly from the 2005 level of $7.51 per thousand cubic feet. The *AEO2007* low price case shows the lowest projection for natural gas wellhead prices in 2015 [*174*], followed by the *AEO2007* reference case. Natural gas wellhead prices in the *AEO2007* reference and low price cases in 2025 are at or below the levels in all the other projections. Among the other organizations, only DB projects a natural gas wellhead price below that in the *AEO2007* reference case for 2030, and only Altos projects a price that exceeds the 2005 price. In the GII and SEER projections, natural gas wellhead prices in 2030 exceed the *AEO2007* reference case projection by less than 2 percent, and the Altos price projection for 2030 exceeds the *AEO2007* reference case projection by 26 percent.

Delivered natural gas price margins [*175*] to electricity generators are consistently the lowest in the *AEO2007* high price case and GII projections. Both are notably lower than the historically high margins in 2005. The margins in the SEER projection exceed those in the *AEO2007* reference case in all years by up to 120 percent. While the industrial sector margins in the other projections exceed those in the *AEO2007* reference case in all years by as much as 120 percent [*176*], the disparity is largely attributable to definitional differences, which can be seen by comparing the 2005 values provided with the other projections. All projections show a decline in industrial margins across the projection period relative to their 2005 values. SEER shows the greatest percentage decline from 2005 to 2025, at 17 percent; EEA shows the smallest decline at 5 percent; and the rest show

Table 22. Comparison of natural gas projections, 2015, 2025, and 2030 (continued)
(trillion cubic feet, except where noted)

Projection	2005	AEO2007			Other projections					
		Reference	Low price	High price	GII [a]	EVA	EEA [b]	DB	SEER	Altos
					2030					
Dry gas production [c]	**18.23**	**20.53**	**20.64**	**20.90**	**18.27**	**NA**	**NA**	**16.32**	**21.17**	**14.33**
Net imports	**3.57**	**5.45**	**8.85**	**3.17**	**5.25**	**NA**	**NA**	**10.72**	**9.08**	**12.67**
Pipeline	3.01	0.92	1.31	0.84	NA	NA	NA	1.25	1.44	0.70
LNG	0.57	4.53	7.54	2.33	NA	NA	NA	9.47	7.64	11.97
Consumption	**21.98**	**26.12**	**29.74**	**24.09**	**23.74**	**NA**	**NA**	**27.20**	**30.26**	30.95 [d]
Residential	4.84	5.31	5.40	5.20	4.92	NA	NA	6.15	5.92	6.34
Commercial	3.05	4.24	4.32	4.01	3.07	NA	NA	4.06	3.56	4.44
Industrial [e]	6.64	8.65	8.34	8.18	6.79	NA	NA	8.34	7.82	7.78 [f]
Electricity generators [g]	5.78	5.92	9.64	4.37	7.27	NA	NA	7.12	10.74	12.39
Other [h]	1.66	2.01	2.05	2.00	1.68	NA	NA	1.52	2.22	NA
Lower 48 wellhead price (2005 dollars per thousand cubic feet) [i]										
	7.51	5.98	5.06	7.63	6.08	NA	NA	5.45	6.07	7.55
End-use prices (2005 dollars per thousand cubic feet)										
Residential	12.80	11.77	10.71	13.52	10.98	NA	NA	NA	11.58	NA
Commercial	11.54	9.58	8.58	11.32	9.81	NA	NA	NA	9.96	NA
Industrial [j]	8.41	6.76	5.82	8.46	7.74	NA	NA	NA	7.58	NA
Electricity generators	8.42	6.51	5.88	8.02	6.65	NA	NA	NA	7.25	NA

NA = not available.

[a]Previously DRI-WEFA. Conversion factors: 1,000 cubic feet = 1.027 million Btu for production, 1.028 million Btu for end-use consumption, 1.019 million Btu for electric power. [b]The EEA projection shows a cyclical price trend; projected values for isolated years may be misleading. [c]Does not include supplemental fuels. [d]Excludes consumption for transportation and pipeline fuels. [e]Includes consumption for industrial CHP plants and a small number of electricity-only plants; excludes consumption by nonutility generators. [f]Includes lease and plant fuel. [g]Includes consumption of energy by electricity-only and CHP plants whose primary business is to sell electricity, or electricity and heat, to the public. Includes electric utilities, small power producers, and exempt wholesale generators. [h]Includes lease, plant, and pipeline fuel and fuel consumed in natural gas vehicles. [i]2005 wellhead natural gas prices for EEA, EVA, and DB are $7.77, $8.84, and $8.36, respectively. [j]The 2005 industrial natural gas prices in other projections are nearly a dollar higher than EIA's.

Sources: **2005 and *AEO2007*:** AEO2007 National Energy Modeling System, runs AEO2007.D112106A (reference case), LP2007.D112106A (low price case), HP2007.D112106A (high price case). **GII:** Global Insight, Inc., *2006 U.S. Energy Outlook* (November 2006). **EVA:** Energy Ventures Analysis, Inc., *FUELCAST: Long-Term Outlook* (August 2006). **EEA:** Energy and Environmental Analysis, Inc., *EEA's Compass Service Base Case* (October 2006). **DB:** Deutsche Bank AG, e-mail from Adam Sieminski on November 27, 2006. **SEER:** Strategic Energy and Economic Research, Inc., *Natural Gas Outlook* (April 2006). **Altos:** Altos Partners North American Regional Gas Model (NARG) Long-Term Base Case (November 2006).

declines of around 13 percent. Residential and commercial sector margins are, on average, about $5.40 and $3.70, respectively, with residential sector margins in the *AEO2007* reference case generally higher than those in the projections from other organizations, and commercial sector margins generally lower.

Petroleum

With significantly lower crude oil prices, the DB projections of U.S. petroleum demand in 2015 and 2030 are only 2 percent higher than those in the *AEO2007* reference case (Table 23). In the IEA reference case, total petroleum consumption in 2015 is within 1 percent of the total petroleum consumption in the *AEO2007* reference case; but in 2030, IEA's total petroleum demand projection is 7 percent lower than in the *AEO2007* reference case. Although the crude oil price is almost $19 per barrel lower than that in the *AEO2007* reference case in 2030, total petroleum demand in the GII projection is lower than in the *AEO2007* reference case throughout the projection period. The GII projection shows the lowest level of petroleum demand among the projections reviewed, lower than the *AEO2007* high price case projection. The *AEO2007* low price case shows the highest levels of total petroleum demand in 2015 and 2030 among all the projections. The *AEO2007* high price case also shows higher petroleum demand than the GII projection in 2030, with projected crude oil prices that are almost $60 per barrel higher. The extent to which the projections from other organizations reviewed above and summarized in Table 24 incorporate expectations of changes in vehicle efficiency standards or other policy actions that could influence petroleum demand is not clear.

The projection of domestic crude oil production in the *AEO2007* reference case differs significantly from the other projections; rising from 5.2 million barrels in 2005 to a peak of 5.9 million barrels per day in 2017 and then declining to 5.4 million barrels per day in 2030. With the exception of the IEA reference case, domestic crude oil production in the other projections declines throughout the projection period to levels more than a million barrels per day lower than in the *AEO2007* reference case. Domestic crude oil production falls to 3.4 million barrels per day in 2025 in the EVA projection and 3.5 million barrels per day in 2030 in the DB projection. In the IEA projection, domestic crude oil production increases until 2010, then declines to 4.0 million barrels per day in 2030.

The higher crude oil prices in the *AEO2007* reference case alone do not fully explain the differences in the projections for domestic crude oil production. For example, crude oil prices in the IEA projection are slightly higher than in the *AEO2007* reference case from 2012 through 2030, but domestic crude oil production in 2030 is more than 1 million barrels per day below domestic crude oil production in the *AEO2007* reference case. The *AEO2007* low price case, with crude oil prices in the mid-$30 per barrel range from 2015 through 2030, shows the same pattern of domestic crude oil production as the *AEO2007* reference case. Production rises from current levels, peaks in 2015, and then gradually declines but still ends up slightly higher in 2030 than the current level of production. The *AEO2007* high price case projects increasing domestic crude oil production, peaking in 2030 at more than 6.0 million barrels per day.

The projections also differ on domestic NGL production. In the *AEO2007* reference case, NGL production increases from current levels to a peak of 1.8 million barrels per day in 2017 before falling back to 1.7 million barrels per day in 2030, about equal to the 2005 level. NGL production is 17 percent lower in 2015 in the DB projection and 37 percent lower in 2030 than in the *AEO2007* reference case. The GII projection is more bullish, with 2030 NGL production slightly higher than in the *AEO2007* reference case.

The differences in domestic crude oil production lead to very different conclusions about U.S. dependence on imported petroleum. In the *AEO2007* reference case, the import share of product supplied decreases from 60 percent in 2005 to below 55 percent in 2009 and then slowly rises back to 61 percent in 2030. The share of imported petroleum increases from 2005 levels in the DB and GII projections throughout the projection period, to 77 percent in 2030 in the DB projection and 75 percent in 2030 in the GII projection. Despite higher petroleum demand in the *AEO2007* low price case, the projected import share rises to only 67 percent in 2030. In the *AEO2007* high price case, the import share is projected to decline to 49 percent in 2030, well below 2005 levels.

Coal

The coal consumption, production, and price projections vary considerably, reflecting uncertainty about environmental regulations and economic growth, among many factors (Table 24). The coal projections from the *AEO2007* cases reflect existing environmental regulations, including CAAA90, CAIR, and CAMR, which restrict SO_2, NO_x, and mercury emissions beginning in 2010. The EVA projection incorporates similar regulations and also includes a carbon tax of \$6 per metric ton CO_2 equivalent beginning in 2013. In addition to differences in environmental assumptions, the *AEO2007*, EVA, and GII projections reflect different assumptions about the outlook for economic growth rates, the natural gas prices, and world oil prices.

All the projections show increases in total coal consumption over their projection periods. Despite early similarities between the projections, total coal consumption in the *AEO2007* reference case after 2015

Table 23. Comparison of petroleum projections, 2015, 2025, and 2030 (million barrels per day, except where noted)

Projection	2005	AEO2007			Other projections			
		Reference	Low price	High price	GII	EVA	DB	IEA
					2015			
Crude oil and NGL production	**6.90**	**7.73**	**8.02**	**7.41**	**6.47**	**6.09**	**6.28**	**NA**
Crude oil	5.18	5.91	6.18	5.67	4.80	4.45	4.78	5.00
Natural gas liquids	1.72	1.82	1.84	1.74	1.67	1.64	1.50	NA
Total net imports	**12.57**	**12.52**	**13.29**	**11.79**	**13.75**	**NA**	**15.37**	**NA**
Crude oil	10.09	10.49	10.62	10.18	NA	NA	NA	NA
Petroleum products	2.48	2.03	2.67	1.61	NA	NA	NA	NA
Petroleum demand	**20.75**	**22.86**	**23.61**	**21.87**	**20.22**	**NA**	**23.26**	**23.10**
Motor gasoline	9.16	10.18	10.45	9.53	NA	NA	10.32	NA
Jet fuel	1.68	2.10	2.12	2.08	NA	NA	1.87	NA
Distillate fuel	4.12	4.86	5.04	4.72	NA	NA	4.81	NA
Residual fuel	0.92	0.82	1.03	0.73	NA	NA	0.78	NA
Other	4.87	4.89	4.96	4.81	NA	NA	5.48	NA
Import share of product supplied (percent)	**60**	**55**	**56**	**54**	**68**	**NA**	**66**	**NA**
					2025			
Crude oil and NGL production	**6.90**	**7.30**	**7.34**	**7.68**	**6.02**	**4.95**	**5.11**	**NA**
Crude oil	5.18	5.58	5.60	5.97	4.27	3.35	3.91	NA
Natural gas liquids	1.72	1.72	1.74	1.70	1.75	1.60	1.20	NA
Total net imports	**12.57**	**14.87**	**16.98**	**11.70**	**17.03**	**NA**	**19.31**	**NA**
Crude oil	10.09	12.20	13.27	10.19	NA	NA	NA	NA
Petroleum products	2.48	2.67	3.71	1.51	NA	NA	NA	NA
Petroleum demand	**20.75**	**25.34**	**26.77**	**23.50**	**23.05**	**NA**	**26.15**	**NA**
Motor gasoline	9.16	11.71	12.26	10.14	NA	NA	11.57	NA
Jet fuel	1.68	2.22	2.24	2.12	NA	NA	2.14	NA
Distillate fuel	4.12	5.48	5.88	5.35	NA	NA	5.47	NA
Residual fuel	0.92	0.82	1.11	0.75	NA	NA	0.83	NA
Other	4.87	5.11	5.29	5.14	NA	NA	6.15	NA
Import share of product supplied (percent)	**60**	**59**	**64**	**50**	**74**	**NA**	**74**	**NA**
					2030			
Crude oil and NGL production	**6.90**	**7.10**	**6.98**	**7.75**	**5.79**	**NA**	**4.62**	**NA**
Crude oil	5.18	5.39	5.25	6.04	4.04	NA	3.53	4.00
Natural gas liquids	1.72	1.72	1.73	1.71	1.75	NA	1.08	NA
Total net imports	**12.57**	**16.37**	**19.31**	**12.04**	**17.03**	**NA**	**21.13**	**NA**
Crude oil	10.09	13.09	14.35	10.59	NA	NA	NA	NA
Petroleum products	2.48	3.28	4.95	1.45	NA	NA	NA	NA
Petroleum demand	**20.75**	**26.95**	**28.84**	**24.58**	**22.82**	**NA**	**27.54**	**25.00**
Motor gasoline	9.16	12.53	13.23	10.47	NA	NA	12.16	NA
Jet fuel	1.68	2.27	2.29	2.06	NA	NA	2.27	NA
Distillate fuel	4.12	5.95	6.64	5.85	NA	NA	5.81	NA
Residual fuel	0.92	0.83	1.16	0.76	NA	NA	0.85	NA
Other	4.87	5.36	5.53	5.45	NA	NA	6.46	NA
Import share of product supplied (percent)	**60**	**61**	**67**	**49**	**75**	**NA**	**77**	**NA**

NA = Not available.

Sources: ***2005 and AEO2007:*** AEO2007 National Energy Modeling System, runs AEO2007.D112106A (reference case), LP2007.D112106A (low price case), HP2007.D112106A (high price case). **GII:** Global Insight, Inc., *2006 U.S. Energy Outlook* (November 2006). **EVA:** Energy Ventures Analysis, Inc., *FUELCAST: Long-Term Outlook* (August 2006). **DB:** Deutsche Bank AG, e-mail from Adam Sieminski on November 27, 2006. **IEA:** International Energy Agency, *World Energy Outlook 2006* (Paris, France, November 2006).

increases more rapidly than in the EVA or GII projections. In the *AEO2007* reference case, total coal consumption grows by 14 percent from 2005 to 2015, to 1,282 million tons in 2015. With more restrictive environmental standards, EVA projects lower levels of total coal consumption (8 percent lower in 2025) than the *AEO2007* reference case. Between 2005 and 2025, coal consumption grows by 2.1 percent per year in the *AEO2007* reference case, which is substantially higher than the 1.3-percent growth rate projected by EVA for the same period. On a Btu basis between 2005 and 2015, GII projects growth in coal consumption similar to that in the *AEO2007* reference case. In 2030, however, coal consumption in the *AEO2007* reference case is 34.1 quadrillion Btu (19 percent) higher than the GII projection of 28.7 quadrillion Btu.

In all the projections, coal consumption in the electricity sector accounts for about 90 percent of total coal use. Coal consumption in the electricity sector in the early years of the EVA and GII projections closely matches that in the *AEO2007* reference case. Both EVA and GII project slower growth in coal consumption for the electric power sector over the entire projection period. EVA projects total coal consumption in the electricity sector at 1,361 million short tons in 2025, 50 million tons less than that in the *AEO2007* reference case. On a Btu basis, the GII projection for coal consumption in the electric power sector is 26.7 quadrillion Btu in 2030, 14 percent less than the 31.1 quadrillion Btu (1,570 million tons) projected for 2030 in the *AEO2007* reference case.

Table 24. Comparison of coal projections, 2015, 2025, and 2030 (million short tons, except where noted)

Projection	2005	AEO2007			Other projections		
		Reference	Low economic growth	High economic growth	GII[a]	EVA	Hill
				2015			
Production	**1,131**	**1,266**	**1,227**	**1,300**	**24.3**	**1,289**	**NA**
Consumption by sector							
Electric power	1,039	1,178	1,151	1,202	22.5	1,179	NA
Coke plants	23	21	20	23	NA	24	NA
Coal-to-liquids	NA	16	6	25	NA	NA	NA
Other industrial/buildings	66	67	66	70	2.1[b]	73	NA
Total	**1,128**	**1,282**	**1,243**	**1,318**	**24.6**	**1,276**	**NA**
Net coal exports	**21.1**	**-4.6**	**-3.6**	**-7.3**	**-0.3**	**2.1**	**NA**
Exports	49.9	37.4	38.4	36.5	NA	44.0	NA
Imports	28.8	42.0	42.0	43.8	NA	41.9	NA
Minemouth price							
(2005 dollars per short ton)	23.34	22.41	22.06	22.82	NA	22.73	19.85[c]
(2005 dollars per million Btu)	1.15	1.11	1.09	1.13	NA	1.12	0.99[c]
Average delivered price to electricity generators							
(2005 dollars per short ton)	30.83	31.84	31.43	32.42	NA	34.02[d]	31.53[c]
(2005 dollars per million Btu)	1.53	1.60	1.58	1.63	1.42	1.67[c]	1.57[c]
				2025			
Production	**1.131**	**1.517**	**1.367**	**1.669**	**26.8**	**1.452**	**NA**
Consumption by sector							
Electric power	1.039	1.411	1.296	1.529	25.1	1.361	NA
Coke plants	23	21	18	24	NA	22	NA
Coal-to-liquids	NA	82	41	110	NA	NA	NA
Other industrial/buildings	66	68	65	71	2.0[b]	68	NA
Total	**1.128**	**1.582**	**1.420**	**1.734**	**27.1**	**1.452**	**NA**
Net coal exports	**21.1**	**-52.4**	**-39.7**	**-50.8**	**-0.3**	**-9.8**	**NA**
Exports	49.9	26.6	35.1	26.6	NA	49.0	NA
Imports	28.8	79.0	74.8	77.4	NA	58.8	NA
Minemouth price							
(2005 dollars per short ton)	23.34	21.55	20.96	22.68	NA	23.77	25.62[c]
(2005 dollars per million Btu)	1.15	1.09	1.06	1.15	NA	1.18	1.28[c]
Average delivered price to electricity generators							
(2005 dollars per short ton)	30.83	32.20	30.92	33.39	NA	33.81[d]	39.08[c]
(2005 dollars per million Btu)	1.53	1.63	1.57	1.69	1.35	1.68[c]	1.96[c]

Btu = British thermal unit. NA = Not available. See notes and sources at end of table.

Comparison with Other Projections

The *AEO2007* reference case includes the introduction of CTL technology by 2011. Coal use at CTL plants increases to 112 million tons in 2030 in the *AEO2007* reference case, representing 6 percent of total coal consumption. CTL production does not appear to be included in any of the other projections.

The *AEO2007* reference case shows relatively constant coal consumption levels for other industrial and buildings uses as well as at coke plants, in contrast to the other projections. In the EVA projection, other industrial/buildings coal consumption declines by 7 percent after 2015 to 68 million tons in 2025, nearly the same as the amount projected for 2015 in the *AEO2007* reference case. The *AEO2007* reference case projection for other industrial/buildings consumption in 2025 increases only slightly from 2015, to 68 million tons. Coal consumption at coke plants peaks in 2010 in the EVA projection at 26 million tons, slightly higher than the 22 million tons in the *AEO2007* reference case, then declines over the balance of the projection period. In 2025, the EVA and *AEO2007* reference case projections of coal consumption are nearly the same for both coke plants and other industrial/buildings. The GII projection for other industrial/buildings includes coal consumption at coke plants. Compared with the *AEO2007* reference case, GII's projection is lower over the entire period, declining after 2010 to 1.9 quadrillion Btu in 2030—8 percent less than projected in the *AEO2007* reference case.

With growing coal demand for electric power generation, most of the projections show an upward trend in minemouth coal prices after 2020; however, Hill & Associates, Inc. (Hill) and EVA project average minemouth coal prices beginning to increase by 2015. Following a 10-year period of declining minemouth coal prices, the *AEO2007* reference case projects prices increasing by 5 percent from 2020 to 2030. Hill projects the lowest minemouth coal price in 2015, but it also projects the highest price in 2025, at $25.62 per short ton (2005 dollars), with the greatest rate of increase over the projection period. Hill also projects the highest delivered coal price to the electric power sector in 2025, at $39.08 per short ton, 21 percent

Table 24. Comparison of coal projections, 2015, 2025, and 2030 (continued)
(million short tons, except where noted)

Projection	2005	AEO2007			Other projections		
		Reference	Low economic growth	High economic growth	GII[a]	EVA	Hill
		2030					
Production	**1.131**	**1.691**	**1.501**	**1.861**	**28.3**	**NA**	**NA**
Consumption by sector							
Electric power	1.039	1.570	1.393	1.712	26.7	NA	NA
Coke plants	23	21	16	25	NA	NA	NA
Coal-to-liquids	NA	112	100	130	NA	NA	NA
Other industrial/buildings	66	69	64	73	1.9[b]	NA	NA
Total	**1.128**	**1.772**	**1.573**	**1.939**	**28.7**	**NA**	**NA**
Net coal exports	**21.1**	**-67.8**	**-59.7**	**-62.9**	**-0.4**	**NA**	**NA**
Exports	49.9	27.2	25.8	26.4	NA	NA	NA
Imports	28.8	94.9	85.4	89.4	NA	NA	NA
Minemouth price							
(2005 dollars per short ton)	23.34	22.60	20.99	23.64	NA	NA	NA
(2005 dollars per million Btu)	1.15	1.15	1.07	1.20	NA	NA	NA
Average delivered price to electricity generators							
(2005 dollars per short ton)	30.83	33.52	31.69	34.57	NA	NA	NA
(2005 dollars per million Btu)	1.53	1.69	1.60	1.74	1.34	NA	NA

Btu = British thermal unit. NA = Not available.

[a]Coal quantities provided in quadrillion Btu.

[b]Includes coal consumption at coke plants.

[c]Converted from 2006 dollars to 2005 dollars to be consistent with *AEO2007*.

[d]Calculated by multiplying the delivered price of coal to the electric power sector in dollars per million Btu by the average heat content of coal delivered to the electric power sector.

Sources: ***2005 and AEO2007:*** AEO2007 National Energy Modeling System, runs AEO2007.D112106A (reference case), LM2007.D112106A (low economic growth case), and HM2007.D112106A (high economic growth case). **GII:** Global Insight, Inc., *Preliminary 2006 U.S. Energy Outlook* (November 2006). **EVA:** Energy Ventures Analysis, Inc., *FUELCAST: Long-Term Outlook* (August 2006). **Hill:** Hill & Associates, Inc., *2006 Outlook for U.S. Steam Coal Long-Term Forecast* (November 2006).

greater than in the *AEO2007* reference case. In contrast to the other projections, GII projects declining delivered coal prices to the electric power sector through 2030, falling from $1.53 per million Btu (2005 dollars) in 2010 to $1.34 per million Btu in 2030—$0.26 per million Btu (16 percent) less than in the *AEO2007* low economic growth case.

Coal demand is met primarily through domestic production in all the projections. Both the *AEO2007* and EVA projections show U.S. coal production increasing at an average rate of just over 1 percent per year from 2005 to 2015. EVA projects coal production in 2025 at 1,452 million short tons, 65 million short tons (4 percent) less than in the *AEO2007* reference case. The *AEO2007* reference case shows the largest increase in coal production over the entire projection period, with output reaching 1,691 million tons in 2030, nearly 50 percent higher than in 2005. The GII projection for coal production in 2030 is 5.1 quadrillion Btu (15 percent) below the *AEO2007* reference case projection, at a level below that in the *AEO2007* low economic growth case.

U.S. coal exports represent a small percentage of domestic coal production in all the projections. EVA projects the highest level of coal exports, 49 million tons in 2025, but in contrast with the other projections shows exports growing after 2010. Coal exports decline to 27 million short tons or less in 2030 in all the *AEO2007* cases. On a Btu basis, coal exports in the GII projection are lower than those in the *AEO2007* reference case in 2030. All the projections expect the United States to become a net importer of coal by 2020, with the *AEO2007* and GII projections anticipating the transition by 2015. U.S. coal imports reach 59 million tons in 2025 in the EVA projection, 20 million tons less than projected in the *AEO2007* reference case. The *AEO2007* reference case projects the highest level of coal imports, more than tripling over the projection period to 95 million tons in 2030.

List of Acronyms

A.B.	Assembly Bill
AEO	*Annual Energy Outlook*
AEO2006	*Annual Energy Outlook2006*
AEO2007	*Annual Energy Outlook 2007*
ANWR	Arctic National Wildlife Refuge
ASTM	American Society for Testing and Materials
B2, B5, B20	Biodiesel (2, 5, and 20 percent)
BLS	Bureau of Labor Statistics
BNSF	BNSF Railway Company
BTC	Baku-Tbilisi-Ceyhan pipeline
Btu	British thermal unit
CAAA90	Clean Air Act Amendments of 1990
CAFE	Corporate Average Fuel Economy
CAIR	Clean Air Interstate Rule
CAMR	Clean Air Mercury Rule
CBO	Congressional Budget Office
CCS	Carbon capture and sequestration
CFL	Compact fluorescent light
CHP	Combined heat and power
CO_2	Carbon dioxide
CPI	Consumer price index
CRP	Conservation Reserve Program
CSX	CSX Transportation
CT	Computed tomography
CTL	Coal-to-liquids
DB	Deutsche Bank AG
DDGS	Dried distillers' grains and solubles
DM&E	Dakota Minnesota & Eastern Railroad
DOE	U.S. Department of Energy
DOT	U.S. Department of Transportation
DVR	Digital video recorder
E&P	Exploration and production
E10, E85	Ethanol (10 percent and 85 percent)
EEA	Economic and Environmental Analysis, Inc.
EIA	Energy Information Administration
EPA	U.S. Environmental Protection Agency
EPACT2005	Energy Policy Act of 2005
EPS	Energy portfolio standard
EVA	Energy Ventures Analysis, Inc.
FERC	Federal Energy Regulatory Commission
FFV	Flex-fuel vehicle
FY	Fiscal year
GDP	Gross domestic product
GII	Global Insights, Inc.
GTL	Gas-to-liquids
GVWR	Gross vehicle weight rating
Hill	Hill & Associates
IEA	International Energy Agency
IGCC	Integrated gasification combined cycle
INFORUM	Interindustry Forecasting at the University of Maryland
IRAC	Imported refiners' acquisition cost of crude oil
IRS	Internal Revenue Service
LFG	Landfill gas
LNG	Liquefied natural gas
MDPV	Medium-duty passenger vehicle
MECS	Manufacturing Energy Consumption Survey (EIA)
MMS	Minerals Management Service
MRI	Magnetic resonance imaging
MSW	Municipal solid waste
MTBE	Methyl tertiary butyl ether
MY	Model year
NAICS	North American Industry Classification System
NEMS	National Energy Modeling System (EIA)
NHTSA	National Highway Traffic Safety Administration
NO_x	Nitrogen oxides
NS	Norfolk Southern
OCS	Outer Continental Shelf
OMB	Office of Management and Budget
OPEC	Organization of the Petroleum Exporting Countries
POLR	Provider of Last Resort
PTC	Production tax credit
PV	Photovoltaic
R&D	Research and development
RFG	Reformulated gasoline
RFS	Renewable Fuels Standard
RGGI	Regional Greenhouse Gas Initiative
RPS	Renewable Portfolio Standard
SAFETEA-LU	2005 Safe, Accountable, Flexible, Efficient Transportation Equity Act: A Legacy for Users
S.B.	Senate Bill
SEER	Strategic Energy and Economic Research, Inc.
SEP	Supplemental energy payment
SO_2	Sulfur dioxide
SOS	Standard Offer Service
SPR	Strategic Petroleum Reserve
SSA	Social Security Administration
STB	Surface Transportation Board
SUV	Sport utility vehicle
TV	Television
ULSD	Ultra-low-sulfur diesel
UP	Union Pacific Railroad Company
USGS	U.S. Geological Survey
WTI	West Texas Intermediate (crude oil)

Notes and Sources

Text Notes

Overview

1. For the complete text of the Energy Policy Act of 2005, see web site http://frwebgate.access.gpo.gov/cgi-bin/getdoc.cgi?dbname=109_cong_public_laws&docid=f:publ058.109.pdf.
2. Full hybrids include an integrated starter-generator that allows improved efficiency by shutting the engine off when the vehicle is idling and an electric motor that provides tractive power to the vehicle when it is moving. Mild hybrids include only an integrated starter.
3. Energy Information Administration, *Impacts of Modeled Recommendations of the National Commission on Energy Policy*, SR/OIAF/2005-02 (Washington, DC, April 2005), web site www.eia.doe.gov/oiaf/servicerpt/bingaman, and *Energy Market Impacts of Alternative Greenhouse Gas Intensity Reduction Goals*, SR/OIAF/2006-01 (Washington, DC, March 2006), web site www.eia.doe.gov/oiaf/service_rpts.htm.

Legislation and Regulations

4. E85 is a fuel containing a blend of 70 to 85 percent ethanol and 30 to 15 percent gasoline by volume.
5. The ethanol tax credit was first established in 1978. It has been extended in 1980, 1983, 1984, 1990, 1998, and 2005.
6. The PTC was subsequently extended in 1999, 2002, 2004, and 2005. Some extensions have included significant modifications, including changes in eligible resources, changes in the value and duration of the credit for certain resources, and changes in the treatment of the credit with respect to the Alternative Minimum Tax.
7. Regional Greenhouse Gas Initiative, "Model Rule & Amended Memorandum of Understanding" (August 2006), web site www.rggi.org/modelrule.htm.
8. Regional Greenhouse Gas Initiative, "Memorandum of Understanding" (December 2005), web site www.rggi.org/agreement.htm.
9. For the complete text of the Energy Policy Act of 2005, see web site http://frwebgate.access.gpo.gov/cgi-bin/getdoc.cgi?dbname=109_cong_public_laws&docid=f:publ058.109.pdf.
10. Energy Information Administration, "EPACT2005 Summary," *Annual Energy Outlook 2006*, DOE/EIA-0383(2006) (Washington, DC, February 2006), pp. 15-22, web site www.eia.doe.gov/oiaf/archive/aeo06.
11. See, for example, web site http://energy.senate.gov/public/_files/PostConferenceBillSummary.doc.
12. U.S. Department of Transportation, National Highway Traffic Safety Administration, "Average Fuel Economy Standards for Light Trucks Model Years 2008-2011," 49 CFR Parts 523, 533 and 537 (Docket No. 2006-24306, RIN 2127-AJ61), web site www.nhtsa.dot.gov/staticfiles/DOT/NHTSA/Rulemaking/Rules/Associated%20Files/2006FinalRule.pdf.
13. Energy Information Administration, "Proposed Revisions to Light Truck Fuel Economy Standards," *Annual Energy Outlook 2006*, DOE/EIA-0383(2006) (Washington, DC, February 2006), pp. 23-24, web site www.eia.doe.gov/oiaf/archive/aeo06.
14. U.S. Environmental Protection Agency, "Standards of Performance for Stationary Compression Ignition Internal Combustion Engines," 40 CFR Parts 60, 85, 89, 94, 1039, 1065, and 1068 (EPA-HQ-OAR-2005-0029, FRL-8190-7, RIN 2060-AM82), web site www.epa.gov/fedrgstr/EPA-AIR/2006/July/Day-11/a5968.htm.
15. See Energy Information Administration, "Clean Air Nonroad Diesel Rule,"*Annual Energy Outlook 2005*, DOE/EIA-0383(2005) (Washington, DC, February 2005), pp. 14-17, web site www.eia.doe.gov/oiaf/archive/aeo05.
16. Transition regulations apply to engines constructed or ordered after July 11, 2005, and manufactured after April 1, 2006.
17. The regulations specify different time tables and limits for emergency and fire pump engines.
18. Alpha-Gamma Technologies, Inc., "Population and Projection of Stationary Engines" (Memorandum, June 20, 2005), p. 3, web site www.epa.gov/ttn/atw/nsps/cinsps/nsps_population_projection4.pdf.
19. Energy Policy Act of 2005, Section 1501. For complete text, see web site http://frwebgate.access.gpo.gov/cgi-bin/getdoc.cgi?dbname=109_cong_public_laws&docid=f:publ058.109.pdf.
20. Energy Information Administration, "Status and Impact of State MTBE Bans" (March 2003), p. 1, web site http://tonto.eia.doe.gov/FTPROOT/service/mtbe.pdf.
21. U.S. Environmental Protection Agency, "State Actions Banning MTBE" (June 2004), web site www.epa.gov/mtbe/420b04009.pdf.
22. The Federal motor fuels excise tax credit for blending ethanol has been in place since 1978, at 51 cents per gallon. Its expiration date has been extended several times by Congress. Most recently, the American Jobs Creation Act of 2004, Section 301, extended the credit through 2010.
23. U.S. Department of Energy, Energy Efficiency and Renewable Energy, "Federal Fleet Requirements" (May 2006), web site www1.eere.energy.gov/vehiclesandfuels/epact/federal.
24. The requirement is actually for 2.7 percent oxygen by weight, which corresponds to 7.7 percent ethanol by volume. See Minnesota Department of Agriculture, "Economic Impact of the Ethanol Industry in Minnesota" (May 2003), p. 8, web site www.mda.state.mn.us/ethanol/economicimpact.pdf.
25. State of Hawaii, "New-Fuel.com—Your Source for Information About Hawaii's Transition to Ethanol-Blended Fuel," web site www.hawaii.gov/dbedt/ert/new-fuel.
26. National Biodiesel Board, "Minnesota Diesel Now Contains Two Percent Biodiesel Statewide" (Press

Release, September 29, 2005), web site www.biodiesel.org/resources/pressreleases/gen/20050929_mn_mandate_implemented.pdf.

27. National Biodiesel Board, "Washington State Adopts Biodiesel Requirement" (Press Release, March 30, 2006), web site www.biodiesel.org/resources/pressreleases/gen/20060330_WA_B2.pdf.

28. Green Car Congress, "Louisiana Legislature Passes 2% Renewable Fuels Standard" (June 9, 2006), web site www.greencarcongress.com/2006/06/louisiana_legis.html.

29. State of Louisiana, "Governor Blanco Signs HB 685" (Press Release, June 12, 2006), web site www.gov.state.la.us/index.cfm?md=newsroom&tmp=detail&articleID=1945.

30. Testimony of Robert Carroll, Deputy Assistant Secretary, Tax Analysis, U.S. Department of the Treasury, before the Subcommittee on Highways, Transit, and Pipelines, Committee on Transportation and Infrastructure, U.S. House of Representatives (April 4, 2006).

31. Testimony of Donald Marron, Acting Director, Congressional Budget Office, before the Subcommittee on Highways, Transit, and Pipelines, Committee on Transportation and Infrastructure, U.S. House of Representatives (April 4, 2006).

32. The U.S. Department of Defense maintains a detailed compilation of Federal, State, and local fuel taxes, inspection fees, and other environmental taxes and fees that apply to the sale of fuel. This compilation, which was last updated in August 2006, can be found at web site www.desc.dla.mil/DCM/Files/Tax compilation 2006-12.doc.

33. The ethanol tax credit was first established in 1978 and has been modified or extended in 1980, 1983, 1984, 1990, 1998, and 2005.

34. D. Nilles, R. Kotrba, J. Williams, and T. Bryan, "A Boost for Biodiesel," *Biodiesel Magazine* (August/September 2005), web site www.biodieselmagazine.com/article.jsp?article_id=376.

35. Eligible wind, poultry litter, geothermal, and certain hydroelectric and biomass facilities can claim the full 1.9-cent tax credit. Facilities burning eligible waste streams, including municipal waste, landfill gases, agricultural and landscaping residues, and forestry and mill wastes receive one-half the value of the credit given to other facilities, currently 0.95 cents.

36. Expiration on June 30, 1999, reauthorized on December 17, 1999; expiration on December 31, 2001, reauthorized on March 9, 2002; expiration on December 31, 2003, reauthorized on October 4, 2004. Reauthorized on August 8, 2005, before scheduled expiration at the end of 2005. Current expiration scheduled for December 31, 2007.

37. Energy Information Administration, "State Renewable Energy Requirements and Goals: Update Through 2005," *Annual Energy Outlook 2006*, DOE/EIA-0383(2006) (Washington, DC, February 2006), pp. 24-27, web site www.eia.doe.gov/oiaf/archive/aeo06.

38. E-mail communication with Ray Williamson, Arizona Corporation Commission, August 28, 2006.

39. Arizona Corporation Commission, "In the Matter of the Proposed Rulemaking for the Renewable Energy Standard and Tariff Rules," Docket No. RE-00000C-05-0030, web site www.cc.state.az.us/utility/electric/RES-03-14-06.pdf (March 14, 2006).

40. California Public Utilities Commission, "Renewable Energy," web site www.cpuc.ca.gov/static/energy/electric/renewableenergy.

41. California Energy Commission, *Renewable Energy Program: Overall Program Guidebook*, CEC-300-2006-008-F (Sacramento, CA, April 2006), web site www.energy.ca.gov/2006publications/CEC-300-2006-008/CEC-300-2006-008-F.pdf.

42. See web sites www.leginfo.ca.gov/pub/bill/sen/sb_0101-0150/sb_107_bill_20060926_history.html and www.leginfo.ca.gov/pub/bill/sen/sb_0101-0150/sb_107_bill_20060926_chaptered.pdf.

43. State of Connecticut, "An Act Concerning Biomass," Substitute Senate Bill No. 212, Public Act No. 06-74, web site www.cga.ct.gov/2006/ACT/PA/2006PA-00074-R00SB-00212-PA.htm.

44. State of Maine, "An Act To Enhance Maine's Energy Independence and Security," Chapter 677, H.P. 1439 - L.D. 2041, web site www.mainelegislature.org/legis/bills/chapdocs/PUBLIC677.doc.

45. Commonwealth of Massachusetts, Office of Consumer Affairs and Business Regulation, "Policy Statement on the RPS Eligibility of Retooled Biomass Plants" (October 27, 2005), web site www.mass.gov/doer/rps/rps-pol-stat-elig-biomass.pdf.

46. Commonwealth of Massachusetts, Office of Consumer Affairs and Business Regulation, "Annual RPS Compliance Report for 2004" (January 9, 2006), web site www.mass.gov/doer/rps/rps-2004annual-rpt.pdf.

47. Public Utilities Commission of Nevada, "Revised Adopted Regulation of the Public Utilities Commission of Nevada," LCB File No. R167-05 (February 23, 2006), web site www.leg.state.nv.us/Register/2005Register/R167-05RA.pdf.

48. New Jersey Board of Public Utilities, New Jersey Administrative Code Title 14, Chapter 8, Subchapter 2, "Renewable Portfolio Standards (RPS) Rules Adoption" (April 13, 2006), web site www.state.nj.us/bpu/wwwroot/secretary/RPS_148.2.pdf.

49. Wisconsin State Legislature, "Engrossed 2005 Senate Bill 459" (February 22, 2006), web site www.legis.state.wi.us/2005/data/SB-459eng.pdf.

50. Note that the *AEO2007* reference case is based on the California legislative target (as of September 1, 2006) of 10 percent by 2017. Legislation accelerating the deadline to 2010 was signed into law too late to be included in the *AEO2007* projections.

51. *Federal Register*, Vol. 70, No. 91 (May 12, 2005), 40 CFR Parts 51, 72, 73, 74, 77, 78, and 96.

52. *Federal Register*, Vol. 70, No. 95 (May 18, 2005), 40 CFR Parts 60, 72, and 75.
53. Energy Information Administration, "Federal Air Emissions Regulations," *Annual Energy Outlook 2006*, DOE/EIA-0383(2006) (Washington, DC, February 2006), pp. 28-29, web site www.eia.doe.gov/oiaf/archive/aeo06.
54. State and Territorial Air Pollution Program Administrators (STAPPA) and the Association of Local Air Pollution Control Officials (ALAPCO), "State Mercury Programs for Utilities" (August/September 2006), web site www.4cleanair.org/StateTable.doc.
55. State of Maine, "States Reach Agreement on Proposed Rules for the Nation's First Cap-and-Trade Program to Address Climate Change" (Press Release, August 15, 2006), web site http://mobile.maine.gov/news/?sid=22966.
56. M. Finnegan and M. Lifsher, "State's Greenhouse Gas Bill Signed," *Los Angeles Times* (September 27, 2006).
57. F. Barringer, "Officials Reach California Deal To Cut Emissions," *New York Times* (August 31, 2006).
58. West Coast Governors' Global Warming Initiative, web site www.ef.org/westcoastclimate.

Issues in Focus

59. The *AEO2007* near-term world oil prices (for 2006 and 2007) are those published in EIA's September 2006 *Short-Term Energy Outlook*, web site www.eia.doe.gov/pub/forecasting/steo/oldsteos/sep06.pdf.
60. The USGS provides three point estimates for undiscovered and inferred resources: the mean, a 5-percent confidence interval, and a 95-percent confidence interval, with no price relationship. *AEO2007* assumes that proven reserves are not subject to much uncertainty.
61. Includes crude oil, condensates, natural gas liquids, and refinery gains.
62. M.S. Crandall, *Energy, Economics, and Politics in the Caspian Region* (Westport, CT: Praeger Security International, 2006), p. 64.
63. International Energy Agency, *Oil Market Report* (October 6, 2006), p. 24, web site http://omrpublic.iea.org.
64. BP, *Statistical Review of World Energy 2006*, web site www.bp.com/productlanding.do?categoryId=6842&contentId=7021390 (October 12, 2006).
65. Global Insight, "Brazil Country Brief" (October 18, 2006).
66. International Energy Agency, *Oil Market Report* (October 6, 2006), p. 22, web site http://omrpublic.iea.org.
67. American Petroleum Institute, *Joint Association Survey on Drilling Costs* (1996-2004), web site www.api.org/statistics/accessapi/surveys/survey-description.cfm.
68. This well category—that is, a natural gas development well drilled between 7,500 feet and 9,999 feet—was chosen because: (1) about 75 percent of the wells drilled in the United States are natural gas wells; (2) more than 18 development wells are drilled for every exploration well; (3) exploration wells can have significant and unforeseen cost overruns, whereas the drilling of development wells is a more routine operation; and (4) the depth category of 7,500 to 9,999 feet accounts for the greatest number and proportion of natural gas development wells drilled. Consequently, this category is reasonably representative of the oil industry overall.
69. ODS-Petrodata, Inc., web site https://www.ods-petrodata.com/RigPoint.
70. *Oil & Gas Journal*, Nelson-Farrar Cost Indices (published quarterly in January, April, July, and October).
71. Source: Bureau of Labor Statistics, Seasonally Adjusted Producer Price Indices (PPI), Series WPS101. In May 2002, the iron and steel PPI was 112.9; by June 2006 it had risen to 194.0 (preliminary).
72. Baker-Hughes, Inc., "North American Rotary Rig Count," web site www.bhinet.com/investor/rig/rig_na.htm.
73. U.S. Census Bureau, web site www.census.gov/industry/ma333f05.xls, product code 3331311 for underground mining machinery.
74. Peabody Energy, Inc., web site www.peabodyenergy.com/pdfs/05_AR_FINANCIALS.pdf.
75. Peabody Energy, Inc., web site www.peabodyenergy.com/pdfs/05_AR_FINANCIALS.pdf.
76. CONSOL Energy, web site http://media.corporate-ir.net/media_files/irol/66/66439/pdfs/AR05.pdf.
77. Massey Energy Company, web site http://library.corporate-ir.net/library/ 10/102/102864/items/195646/2005AR.pdf.
78. The subsidiaries of Joy Global manufacture mining machinery that includes continuous miners, longwall shearers, powered roof supports, armored face conveyors, shuttle cars, flexible conveyor trains, roof bolters, electric mining shovels, rotary blasthole drills, and walking draglines.
79. R. Barrett, "Mining Boom Brightens Joy Global Profits, Developments in China, *Milwaukee Journal Sentinel* (February 25, 2005).
80. BNSF Railway, 2005 Annual Report, web site www.bnsf.com/investors/annualreports/2005annrpt.pdf.
81. BNSF Railway, 2005 Annual Report, web site www.bnsf.com/investors/annualreports/2005annrpt.pdf.
82. Toby Kolstad, Rail Theory Forecasts, LLC, personal communication, October 22, 2006.
83. "Acquisitions by Railroads and Others," *Progressive Railroading*, Vol. 48, No. 5 (May 2005), p. 49.
84. Imputed from the number and value of backlogged orders reported in Freight Car America's 2005 Annual Report, web site www.johnstownamerica.com.
85. Freight Car America, 2005 Annual Report, web site www.johnstownamerica.com.
86. P. Foran, "Powergrabs: Traffic Volumes Prompt Railroads To Pursue Locomotive Power—and Lots of It,"

Notes and Sources

Progressive Railroading, Vol. 48, No. 5 (May 2005), pp. 50-51.

87. BNSF Railway, 2005 Annual Report, web site www.bnsf.com/investors/annualreports/2005annrpt.pdf.

88. Surface Transportation Board, U.S. Department of Transportation, "Statistics of Class I Freight Railroads in the United States," web site www.stb.dot.gov/econdata.nsf.

89. BNSF Railway, 2005 Annual Report, web site www.bnsf.com/investors/annualreports/2005annrpt.pdf.

90. Reviews of energy demand elasticities include C. Dahl, *A Survey of Energy Demand Elasticities in Support of the Development of the NEMS*, Contract Number DE-AP01-93EI23499 (Washington, DC, October 1993).

91. Federal Energy Regulatory Commission, *Assessment of Demand Response and Advanced Metering*, Staff Report, Docket AD-06-2-000 (Washington, DC, August 2006), web site www.ferc.gov/legal/staff-reports/demand-response.pdf.

92. Energy Information Administration, Commercial Buildings Energy Consumption Survey (2003 data), web site www.eia.doe.gov/emeu/cbecs.

93. L.J. Sandahl, T.L. Gilbride, M.R. Ledbetter, H.E. Steward, and C. Calwell, *Compact Fluorescent Lighting in America: Lessons Learned on the Way to Market*, PNNL-15730 (Richland, WA: Pacific Northwest National Laboratory, June 2006), web site www.eere.energy.gov/buildings/info/documents/pdfs/cfl_lessons_learned_web.pdf.

94. Web sites www.energystar.gov and www.eere.energy.gov/buildings/building_america.

95. K.A. Hanson, "Seasonality of Moves and Duration of Residence," U.S. Census Bureau, Current Population Reports, Household Economic Studies (October 1998), web site www.census.gov/prod/3/98pubs/p70-66.pdf.

96. R. Nevin and G. Watson, "Evidence of Rational Market Valuations for Home Energy Efficiency," *Appraisal Journal* (October 1998).

97. Apparently, some very low-cost projects are never undertaken. See S.T. Anderson and R.G. Newell, "Information Programs for Technology Adoption: The Case of Energy-Efficiency Audits," *Resource and Energy Economics*, Vol. 26, No. 1 (March 2004), pp. 27-50. See web site www.rff.org/~newell/ResEnergyEcon.pdf.

98. R.B. Lung, A. McKane, and M. Olszewski, "Industrial Motor System Optimization Projects in the US: An Impact Study," TAPPI, web site www.tappi.org/s_tappi/sec_publications.asp?CID=4583&DID=510344; and U.S. Department of Energy, Office of Industrial Technologies, "Georgia Pacific's Insulation Upgrade Leads to Reduced Fuel Costs and Increased Process Efficiency," web site www1.eere.energy.gov/industry/bestpractices/pdfs/insulation.pdf.

99. Based on data from the U.S. Census Bureau's 2004 Annual Survey of Manufactures, published in *Statistics for Industry Groups and Industries: 2004*, M04(AS)-1 (Washington, DC, December 2005), web site www.census.gov/mcd/asm-as1.html.

100. Based on data from EIA's 2002 Manufacturing Energy Consumption Survey (MECS), Tables 10.2 to 10.13, web site www.eia.doe.gov/emeu/mecs/mecs2002/data02/shelltables.html. Petrochemical feedstocks are not reported separately in MECS but are included in this calculation.

101. See W.A. Pizer, W. Harrington, R.J. Kopp, R.D. Morgenstern, and J-S Shih, "Technology Adoption and Aggregate Energy Efficiency," Discussion Paper 02-52 (Washington, DC: Resources for the Future, December 2002), web site www.rff.org/Documents/RFF-DP-02-52.pdf.

102. Electricity demand for commercial office equipment is an end-use category separate from miscellaneous consumption.

103. TIAX LLC, *Commercial and Residential Sector Miscellaneous Electricity Consumption: Y2005 and Projections to 2030* , EIA contract report (September 2006).

104. The decline in electricity use per unit is driven by the new California standard that requires less electricity use while in standby mode, which accounts for more than 99 percent of hours of use over the course of a year. See California Energy Commission, *2006 Appliance Efficiency Regulations*, CEC-400-2006-002 (Sacramento, CA, January 2006).

105. Alaska's former governor, Frank Murkowski, supported 20-percent equity ownership for the State, which would also bear 20 percent of the construction costs.

106. The Northern Pipeline Act awarded Foothills Pipe Lines Ltd. the certificate for construction of the Canadian portion of the Alaska Natural Gas Transportation System along the Alaska Highway. After the passage of the Northern Pipeline Act, TransCanada purchased Foothills Pipe Line.

107. BP, ExxonMobil, and ConocoPhillips, "Alaska Producer Pipeline Update" (May 2002), Slide 15, web site www.arcticgaspipeline.com/Reference/Documents&Presentations/ProducerInformation/5-02Alaska ProducerUpdate.ppt.

108. U.S. Department of Labor, Bureau of Labor Statistics, Seasonally Adjusted Producer Price Indices (PPI), Series WPS101. In May 2002 the PPI for iron and steel was 112.9; in June 2006 it was 194.0.

109. Energy Information Administration, "Coal Distribution Back Issues," web site www.eia.doe.gov/cneaf/coal/page/coaldistrib/coal_distributions.html.

110. EIA's coal distribution data for 2004 indicate that the mode of transportation for 28 million tons (3 percent) of domestic coal shipments was "unknown."

111. Energy Information Administration, "Coal Transportation: Rates and Trends in the United States, 1979-2001 (with Supplementary Data to 2002)," Table 2.06, web site www.eia.doe.gov/cneaf/coal/page/trans/ratesntrends.html.

112. U.S. Senate, Committee on Energy and Natural Resources, "Coal-Based Generation Reliability," Statement of David Wilks, President of Energy Supply, Xcel Energy, on behalf of the Edison Electric Institute and Consumers United for Rail Equity (May 25, 2006).

113. Average transportation rates are imputed from the difference between average delivered prices, excluding imports, and average minemouth prices.

114. BNSF Railway Company, Quarterly Report (June 30, 2006) web site www.bnsf.com/investors/secfilings/10Q_Railway_2Q_2006.pdf.

115. U.S. Government Accountability Office, *Freight Railroads: Industry Health Has Improved, But Concerns about Competition and Capacity Should Be Addressed*, GAO-07-94 (Washington, DC, October 2006), web site www.gao.gov/new.items/d0794.pdf.

116. U.S. Senate, Subcommittee on Surface Transportation and Merchant Marine, "Economics, Service and Capacity in the Freight Railroad Industry," Statement of W. Douglas Buttrey, Chairman, Surface Transportation Board (June 21, 2006).

117. Surface Transportation Board, web site www.stb.dot.gov.

118. N. Carey, "Margins, Fuel Charges Boost Union Pacific 4th Qtr," Reuters.com (January 2006).

119. Norfolk Southern, web site www.nscorp.com/nscorphtml/marketing/pricing/NS8003_terms.html.

120. CSX, web site http://shipcsx.com/public/ec.shipcsx public/Main?module_url=/ec.pricingpublic/About.

121. Surface Transportation Board Decision, *Rail Fuel Surcharges*, Ex Parte No. 661 (August 3, 2006), web site www.stb.dot.gov.

122. Surface Transportation Board, *Rail Fuel Surcharge*, Comments of BNSF Railway (October 2, 2006).

123. "CSX STB Fuel Surcharge Filing" (October 3, 2006) web site www.csx.com.

124. U.S. Senate, Committee on Energy and Natural Resources, "Coal-Based Generation Reliability," Statement of Howard Gruenspecht, Deputy Administrator, Energy Information Administration (May 25, 2006).

125. R. Smith and D. Machalaba, "As Utilities Seek More Coal, Railroads Struggle To Deliver," *Wall Street Journal* (March 15, 2006), p. A1; and "Coal Stocks Outperform S&P, Metals Securities," *Platts Coal Trader* (November 15, 2005), pp. 2-3.

126. S. Bobb, Group Vice President of Coal Marketing, BNSF Railway, "June 2006 Coal Update," paper presented to staff at the Energy Information Administration (Washington, DC, June 27, 2006).

127. U.S. Senate, Committee on Energy and Natural Resources, "Coal-Based Generation Reliability," Statement of Edward Hamberger, President and Chief Executive Officer, Association of American Railroads (May 15, 2006); and S. Bobb, Group Vice President of Coal Marketing, BNSF Railway, "June 2006 Coal Update," paper presented to staff at the Energy Information Administration (Washington, DC, June 27, 2006).

128. A.J. Cebula, Vice President of Planning & Engineering, CANAC Inc., "Southern Powder River Basin, BNSF/UP Joint Line, Towards Sustainable Operations of 500 mmT," paper presented at the National Coal Transportation Association Annual Fall Meeting & Conference (Denver, CO, September 13, 2006).

129. "CANAC Sees 50 Percent Jump in SPRB Output by 2012," *Argus Coal Transportation,* Vol. 25, No. 17 (September 14, 2006), p. 1.

130. Personal communication to EIA from Doug Conway, General Manager–Coal, DM&E and IC&E Railroads (November 16, 2006).

131. O. Port, "Not Your Father's Ethanol," *Business Week* (February 21, 2005), web site www.businessweek.com/magazine/content/05_08/b3921117.htm.

132. National Renewable Energy Laboratory, *Lignocellulosic Biomass to Ethanol Process Design and Economics Utilizing Co-Current Dilute Acid Prehydrolysis and Enzymatic Hydrolysis for Corn Stover*, NREL/TP-510-32438 (Golden, CO, June 2002), web site www.nrel.gov/docs/fy02osti/32438.pdf.

133. National Biodiesel Board, "Biodiesel—Commonly Asked Questions," web site www.nbb.org/pdf_files/fuelfactsheets/CommonlyAsked.PDF.

134. U.S. Department of Energy, Energy Efficiency and Renewable Energy, Alternative Fuels Data Center, "State and Federal Incentives and Laws," web site www.eere.energy.gov/afdc/progs/view_ind_mtx.cgi?in/TAX/US/0.

135. U.S. Department of Agriculture, *The Economic Feasibility of Ethanol Production from Sugar in the United States* (Washington, DC, July 2006), Table 19, web site www.usda.gov/oce/EthanolSugarFeasibilityReport3.pdf.

136. U.S. Department of Agriculture, *USDA's 2002 Ethanol Cost-of-Production Survey* (Washington, DC, July 2005), Table 3, web site www.usda.gov/oce/reports/energy/USDA_2002_ETHANOL.pdf.

137. Promar International, *Evaluation and Analysis of Vegetable Oil Market: The Implication of Increased Demand for Industrial Uses on Markets & USB Strategy* (November 2005), pp. 37-41, web site www.nbb.org/resources/reportsdatabase/reports/gen/20051101_gen-368.pdf.

138. National Renewable Energy Laboratory, *Biomass Oil Analysis: Research Needs and Recommendations*, NREL/TP-510-34796 (June 2004), Table 23, web site www.nrel.gov/docs/fy04osti/34796.pdf.

139. In 2005, the wholesale price of gasoline was imputed to be $1.74 per gallon and the wholesale price of diesel $1.78 per gallon. In 2012, the average U.S. wholesale prices are projected to be $1.52 and $ 1.48 per gallon, respectively, as some of the unusual margins of the refinery market are projected to return to the historical pattern.

140. U.S. Department of Agriculture, *USDA's 2002 Ethanol Cost-of-Production Survey* (Washington, DC, July

2005), Table 3, web site www.usda.gov/oce/reports/energy/USDA_2002_ETHANOL.pdf.

141. S. Howell, "Time To Take the Biodiesel Plunge?" *Render Magazine* (February 2005), pp. 10-14.

142. R. Pruszko, "Rendered Fats and Oils as a Biodiesel Feedstock: The Opportunities and Challenges," *Render Magazine* (February 2006), pp. 10-12.

143. K. Collins, Chief Economist, U.S. Department of Agriculture, "Statement Before the U.S. Senate Committee on Agriculture, Nutrition, and Forestry" (January 10, 2007), p. 2, web site www.usda.gov/oce/newsroom/congressional_testimony/Collins_011007.pdf.

144. A. Baker and S. Zahniser, "Ethanol Reshapes the Corn Market," *Amber Waves Magazine*, Vol. 4, No. 2 (2006), pp. 30-35, web site www.ers.usda.gov/AmberWaves/April06/Features/Ethanol.htm.

145. Promar International, *Evaluation and Analysis of Vegetable Oil Market: The Implication of Increased Demand for Industrial Uses on Markets & USB Strategy* (November 2005), pp. 25-35, web site www.nbb.org/resources/reportsdatabase/reports/gen/20051101_gen-368.pdf.

146. U.S. Department of Agriculture, *2002 Census of Agriculture* (Washington, DC, June 2004), Vol. 1, Chapter 1, "U.S. National Level Data," Table 8, "Land: 2002 and 1997," web site www.nass.usda.gov/census/census02/volume1/us/st99_1_008_008.pdf.

147. U.S. Department of Agriculture, *Summary Report: 1997 National Resources Inventory* (Washington, DC, December 2000), Figure 1, "How Our Land Is Used, 1997," web site www.nrcs.usda.gov/Technical/NRI/1997/summary_report/figures.html#figure1.

148. For a description of the Conservation Reserve Program, see web site www.nrcs.usda.gov/programs/crp.

149. U.S. Department of Agriculture, Baseline Projection (Washington, DC, February 2006), Tables 8 and 13, "Planted Acres, 2005/06 Marketing Year."

150. Historic crop yields tabulated by U.S. Department of Agriculture, National Agriculture Statistics Service, web site www.nass.usda.gov/Data_and_Statistics/Quick_Stats/index.asp.

151. Average yield increase calculated from 30-year historic data supplied by U.S. Department of Agriculture, National Agriculture Statistics Service.

152. U.S. Department of Agriculture and U.S. Department of Energy, *Biomass as a Feedstock for a Bioenergy and Bioproducts Industry: The Technical Feasibility of a Billion-Ton Annual Supply*, DOE/GO-102995-2135 (Washington, DC, April 2005), web site http://feedstockreview.ornl.gov/pdf/billion_ton_vision.pdf.

153. Renewable Fuels Association, *From Niche to Nation: Ethanol Industry Outlook 2006* (Washington, DC, February 2006), pp. 14-15, web site www.ethanolrfa.org/objects/pdf/outlook/outlook_2006.pdf.

154. Promar International, *Evaluation and Analysis of Vegetable Oil Market: The Implication of Increased Demand for Industrial Uses on Markets & USB Strategy* (November 2005), pp. 25-35, web site www.nbb.org/resources/reportsdatabase/reports/gen/20051101_gen-368.pdf.

155. National Biodiesel Board, "Biodiesel Production," web site www.nbb.org/pdf_files/fuelfactsheets/Production.pdf.

156. F. Luxem, "Glycerin as a C-3 Raw Material," presented at the 2006 National Biodiesel Conference.

157. National Corn Grower's Association, "How Much Ethanol Can Come From Corn?" (May 2006), web site www.ncga.com/ethanol/pdfs/2006/HowMuchEthanol Can ComeFromCorn.v.2.pdf.

158. U.S. Department of Agriculture, "Agricultural Baseline Projections 2006-2015," web site www.ers.usda.gov/Briefing/Baseline/crops.htm.

159. National Corn Growers Association, "U.S. Corn Usage by Segment, 2005" (pie chart), web site www.ncga.com/WorldOfCorn/main/consumption1.asp.

160. U.S. Department of Energy, Energy Efficiency and Renewable Energy, Alternative Fuels Data Center, web site www.eere.energy.gov/afdc/progs/reg_matrx.cgi.

161. K. Collins, Chief Economist, U.S. Department of Agriculture, "Statement before the U.S. Senate Committee on Environment and Public Works" (September 6, 2006), p. 8, web site www.usda.gov/oce/newsroom/congressional_testimony/Biofuels Testimony 9-6-2006.doc.

Market Trends

162. The year 2004 was used as the end point (as opposed to 2005, which is the base year of the *AEO2007* projections) because of the precipitous drop in industrial energy consumption between 2004 and 2005 caused by the impact of hurricanes Katrina and Rita.

163. When the reference case industrial energy intensity projections are decomposed using the Divisia index, structural change accounts for 61 percent of the projected change in energy intensity. A discussion of the index can be found in Boyd et al., "Separating the Changing Effects of U.S. Manufacturing Production from Energy Efficiency Improvements," *Energy Journal*, Vol. 8, No. 2 (1987).

164. The refining portion of the industrial sector is projected to become more energy-intensive over time. Its energy intensity is projected to increase as a result of declining crude oil quality, higher quality products, and the use of alternative inputs and technologies to produce liquid fuels. Coal-to-liquids and biofuel techniques are more energy-intensive than standard refining processes.

165. The alternative technology cases change technology characterizations only for sectors represented in the NEMS industrial model. Consequently, in the technology cases portrayed in Figure 48, refining values are unchanged from those in the reference case projections. The petroleum refining industry displays a range of intensity changes in other alternative *AEO-2007* cases but responds differently from the other

industrial subsectors. For example, because of increased CTL production in the high price case, energy intensity in the petroleum refining industry is higher than in the reference case. In all the other industrial subsectors, energy intensity is lower in the high price case.

166. S.C. Davis and S.W. Diegel, *Transportation Energy Data Book: Edition 25*, ORNL-6974 (Oak Ridge, TN, May 2006), Chapter 3, "All Highway Vehicles and Characteristics," web site http://cta.ornl.gov/data/chapter3.shtml.

167. Unless otherwise noted, the term "capacity" in the discussion of electricity generation indicates utility, nonutility, and combined heat and power capacity. Costs reflect the weighted average of regional costs.

168. Does not include off-grid photovoltaics (PV). Based on annual PV shipments from 1989 through 2004, EIA estimates that as much as 167 megawatts of remote PV applications for electricity generation (off-grid power systems) was in service in 2004, plus an additional 447 megawatts in communications, transportation, and assorted other non-grid-connected, specialized applications. See Energy Information Administration, *Annual Energy Review 2005*, DOE/EIA-0384(2005) (Washington, DC, July 2006), Table 10.6 (annual PV shipments, 1989-2004). The approach used to develop the estimate, based on shipment data, provides an upper estimate of the size of the PV stock, including both grid-based and off-grid PV. It overestimates the size of the stock, because shipments include a substantial number of units that are exported, and each year some of the PV units installed earlier are retired from service or abandoned.

169. Avoided cost estimates the incremental cost of fuel and capacity displaced by a unit of the specified resource and more accurately reflects its as-dispatched energy value than comparison to the levelized cost of other individual technologies. It does not reflect system reliability cost, nor does it necessarily indicate the lowest cost alternative for meeting system energy and capacity needs.

170. Although cellulosic ethanol technology currently is not a commercially proven process, researchers and developers are vigorously pursuing cost reduction goals in the technology and production processes that would substantially exceed those considered in the *AEO2007* "lower cost" cases. These even lower production cost goals may be possible, but it is uncertain at present whether, and when, the technology advances necessary to achieve the lowest of the production cost goals will occur. Nevertheless, even the relatively modest reductions in production costs assumed in the *AEO-2007* "lower cost" cases can be seen to result in a significant increase in cellulosic ethanol production.

171. CAIR mandates SO_2 emissions caps in 28 eastern and midwestern States and the District of Columbia. The first compliance period begins in 2010, and a second, more stringent cap takes effect in 2015.

172. The first milestone for reducing NO_x emissions from electric power generation becomes effective in 2009. A lower limit is mandated for 2015.

173. The Phase I mercury cap is 38 short tons, beginning in 2010. The Phase II cap is 15 short tons, beginning in 2018.

Comparison with Other Projections

174. Because EVA reports a 2005 price of $8.84 (2005 dollars per thousand cubic feet), its projection actually shows a greater decline relative to the reported 2005 price than does the *AEO2007* reference case.

175. A delivered natural gas price margin equals the end-use sector natural gas price minus the wellhead natural gas price.

176. It should be noted that the 2005 industrial price reported by the other organizations is about a dollar higher than that reported in *AEO2007*.

Table Notes and Sources

Note: Tables indicated as sources in these notes refer to the tables in Appendixes A, B, C, and D of this report.

Table 1. Total energy supply and disposition in the *AEO2007* and *AEO2006* reference cases, 2005-2030: ***AEO2006***: AEO2006 National Energy Modeling System , run AEO2006.D111905A. ***AEO2007***: AEO2007 National Energy Modeling System, run AEO2007.D112106A. **Notes:** Quantities are derived from historical volumes and assumed thermal conversion factors. Other production includes liquid hydrogen, methanol, and some inputs to refineries. Net imports of petroleum include crude oil, petroleum products, unfinished oils, alcohols, ethers, and blending components. Other net imports include coal coke and electricity.

Table 2. Changes in Standard Offer Supply price determinations by supply region and State: Brubaker and Associates, Inc., "Customer Choice Update: State by State Summary of Restructuring Activities Around the Nation," in *Retail Competition Struggles As Fuel Costs Rise And Rate Caps Expire* (Spring 2006), web site www.consultbai.com/publications/newsletter/spring06_newsletter.pdf. Energy Information Administration Survey Data: 2005 preliminary f-861 database for the Arizona, Connecticut, Delaware, District of Columbia, Illinois, Montana, Nevada, New Hampshire, New York, Ohio, Oregon, Pennsylvania, Rhode Island, and Virginia percent of competitive load. Energy Information Administration, "Status of State Electric Industry Restructuring Activity—as of February 2003," web site www.eia.doe.gov/cneaf/electricity/chg_str/restructure.pdf. Connecticut Department of Public Utility Control, "Decision: Docket No. 06-01-08PH01, DPUC Development and Review of Standard Service and Supplier of Last Report Service—Phase I" (June 21, 2006). District of Columbia Public Service Commission, "Questions And Answers Regarding Pepco's 2004 Request for Proposals for Standard Offer Service, web site www.dcpsc.org/pdf_files/hottopics/RisingElectricityPrices_FAQ.pdf. Illinois Commerce Commission, "Electric Switching Statistics" (April

2006), web site www.icc.illinois.gov/en/switchstats.aspx. Maine State Legislature, Office of the Revisor of Statutes, "Maine Public Utilities/Standard Offer Law, Title 35-A, Public Utilities, Part 3, Electric Power, Chapter 32, Electric Industry Restructuring, §3212," web site http://janus.state.me.us/legis/statutes/35-A/title35-Asec3212.html. Maine Public Utility Commission, Standard Offer Service Rule, Chapter 301, web site http://Mainegov-Images.Informe.Org/Mpuc/Doing_Business/Rules/Part_3/Ch-301.pdf. Maine Public Utilities Commission, "Maine's Load Served by Competitive Providers as of June 1, 2006," web site http://mainegov-images.informe.org/mpuc/industries/electricity/electric%20restructuring/Migration_Rates_Monthly-June06.pdf. Maryland Public Service Commission, Case No. 8908, "In the Matter of the Commission's Inquiry into the Competitive Selection of Electricity Supplier/Standard Offer Service, The Commission Staff's Report/Observations on the Standard Offer Service Bidding Process and Results" (March 7, 2006). Maryland Public Service Commission, "Electric Choice Enrollment Monthly Report" (July 19, 2006), web site www.psc.state.md.us/psc/electric/enrollmentrpt.htm. Massachusetts Department of Technology and Energy, Electric Restructuring in Massachusetts, "Default Service" (2006), web site www.mass.gov/dte/restruct/competition/defaultservice.htm. Massachusetts Division of Energy Resources, "2006 Electric Power Customer Migration Data" (May 2006), web site www.mass.gov/Eoca/docs/doer/2006migrate.pdf. Michigan Public Service Commission, "Electric Customer Choice, Program Status: MPSC Reports on the Status of Electric Competition in Michigan, as of July 31, 2006," web site www.dleg.state.mi.us/mpsc/electric/restruct/status.htm. Public Utilities Company of Ohio, "Electric Rate Stabilization Plans: Ensuring Rate Certainty in Ohio," web site www.puco.ohio.gov/PUCO/Consumer/information.cfm?doc_id=1449. New Hampshire Public Utilities Commission, Electric Division, web site www.puc.state.nh.us/Electric/electric.htm. New York Public Service Commission, "Mandatory Hourly Pricing," web site www.dps.state.ny.us/Mandatory_Hourly_Pricing.html. New York Public Service Commission, "Retail Access Plans," web site www.dps.state.ny.us/Retail_Access_Plans.html. New York Public Service Commission, "Staff Report on the State of Competitive Energy Markets: Progress to Date and Future Opportunities" (March 2, 2006), web site www.dps.state.ny.us/StaffReportCompetition.pdf. New York Public Service Commission, Case 00-M-0504, "Proceeding on Motion of the Commission Regarding Provider of Last Resort Responsibilities: The Role of Utilities in Competitive Energy Markets and Fostering Development of Retail Competitive Opportunities. Statement of Policy on Further Steps Toward Competition in Retail Energy Markets" (August 25, 2004) web site http://www3.dps.state.ny.us/pscweb/WebFileRoom.nsf/ArticlesByCategory/F4746B665D1C642685256EFB00622E91/$File/201a.00m0504.pdf?OpenElement. Ohio Public Utilities Commission, "The End of the Market Development Period for Retail Electric Choice," web site www.puco.ohio.gov/puco/mediaroom/annualpublications.cfm?doc_id=1679#Electric. Oregon Public Utility Commission, "Competitive Power Market for Oregon Business Customers," web site www.puc.state.or.us/PUC/electric_restruc/ORDA03-05.pdf. Oregon Public Utility Commission, "Restructuring for Residential Consumers," web site www.oregon.gov/PUC/electric_restruc/consumer/resident.shtml. Oregon Public Utility Commission, "Status Report: Oregon Electric Industry Restructuring," (September, 2006), web site www.puc.state.or.us/PUC/electric_restruc/statrpt/2006/092006_status_report.pdf. Pennsylvania Public Utility Commission, "Metropolitan Edison Company & Pennsylvania Electric Company Rate Filing," "Duquesne Light Company Rate Filing," and "Penn Power's POLR Filing," web site www.puc.state.pa.us/electric/electric_index.aspx. Pennsylvania Public Utility Commission, "Provider of Last Resort," web site www.puc.state.pa.us/electric/electric_last_resort.aspx. Pennsylvania Public Utility Commission, "Rulemaking Re Electric Distribution Companies," Docket No. L-00040169, and "Obligation to Serve Retail Customers at the Conclusion of the Transition Period Pursuant to 66 Pa. C.S. §2807(e)(2), Provider of Last Resort Roundtable, Docket No. M-00041792, Proposed Rulemaking Order" (December 14, 2004), web site www.puc.state.pa.us/PcDocs/514337.doc. Texas Public Utility Commission, "Report to the 79th Texas Legislature: Scope of Competition in Electric Markets in Texas" (January 2005), web site www.puc.state.tx.us/electric/reports/scope/2005/2005scope_elec.pdf.

Table 3. OPEC and non-OPEC oil production in three *AEO2007* world oil price cases, 2005-2030: AEO2007 National Energy Modeling System, runs AEO2007.D112106A, HP2007.D112106A, and LP2007.D112106A; and Energy Information Administration, *Short-Term Energy Outlook* (September 2006), web site www.eia.doe.gov/pub/forecasting/steo/oldsteos/sep06.pdf.

Table 4. Changes in surface coal mining equipment costs, 2002-2005: 2002 and 2003: Previous editions, web site www.census.gov/cir/www. **2004 and 2005:** U.S. Department of Commerce, Economics and Statistics Administration, U.S. Census Bureau, Current Industrial Reports, *Construction Machinery: 2005*, MA333D(05)-1 (Washington, DC, July 2006).

Table 5. Miscellaneous electricity uses in the residential sector, 2005, 2015, and 2030: TIAX LLC, "Commercial and Residential Sector Miscellaneous Electricity Consumption: Y2005 and Projections to 2030" (September 2006); and AEO2007 National Energy Modeling System, run AEO2007.D112106A.

Table 6. Electricity use and market share for televisions by type, 2005 and 2015: TIAX LLC, "Commercial and Residential Sector Miscellaneous Electricity Consumption: Y2005 and Projections to 2030" (September 2006).

Table 7. Miscellaneous electricity uses in the commercial sector, 2005, 2015, and 2030: TIAX LLC, "Commercial and Residential Sector Miscellaneous Electricity Consumption: Y2005 and Projections to 2030" (September 2006); and AEO2007 National Energy Modeling System, run AEO2007.D112106A.

Table 8. Revised subgroups for the non-energy-intensive manufacturing industries in *AEO2007*: energy demand and value of shipments, 2002: Office of Management and Budget, *North American Industry Classification System—United States* (Springfield, VA, 2002). **Energy:** Energy Information Administration, Manufacturing Energy Consumption Survey 2002. **Value of Shipments:** Global Insight, Inc., *2006 U.S. Energy Outlook* (November 2006).

Table 9. Effects of DOE's loan guarantee program on the economics of electric power plant generating

technologies, 2015: AEO2007 National Energy Modeling System, runs AEO2007.D112106A and NUCLOAN. D020207A.

Table 10. Technically recoverable undiscovered oil and natural gas resources in the lower 48 Outer Continental Shelf as of January 1, 2003: U.S. Department of the Interior, Minerals Management Service, *Report to Congress: Comprehensive Inventory of U.S. OCS Oil and Natural Gas Resources* (February 2006), web site www.mms.gov/revaldiv/PDFs/FinalInvRptToCongress050106.pdf.

Table 11. U.S. motor fuels consumption, 2000-2005: Energy Information Administration, *Annual Energy Review 2005*, DOE/EIA-0384(2005) (Washington, DC, July 2006), Table 5.13c, web site www.eia.doe.gov/emeu/aer; Renewable Fuels Association, "Industry Statistics 2006," web site www.ethanolrfa.org/industry/statistics; National Biodiesel Board, "Estimated U.S. Biodiesel Production," web site www.biodiesel.org/pdf_files/fuelfactsheets/Production_Graph_Slide.pdf.

Table 12. Energy content of biofuels: Energy Information Administration, *Annual Energy Review 2005*, DOE/EIA-0384(2005) (Washington, DC, July 2006), Table A.1, web site www.eia.doe.gov/emeu/aer; U.S. Environmental Protection Agency, *A Comprehensive Analysis of Biodiesel Impacts on Exhaust Emissions*, EPA420-P-02-001 (Washington, DC, October 2002), web site http://65.166.250.34/resources/reportsdatabase/reports/gen/20021001_gen-323.pdf; Argonne National Laboratory, Center for Transportation Research, *GREET 1.5 Transportation Fuel Cycle Model: Volume 1, Methodology, Development Use and Results*," ANL/ESD-39 (Argonne, IL, August 1999), Table 3.3, web site www.transportation.anl.gov/software/GREET/pdfs/esd_39v1.pdf. **Note:** E85 is rarely blended as 85 percent ethanol and 15 percent gasoline. The more typical blend is 77 percent ethanol and 23 percent gasoline, but the term E85 is still used.

Table 13. U.S. production and values of biofuel coproducts: U.S. Department of Agriculture, "USDA's 2002 Ethanol Cost-of-Production Survey" (Washington, DC, July 2005), web site www.usda.gov/oce/reports/energy/USDA_2002_ETHANOL.pdf; USDA NASS Statistical Database, web site www.nass.usda.gov/index.asp; Renewable Fuels Association, *2005 Annual Industry Outlook*, web site www.ethanolrfa.org/industry/outlook.

Table 14. Vehicle fueling stations in the United States as of July 2006: National Ethanol Vehicle Coalition E85 Retails Locations Database, web site www.e85refueling.org; National Biodiesel Board Biodiesel Retail Locations, web site www.biodiesel.org/buying biodiesel/retailfuelingsites/showall.asp.

Table 15. Potential U.S. market for biofuel blends, 2005: Energy Information Administration, Office of Integrated Analysis and Forecasting.

Table 16. Costs of producing electricity from new plants, 2015 and 2030: AEO2007 National Energy Modeling System, run AEO2007.D112106A.

Table 17. Technically recoverable U.S. natural gas resources as of January 1, 2005: Energy Information Administration, Office of Integrated Analysis and Forecasting.

Figure Notes and Sources

Note: Tables indicated as sources in these notes refer to the tables in Appendixes A, B, C, and D of this report.

Figure 1. Energy prices, 1980-2030: History: Energy Information Administration, *Annual Energy Review 2005*, DOE/EIA-0384(2005) (Washington, DC, July 2006). **Projections:** Table A1.

Figure 2. Delivered energy consumption by sector, 1980-2030: History: Energy Information Administration, *Annual Energy Review 2005*, DOE/EIA-0384(2005) (Washington, DC, July 2006). **Projections:** Table A2.

Figure 3. Energy consumption by fuel, 1980-2030: History: Energy Information Administration, *Annual Energy Review 2005*, DOE/EIA-0384(2005) (Washington, DC, July 2006). **Projections:** Tables A1 and A17.

Figure 4. Energy use per capita and per dollar of gross domestic product, 1980-2030: History: Energy Information Administration, *Annual Energy Review 2005*, DOE/EIA-0384(2005) (Washington, DC, July 2006). **Projections: Energy use per capita:** Calculated from data in Table A2. **Energy use per dollar of GDP:** Table A19.

Figure 5. Electricity generation by fuel, 1980-2030: History: Energy Information Administration (EIA), Form EIA-860B, "Annual Electric Generator Report—Nonutility"; EIA, *Annual Energy Review 2005*, DOE/EIA-0384(2005) (Washington, DC, July 2006); and Edison Electric Institute. **Projections:** Table A8.

Figure 6. Total energy production and consumption, 1980-2030: History: Energy Information Administration, *Annual Energy Review 2005*, DOE/EIA-0384(2005) (Washington, DC, July 2006). **Projections:** Table A1.

Figure 7. Energy production by fuel, 1980-2030: History: Energy Information Administration, *Annual Energy Review 2005*, DOE/EIA-0384(2005) (Washington, DC, July 2006). **Projections:** Tables A1 and A17.

Figure 8. U.S. carbon dioxide emissions by sector and fuel, 1990-2030: History: Energy Information Administration, *Emissions of Greenhouse Gases in the United States 2005*, DOE/EIA-0573(2005) (Washington, DC, November 2006). **Projections:** Table A18.

Figure 9. Reformed CAFE standards for light trucks, by model year and vehicle footprint: U.S. Department of Transportation, National Highway Traffic Safety Administration, 49 CFR Part 523, 533 and 537 [Docket No. 2006-24306] RIN 2127-AJ61, "Average Fuel Economy Standards for Light Trucks, Model Years 2008-2011."

Figure 10. World oil prices in three *AEO2007* cases, 1990-2030: History: Energy Information Administration, *Annual Energy Review 2005*, DOE/EIA-0384(2005) (Washington, DC, July 2006). **Projections:** Table C1.

Figure 11. Changes in construction commodity costs, 1973-2006: U.S. Department of Labor, Bureau of Labor Statistics, Producer Price Index for WPU112, WPU101, WPU133, and WPU1322.

Figure 12. Drilling costs for onshore natural gas development wells at depths of 7,500 to 9,999 feet, 1996-2004: American Petroleum Institute, 2004 Joint Association Survey of Drilling Costs.

Figure 13. Changes in iron and steel, mining equipment and machinery, and railroad equipment costs, 1973-2006: U.S. Department of Labor, Bureau of Labor Statistics, Producer Price Indexes for Iron and Steel, Series ID WPU101; Mining Machinery and Equipment Manufacturing, Series ID PCU333131333131; and Railroad Equipment, Series ID WPU144.

Figure 14. Changes in construction commodity costs and electric utility construction costs, 1973-2006: *Handy-Whitman Bulletin*, No. 163, "Cost Trends of Electric Utility Construction"; and U.S. Department of Labor, Bureau of Labor Statistics, Producer Price Index, Series ID WPU112.

Figure 15. Additions to electricity generation capacity in the electric power sector, 1990-2030: History: Energy Information Administration, Form EIA-860, "Annual Electric Generator Report." **Projections:** AEO2007 National Energy Modeling System, run AEO2007.D112106A.

Figure 16. Energy intensity of industry subgroups in the metal-based durables group of non-energy-intensive manufacturing industries, 2002: Computed from Energy Information Administration, Manufacturing Energy Consumption Survey 2002, and Global Insight, Inc., *2006 U.S. Energy Outlook* (November 2006).

Figure 17. Average annual growth rates of value of shipments for metal-based durables industries in the *AEO2006* and *AEO2007* reference case projections, 2005-2030: *AEO2006*: AEO2006 National Energy Modeling System, run AEO2006.D111905A. ***AEO2007*:** AEO2007 National Energy Modeling System, run AEO2007.D112106A.

Figure 18. Average annual increases in energy demand for metal-based durables industries in the *AEO2006* and *AEO2007* reference case projections, 2005-2030: *AEO2006*: AEO2006 National Energy Modeling System, run AEO2006.D111905A. ***AEO2007*:** AEO2007 National Energy Modeling System, run AEO2007.D112106A.

Figure 19. Annual delivered energy demand for the non-energy-intensive manufacturing industry groups in the *AEO2006* and *AEO2007* reference case projections, 2005-2030: *AEO2006*: AEO2006 National Energy Modeling System, run AEO2006.D111905A. ***AEO-2007*:** National Energy Modeling System, run AEO2007.D112106A.

Figure 20. Lower 48 offshore crude oil production in two cases, 1990-2030: History: Energy Information Administration, Office of Integrated Analysis and Forecasting. **Projections:** AEO2007 National Energy Modeling System, runs AEO2007.D112106A and OCSACC.D112706A.

Figure 21. Lower 48 offshore natural gas production in two cases, 1990-2030: History: Energy Information Administration, Office of Integrated Analysis and Forecasting. **Projections:** AEO2007 National Energy Modeling System, runs AEO2007.D112106A and OCSACC.D112706A.

Figure 22. U.S. ethanol production and production capacity, 1999-2007: Renewable Fuels Association, "Industry Statistics 2006," web site www.ethanolrfa.org/industry/statistics. **Note:** In some years, the production capacity is less than the actual amount of consumption, because the industry as a whole operated above reported nameplate capacity.

Figure 23. Average U.S. prices for ethanol and gasoline, 2003-2006: Bloomberg Data Service, Regular Gasoline Average Rack Price (point of first sale) and Ethanol Average Rack Prices (accessed 8/20/06). **Notes:** The large price spike in June 2006 resulted from supply bottlenecks during transition from winter to summer reformulated gasoline blends that incorporated ethanol for the first time as a result of EPACT2005. Corn-based ethanol prices are expected to remain competitive with gasoline over the long term.

Figure 24. Average annual growth rates of real GDP, labor force, and productivity, 2005-2030: Table B4.

Figure 25. Average annual inflation, interest, and unemployment rates, 2005-2030: Table B4.

Figure 26. Sectoral composition of industrial output growth rates, 2005-2030: AEO2007 National Energy Modeling System, runs AEO2007.D112106A, HM2007.D112106A, and LM2007.D112106A.

Figure 27. Energy expenditures in the U.S. economy, 1990-2030: History: Energy Information Administration, *Annual Energy Review 2005*, DOE/EIA-0384(2005) (Washington, DC, July 2006). **Projections:** AEO2007 National Energy Modeling System, run AEO2007.D112106A.

Figure 28. Energy expenditures as share of gross domestic product, 1970-2030: History: U.S. Department of Commerce, Bureau of Economic Analysis; and Energy Information Administration, *Annual Energy Review 2005*, DOE/EIA-0384(2005) (Washington, DC, July 2006). **Projections:** AEO2007 National Energy Modeling System, run AEO2007.D112106A.

Figure 29. World oil prices, 1980-2030: History: Energy Information Administration, *Annual Energy Review 2005*, DOE/EIA-0384(2005) (Washington, DC, July 2006). **Projections:** Table C1.

Figure 30. U.S. gross petroleum imports by source, 2005-2030: AEO2007 National Energy Modeling System, run AEO2007.D112106A.

Figure 31. Unconventional resources as a share of the world liquids market, 1990-2030: History: Derived from Energy Information Administration, *International Energy Annual 2004*, DOE/EIA-0219(2004) (Washington, DC, July 2006), Table G.4. **Projections:** Table A20. **Note:** Data from Table G.4 are used as a proxy for historical unconventional oil production, because international data are limited. In addition, estimates of historical production from Canadian oil sands and Venezuelan ultra-heavy oil were added to Table G.4. Assumptions about future unconventional oil production are based on current investment reports, published production targets, resource availabilities, and marketplace competition.

Figure 32. World liquids production shares by region, 2005 and 2030: AEO2007 National Energy Modeling System, runs AEO2007.D112106A, HP2007.D112106A, and LP2007.D112106A.

Figure 33. Energy use per capita and per dollar of gross domestic product, 1980-2030: History: Energy Information Administration, *Annual Energy Review 2005*, DOE/EIA-0384(2005) (Washington, DC, July 2006). **Projections: Energy use per capita:** Calculated from data in Table A2. **Energy use per dollar of GDP:** Table A19.

Figure 34. Primary energy use by fuel, 2005-2030: History: Energy Information Administration, *Annual Energy Review 2005*, DOE/EIA-0384(2005) (Washington, DC, July 2006). **Projections:** Tables A1 and A17.

Figure 35. Delivered energy use by fuel, 1980-2030: History: Energy Information Administration, *Annual Energy Review 2005*, DOE/EIA-0384(2005) (Washington, DC, July 2006). **Projections:** Table A2.

Figure 36. Primary energy consumption by sector, 1980-2030: History: Energy Information Administration, *Annual Energy Review 2005*, DOE/EIA-0384(2005) (Washington, DC, July 2006). **Projections:** Table A2.

Figure 37. Residential delivered energy consumption per capita, 1990-2030: History: Energy Information Administration, "State Energy Consumption, Price, and Expenditure Estimates (SEDS)," (Washington, DC, October 2006), web site www.eia.doe.gov/emeu/states/_seds.html, and *Annual Energy Review 2005*, DOE/EIA-0384 (2005) (Washington, DC, July 2006). **Projections:** AEO2007 National Energy Modeling System, runs AEO2007.D112106A, BLDFRZN.D112206A, and BLDHIGH.D112206A.

Figure 38. Residential delivered energy consumption by fuel, 2005, 2015, and 2030: AEO2007 National Energy Modeling System, run AEO2007.D112106A.

Figure 39. Efficiency indicators for selected residential appliances, 2005 and 2030: Energy Information Administration, *Technology Forecast Updates—Residential and Commercial Building Technologies—Advanced Adoption Case* (Navigant Consulting, Inc., September 2004); and AEO2007 National Energy Modeling System, runs AEO2007.D112106A, BLDFRZN.D112206A, and BLDBEST.D112206A.

Figure 40. Commercial delivered energy consumption per capita, 1980-2030: History: Energy Information Administration, "State Energy Consumption, Price, and Expenditure Estimates (SEDS)" (Washington, DC, October 2006), web site www.eia.doe.gov/emeu/states/_seds.html, and *Annual Energy Review 2005*, DOE/EIA-0384(2005) (Washington, DC, July 2006). **Projections:** AEO2007 National Energy Modeling System, runs AEO2007.D112106A, BLDFRZN.D112206A, and BLDHIGH.D112206A.

Figure 41. Commercial delivered energy consumption by fuel, 2005, 2015, and 2030: AEO2007 National Energy Modeling System, run AEO2007.D112106A.

Figure 42. Efficiency indicators for selected commercial energy end uses, 2005 and 2030: Energy Information Administration, *Technology Forecast Updates—Residential and Commercial Building Technologies—Advanced Adoption Case* (Navigant Consulting, Inc., September 2004); and AEO2007 National Energy Modeling System, runs AEO2007.D112106A, BLDFRZN.D112206A, and BLDBEST.D112206A.

Figure 43. Buildings sector electricity generation from advanced technologies, 2030: AEO2007 National Energy Modeling System, runs AEO2007.D112106A, BLDHIGH.D112206A, and BLDBEST.D112206A.

Figure 44. Industrial delivered energy consumption, 1980-2030: History: Energy Information Administration, *Annual Energy Review 2005*, DOE/ EIA-0384(2005) (Washington, DC, July 2006). **Projections:** AEO2007 National Energy Model System, runs AEO2007.D112106A, HM2007.D112106A, and LM2007. D112106A.

Figure 45. Average output growth in the manufacturing subsectors, 2005-2030: AEO2007 National Energy Model System, runs AEO2007.D112106A, HM2007.D112106A, and LM2007.D112106A.

Figure 46. Average growth of delivered energy consumption in the manufacturing subsectors, 2005-2030: AEO2007 National Energy Model System, runs AEO2007.D112106A, HM2007.D112106A, and LM2007.D112106A.

Figure 47. Industrial delivered energy intensity, 1980-2030: History: Energy Information Administration, *Annual Energy Review 2005*, DOE/EIA-0384 (2005) (Washington, DC, July 2006); and Global Insight, Inc., *2006 U.S. Energy Outlook* (November 2006). **Projections:** AEO2007 National Energy Model System, runs AEO2007.D112106A, INDHIGH.D112406A, and INDFRZN.D112406A.

Figure 48. Average change in energy intensity in the manufacturing subsectors, 2005-2030: AEO2007 National Energy Model System, runs AEO2007.D112106A, INDHIGH.D112406A, and INDFRZN.D112406A.

Figure 49. Delivered energy consumption for transportation, 1980-2030: History: Energy Information Administration, *Annual Energy Review 2005*, DOE/EIA-0384 (2005) (Washington, DC, July 2006). **Projections:** AEO2007 National Energy Modeling System, runs AEO2007.D112106A, HM2007.D112106A, LM2007.D112106A, HP2007.D112106A, and LP2007.D112106A.

Figure 50. Delivered energy consumption in light-duty vehicles, 1980-2030: History: S.C. Davis and S.W. Diegel, *Transportation Energy Data Book: Edition 25*, ORNL-6974 (Oak Ridge, TN, May 2006), Chapter 2, Table 2.6, p. 2-8. **Projections:** AEO2007 National Energy Modeling System, runs AEO2007.D112106A, HM2007.D112106A, LM2007.D112106A, HP2007.D112106A, and LP2007.D112106A.

Figure 51. Average fuel economy of new light-duty vehicles, 1980-2030: History: U.S. Department of Transportation, National Highway Traffic Safety Administration, *Summary of Fuel Economy Performance* (Washington, DC, March 2005), web site www. nhtsa.dot.gov/staticfiles/DOT/NHTSA/Vehicle%20Safety/Articles/Associated%20Files/SummaryFuelEconomyPerformance-2005.pdf. **Projections:** AEO2007 National Energy Modeling System, runs AEO2007.D112106A, TRNFRZN.D120806A, TRNHIGH.D120806A, HP2007.D112106A, and LP2007.D112106A.

Figure 52. Sales of unconventional light-duty vehicles by fuel type, 2005, 2015, and 2030: AEO2007 National Energy Modeling System, run AEO2007.D112106A.

Figure 53. Annual electricity sales by sector, 1980-2030: History: Energy Information Administration, *Annual Energy Review 2005*, DOE/EIA-0384(2005) (Washington, DC, July 2006). **Projections:** Table A8.

Figure 54. Electricity generation by fuel, 2005 and 2030: Table A8.

Figure 55. Electricity generation capacity additions by fuel type, including combined heat and power, 2006-2030: Table A9.

Figure 56. Levelized electricity costs for new plants, 2015 and 2030: AEO2007 National Energy Modeling System, run AEO2007.D112106A.

Figure 57. Electricity generation capacity additions, including combined heat and power, by region and fuel, 2006-2030: AEO2007 National Energy Modeling System, run AEO2007.D112106A.

Figure 58. Electricity generation from nuclear power, 1973-2030: History: Energy Information Administration, *Annual Energy Review 2005*, DOE/EIA-0384 (2005) (Washington, DC, July 2006). **Projections:** AEO2007 National Energy Modeling System, runs AEO2007.D112106A, HM2007.D112106A, LM2007.D112106A, HP2007.D112106A, and LP2007.D112106A.

Figure 59. Levelized electricity costs for new plants by fuel type, 2015 and 2030: AEO2007 National Energy Modeling System, runs AEO2007.D112106A, LONUC07.D112706A, and ADVNUC07.D112906A. **Note:** Includes generation and interconnection costs.

Figure 60. Nonhydroelectric renewable electricity generation by energy source, 2005-2030: Table A16.

Figure 61. Grid-connected electricity generation from renewable energy sources, 1990-2030: History: Energy Information Administration, *Annual Energy Review 2005*, DOE/EIA-0384(2005) (Washington, DC, July 2006). **Projections:** Table A16. **Note:** Data for nonutility producers are not available before 1989.

Figure 62. Levelized and avoided costs for new renewable plants in the Northwest, 2030: AEO2007 National Energy Modeling System, runs AEO2007.D112106A, LOREN07.D120806A, and HIREN07.D120806A.

Figure 63. Renewable electricity generation, 2005-2030: AEO2007 National Energy Modeling System, runs AEO2007.D112106A and RGRPS07.D121206C.

Figure 64. Cumulative new generating capacity by technology type, 2006-2030: Table D6.

Figure 65. Fuel prices to electricity generators, 1995-2030: History: Energy Information Administration, *Annual Energy Review 2005*, DOE/EIA-0384(2005) (Washington, DC, July 2006). **Projections:** AEO2007 National Energy Modeling System, run AEO2007.D112106A.

Figure 66. Average U.S. retail electricity prices, 1970-2030: History: Energy Information Administration, *Annual Energy Review 2005*, DOE/EIA-0384(2005) (Washington, DC, July 2006). **Projections:** Table A8.

Figure 67. Natural gas consumption by sector, 1990-2030: History: Energy Information Administration, *Annual Energy Review 2005*, DOE/EIA-0384 (2004) (Washington, DC, July 2006). **Projections:** AEO2007 National Energy Modeling System, run AEO2007.D112106A.

Figure 68. Total natural gas consumption, 1990-2030: History: Energy Information Administration, *Annual Energy Review 2005*, DOE/EIA-0384(2005) (Washington, DC, July 2006). **Projections:** AEO2007 National Energy Modeling System, runs AEO2007.D112106A, HP2007.D112106A, LP2007.D112106A, HM2007.D112106A, and LM2007.D112106A.

Figure 69. Natural gas consumption in the electric power and other end-use sectors in alternative price cases, 1990-2030: History: Energy Information Administration, *Annual Energy Review 2005*, DOE/EIA-0384(2005) (Washington, DC, July 2006). **Projections:** AEO2007 National Energy Modeling System, runs AEO2007.D112106A, HP2007.D112106A, and LP2007.D112106A.

Figure 70. Natural gas consumption in the electric power and other end-use sectors in alternative growth cases, 1990-2030: History: Energy Information Administration, *Annual Energy Review 2005*, DOE/EIA-0384(2005) (Washington, DC, July 2006). **Projections:** AEO2007 National Energy Modeling System, runs AEO2007.D112106A, HM2007.D112106A, and LM2007.D112106A.

Figure 71. Lower 48 wellhead and Henry Hub spot market prices for natural gas, 1990-2030: History: For lower 48 wellhead prices: Energy Information Administration, *Annual Energy Review 2005*, DOE/EIA-0384(2005) (Washington, DC, July 2006). For Henry Hub natural gas prices: Energy Information Administration, *Short-Term Energy Outlook* Query System, Monthly Natural Gas Data, Variable NGHHMCF. **Projections:** AEO2007 National Energy Modeling System, run AEO2007.D112106A.

Figure 72. Lower 48 wellhead natural gas prices, 1990-2030: History: Energy Information Administration, *Annual Energy Review 2005*, DOE/EIA-0384(2005) (Washington, DC, July 2006). **Projections:** AEO2007 National Energy Modeling System, runs AEO2007.D112106A, HP2007.D112106A, LP2007.D112106A, OGHTEC07.D112706A, and OGLTEC07.D112706A.

Figure 73. Natural gas prices by end-use sector, 1990-2030: History: Energy Information Administration, *Annual Energy Review 2005*, DOE/EIA-0384(2005) (Washington, DC, July 2006). **Projections:** AEO2007 National Energy Modeling System, run AEO2007.D112106A.

Figure 74. Average natural gas transmission and distribution margins, 1990-2030: History: Calculated as the difference between natural gas end-use prices and lower 48 wellhead natural gas prices; Energy Information Administration, *Annual Energy Review 2005*, DOE/EIA-0384 (2005) (Washington, DC, July 2006). **Projections:** AEO2007 National Energy Modeling System, runs AEO2007.D112106A, HP2007.D112106A, and LP2007. D112106A.

Figure 75. Natural gas production by source, 1990-2030: History: Energy Information Administration, Office of Integrated Analysis and Forecasting. **Projections:** AEO2007 National Energy Modeling System, run AEO2007.D112106A.

Figure 76. Total U.S. natural gas production, 1990-2030: History: Energy Information Administration, *Annual Energy Review 2005*, DOE/ EIA-0384(2005) (Washington, DC, July 2006). **Projections:** AEO2007 National Energy Modeling System, runs AEO2007.D112106A, HP2007.D112106A, LP2007.D112106A, OGHTEC07.D112706A, and OGLTEC07.D112706A.

Figure 77. Net U.S. imports of natural gas by source, 1990-2030: History: Energy Information Administration, *Annual Energy Review 2005*, DOE/EIA-0384(2005) (Washington, DC, July 2006). **Projections:** AEO2007 National Energy Modeling System, run AEO2007.D112106A.

Figure 78. Net U.S. imports of liquefied natural gas, 1990-2030: History: Energy Information Administration, *Annual Energy Review 2005*, DOE/ EIA-0384(2005) (Washington, DC, July 2006). **Projections:** AEO2007 National Energy Modeling System, runs AEO2007.D112106A, LP2007.D112106A, HP2007.D112106A, HM2007.D112106A, and LM2007.D112106A.

Figure 79. Domestic crude oil production by source, 1990-2030: History: Energy Information Administration, Office of Integrated Analysis and Forecasting. **Projections:** AEO2007 National Energy Modeling System, run AEO2007.D112106A.

Figure 80. Total U.S. crude oil production, 1990-2030: History: Energy Information Administration, *Annual Energy Review 2005*, DOE/ EIA-0384(2005) (Washington, DC, July 2006). **Projections:** AEO2007 National Energy Modeling System, runs AEO2007.D112106A, OGHTEC07.D112706A, and OGLTEC07.D112706A.

Figure 81. Total U.S. unconventional oil production, 2005-2030: AEO2007 National Energy Modeling System, runs AEO2007.D112106A and HP2007.D112106A.

Figure 82. Liquid fuels consumption by sector, 1990-2030: History: Energy Information Administration, *Annual Energy Review 2005*, DOE/EIA-0384(2005) (Washington, DC, July 2006). **Projections:** AEO2007 National Energy Modeling System, run AEO2007.D112106A.

Figure 83. Net import share of U.S. liquid fuels consumption, 1990-2030: History: Energy Information Administration, *Annual Energy Review 2005*, DOE/EIA-0384 (2005) (Washington, DC, July 2006). **Projections:** AEO-2007 National Energy Modeling System, runs AEO2007. D112106A, LP2007.D112106A, HP2007.D112106A, HM2007.D112106A, and LM2007.D112106A.

Figure 84. Average U.S. delivered prices for motor gasoline, 1990-2030: History: Energy Information Administration, *Annual Energy Review 2005*, DOE/EIA-0384(2005) (Washington, DC, July 2006). **Projections:** AEO2007 National Energy Modeling System, runs AEO2007.D112106A, LP2007.D112106A, and HP2007. D112106A.

Figure 85. Cellulose ethanol production, 2005-2030: AEO2007 National Energy Modeling System, runs HP2007.D112106A, CT_80PCT_7l.D120406A, and CT_80 PCT_7l_RF.D120406A.

Figure 86. Coal production by region, 1970-2030: History (short tons): 1970-1990: Energy Information Administration (EIA), *The U.S. Coal Industry, 1970-1990: Two Decades of Change*, DOE/EIA-0559 (Washington, DC, November 2002). **1991-2000:** EIA, *Coal Industry Annual*, DOE/EIA-0584 (various years). **2001-2005:** EIA, *Annual Coal Report 2005*, DOE/EIA-0584(2005) (Washington, DC, October 2006), and previous issues. **History (conversion to quadrillion Btu): 1970-2005: Estimation Procedure:** EIA, Office of Integrated Analysis and Forecasting. Estimates of average heat content by region and year are based on coal quality data for 2005, collected in various energy surveys (see sources), and national-level estimates of U.S. coal production by year in units of quadrillion Btu, published in EIA's *Annual Energy Review*. **Sources:** EIA, *Annual Energy Review 2005*, DOE/EIA-0384(2005) (Washington, DC, July 2006), Table 1.2; Form EIA-3, "Quarterly Coal Consumption and Quality Report, Manufacturing Plants"; Form EIA-5, "Quarterly Coal Consumption and Quality Report, Coke Plants"; Form EIA-6A, "Coal Distribution Report"; Form EIA-7A, "Coal Production Report"; Form EIA-423, "Monthly Cost and Quality of Fuels for Electric Plants Report"; Form EIA-906, "Power Plant Report"; Form EIA-920, "Combined Heat and Power Plant Report"; U.S. Department of Commerce, Bureau of the Census, "Monthly Report EM 545"; and Federal Energy Regulatory Commission, Form 423, "Monthly Report of Cost and Quality of Fuels for Electric Plants." **Projections:** AEO2007 National Energy Modeling System, run AEO2007. D112106A. **Note:** For 2004-2030, the data for "Total Coal Production" include waste coal.

Figure 87. Distribution of coal to domestic markets by supply and demand regions, including imports, 2005 and 2030: 2005: Energy Information Administration (EIA), Form EIA-3, "Quarterly Coal Consumption and Quality Report, Manufacturing Plants"; Form EIA-5, "Quarterly Coal Consumption and Quality Report, Coke Plants"; Form EIA-6A, "Coal Distribution Report"; Form EIA-7A, "Coal Production Report"; Form EIA-423, "Monthly Cost and Quality of Fuels for Electric Plants Report"; Form EIA-906, "Power Plant Report"; Form EIA-920, "Combined Heat and Power Plant Report"; U.S. Department of Commerce, Bureau of the Census, "Monthly Report IM 145"; and Federal Energy Regulatory Commission, Form 423, "Monthly Report of Cost and Quality of Fuels for Electric Plants." **Projections:** AEO2007 National Energy Modeling System, run AEO2007.D112106A. **Note:** The region labeled "Demand east of the Mississippi River" includes the New England, Middle Atlantic, South Atlantic, East North Central, and East South Central Census divisions. The region labeled "Demand west of the Mississippi River" includes the West North Central, West South Central, Mountain, and Pacific Census divisions.

Figure 88. U.S. coal production, 2005, 2015, and 2030: AEO2007 National Energy Modeling System, runs HCCST07.D112906A; LP2007.D112106A, LM2007. D112106A, AEO2007.D112106A, HM2007.D112106A, HP2007.D112106A, and LCCST07.D112906A. **Note:** Coal production totals include waste coal.

Figure 89. Average minemouth price of coal by region, 1990-2030: History (dollars per short ton): 1990-2000: Energy Information Administration (EIA), *Coal Industry Annual*, DOE/EIA-0584 (various years). **2001-2005:** EIA, *Annual Coal Report 2005*, DOE/EIA-0584 (2005) (Washington, DC, October 2006), and previous issues. **History (conversion to dollars per million Btu): 1990-2005: Estimation Procedure:** EIA, Office of Integrated Analysis and Forecasting. Estimates of average heat content by region and year are based on coal quality data for 2005, collected in various energy surveys (see sources), and national-level estimates of U.S. coal production by year in units of quadrillion Btu published in EIA's *Annual Energy Review*. **Sources:** EIA, *Annual Energy Review 2005*, DOE/EIA-0384(2005) (Washington, DC, July 2006), Table 1.2; Form EIA-3, "Quarterly Coal Consumption and Quality Report, Manufacturing Plants"; Form EIA-5, "Quarterly Coal Consumption and Quality Report, Coke Plants"; Form EIA-6A, "Coal Distribution Report"; Form EIA-7A, "Coal Production Report"; Form EIA-423, "Monthly Cost and Quality of Fuels for Electric Plants Report"; Form EIA-906, "Power Plant Report"; Form EIA-920, "Combined Heat and Power Plant Report"; U.S. Department of Commerce, Bureau of the Census, "Monthly Report EM 545"; and Federal Energy Regulatory Commission, Form 423, "Monthly Report of Cost and Quality of Fuels for Electric Plants." **Projections:** AEO2007 National Energy Modeling System, run AEO2007.D112106A. **Note:** Includes reported prices for both open market and captive mines.

Figure 90. Average delivered coal prices, 1990-2030: History: Energy Information Administration (EIA), *Quarterly Coal Report, October-December 2005*, DOE/EIA-0121 (2005/4Q) (Washington, DC, March 2006), and previous issues; EIA, *Electric Power Monthly*, October 2006, DOE/EIA-0226(2006/10) (Washington, DC, October 2006); and EIA, *Annual Energy Review 2005*, DOE/EIA-0384(2005) (Washington, DC, July 2006). **Projections:** AEO2007 National Energy Modeling System, runs AEO2007.D112106A, LP2007.D112106A, HP2007.D112106A, LM2007.D112106A, HM2007.D112106A, LCCST07.D112906A, and HCCST07.D112906A. **Note:** Historical prices are weighted by consumption but exclude residential/commercial prices and export free-alongside-ship (f.a.s.) prices.

Figure 91. Coal consumption in the industrial and buildings sectors and at coal-to-liquids plants, 2005, 2015, and 2030: AEO2007 National Energy Modeling System, run AEO2007.D112106A.

Figure 92. Carbon dioxide emissions by sector and fuel, 2005 and 2030: 2005: Energy Information Administration, *Emissions of Greenhouse Gases in the United States 2005*, DOE/EIA-0573(2005) (Washington, DC, November 2006). **Projections:** Table A18.

Figure 93. Carbon dioxide emissions, 1990-2030: History: Energy Information Administration, *Emissions of Greenhouse Gases in the United States 2005*, DOE/EIA-0573(2005) (Washington, DC, November 2006). **Projections:** Table B2.

Figure 94. Sulfur dioxide emissions from electricity generation, 1995-2030: History: 1995: U.S. Environmental Protection Agency, *National Air Pollutant Emissions Trends, 1990-1998*, EPA-454/R-00-002 (Washington, DC, March 2000). **2000:** U.S. Environmental Protection Agency, *Acid Rain Program Preliminary Summary Emissions Report, Fourth Quarter 2004*, web site www.epa.gov/airmarkets/emissions/prelimarp/index.html. **Projections:** AEO2007 National Energy Modeling System, run AEO2007.D112106A.

Figure 95. Nitrogen oxide emissions from electricity generation, 1995-2030: History: 1995: U.S. Environmental Protection Agency, *National Air Pollutant Emissions Trends, 1990-1998*, EPA-454/R-00-002 (Washington, DC, March 2000). **2000:** U.S. Environmental Protection Agency, *Acid Rain Program Preliminary Summary Emissions Report, Fourth Quarter 2004*, web site www.epa.gov/airmarkets/emissions/prelimarp/index.html. **Projections:** AEO2007 National Energy Modeling System, run AEO2007.D112106A.

Figure 96. Mercury emissions from electricity generation, 1995-2030: History: 1995, 2000, and 2005: Energy Information Administration, Office of Integrated Analysis and Forecasting. **Projections:** AEO2007 National Energy Modeling System, run AEO2007.D112106A.

Appendixes

Table A1. Total Energy Supply and Disposition Summary
(Quadrillion Btu per Year, Unless Otherwise Noted)

Supply, Disposition, and Prices	Reference Case							Annual Growth 2005-2030 (percent)
	2004	2005	2010	2015	2020	2025	2030	
Production								
Crude Oil and Lease Condensate	11.58	10.96	11.99	12.52	12.48	11.82	11.40	0.2%
Natural Gas Plant Liquids	2.46	2.33	2.43	2.45	2.38	2.32	2.31	-0.0%
Dry Natural Gas	19.32	18.77	19.93	20.19	21.41	21.21	21.15	0.5%
Coal[1]	22.85	23.20	24.47	25.74	26.61	30.09	33.52	1.5%
Nuclear Power	8.22	8.13	8.23	8.47	9.23	9.23	9.33	0.6%
Hydropower	2.71	2.71	3.02	3.07	3.08	3.09	3.09	0.5%
Biomass[2]	2.81	2.71	4.22	4.45	4.69	5.04	5.26	2.7%
Other Renewable Energy[3]	0.74	0.76	1.18	1.26	1.33	1.37	1.44	2.6%
Other[4]	0.29	0.22	0.67	0.98	0.89	0.89	1.12	6.8%
Total	**70.98**	**69.80**	**76.13**	**79.12**	**82.09**	**85.06**	**88.63**	**1.0%**
Imports								
Crude Oil[5]	22.02	22.09	21.88	22.96	24.72	26.70	28.63	1.0%
Liquid Fuels and Other Petroleum[6]	6.11	7.16	6.02	6.56	7.05	7.81	9.02	0.9%
Natural Gas	4.36	4.42	5.36	6.43	6.17	6.53	6.47	1.5%
Other Imports[7]	0.82	0.85	0.92	1.02	1.73	1.89	2.26	4.0%
Total	**33.30**	**34.52**	**34.18**	**36.97**	**39.66**	**42.93**	**46.37**	**1.2%**
Exports								
Petroleum[8]	2.07	2.31	2.71	2.77	2.84	2.85	2.90	0.9%
Natural Gas	0.87	0.75	0.69	0.66	0.69	0.80	0.87	0.6%
Coal	1.25	1.27	1.12	0.96	0.80	0.67	0.69	-2.4%
Total	**4.19**	**4.33**	**4.52**	**4.39**	**4.33**	**4.32**	**4.47**	**0.1%**
Discrepancy[9]	**-0.58**	**-0.20**	**-0.70**	**-0.58**	**-0.74**	**-0.73**	**-0.63**	**N/A**
Consumption								
Liquid Fuels and Other Petroleum[10]	40.79	40.61	41.76	44.26	46.52	49.05	52.17	1.0%
Natural Gas	23.05	22.63	24.73	26.07	27.04	27.08	26.89	0.7%
Coal	22.60	22.87	24.24	25.64	27.29	30.62	34.14	1.6%
Nuclear Power	8.22	8.13	8.23	8.47	9.23	9.23	9.33	0.6%
Hydropower	2.71	2.71	3.02	3.07	3.08	3.09	3.09	0.5%
Biomass[11]	2.53	2.38	3.30	3.48	3.64	3.91	4.06	2.2%
Other Renewable Energy[3]	0.74	0.76	1.18	1.26	1.33	1.37	1.44	2.6%
Other[12]	0.04	0.08	0.04	0.03	0.04	0.04	0.04	-2.6%
Total	**100.67**	**100.19**	**106.50**	**112.28**	**118.16**	**124.39**	**131.16**	**1.1%**

Table A1. Total Energy Supply and Disposition Summary (Continued)
(Quadrillion Btu per Year, Unless Otherwise Noted)

Supply, Disposition, and Prices	Reference Case							Annual Growth 2005-2030 (percent)
	2004	2005	2010	2015	2020	2025	2030	
Prices (2005 dollars per unit)								
Petroleum (dollars per barrel)								
Imported Low Sulfur Light Crude Oil Price[13]	42.87	56.76	57.47	49.87	52.04	56.37	59.12	0.2%
Imported Crude Oil Price[13]	37.09	49.19	51.20	44.61	46.47	49.57	51.63	0.2%
Natural Gas (dollars per million Btu)								
Price at Henry Hub	6.08	8.60	6.28	5.46	5.71	6.14	6.52	-1.1%
Wellhead Price[14]	5.63	7.29	5.59	4.84	5.07	5.46	5.80	-0.9%
Natural Gas (dollars per thousand cubic feet)								
Wellhead Price[14]	5.80	7.51	5.76	4.99	5.22	5.62	5.98	-0.9%
Coal (dollars per ton)								
Minemouth Price[15]	20.68	23.34	24.20	22.41	21.58	21.55	22.60	-0.1%
Coal (dollars per million Btu)								
Minemouth Price[15]	1.01	1.15	1.18	1.11	1.08	1.09	1.15	-0.0%
Average Delivered Price[16]	1.45	1.61	1.77	1.65	1.62	1.66	1.71	0.2%
Average Electricity Price (cents per kilowatthour)	7.9	8.1	8.1	7.7	7.9	8.0	8.1	-0.0%

[1]Includes waste coal.
[2]Includes grid-connected electricity from wood and waste; biomass, such as corn, used for liquid fuels production; and non-electric energy demand from wood. Refer to Table A17 for details.
[3]Includes grid-connected electricity from landfill gas; municipal solid waste; wind; photovoltaic and solar thermal sources; and non-electric energy from renewable sources, such as active and passive solar systems. Excludes electricity imports using renewable sources and nonmarketed renewable energy. See Table A17 for selected nonmarketed residential and commercial renewable energy.
[4]Includes liquid hydrogen, methanol, and some domestic inputs to refineries.
[5]Includes imports of crude oil for the Strategic Petroleum Reserve.
[6]Includes imports of finished petroleum products, unfinished oils, alcohols, ethers, blending components, and renewable fuels such as ethanol.
[7]Includes coal, coal coke (net), and electricity (net).
[8]Includes crude oil and petroleum products.
[9]Balancing item. Includes unaccounted for supply, losses, gains, and net storage withdrawals.
[10]Includes petroleum-derived fuels and non-petroleum derived fuels, such as ethanol and biodiesel. Petroleum coke, which is a solid, is included. Also included are natural gas plant liquids, crude oil consumed as a fuel, and liquid hydrogen. Refer to Table A17 for detailed renewable liquid fuels consumption.
[11]Includes grid-connected electricity from wood and wood waste, non-electric energy from wood, and biofuels heat and coproducts used in the production of liquid fuels, but excludes the energy content of the liquid fuels.
[12]Includes net electricity imports.
[13]Weighted average price delivered to U.S. refiners.
[14]Represents lower 48 onshore and offshore supplies.
[15]Includes reported prices for both open market and captive mines.
[16]Prices weighted by consumption; weighted average excludes residential and commercial prices, and export free-alongside-ship (f.a.s.) prices.
Btu = British thermal unit.
N/A = Not applicable.
Note: Totals may not equal sum of components due to independent rounding. Data for 2004 and 2005 are model results and may differ slightly from official EIA data reports.
Sources: 2004 natural gas supply values: Energy Information Administration (EIA), *Natural Gas Annual 2004*, DOE/EIA-0131(2004) (Washington, DC, December 2005). 2005 natural gas supply values and natural gas wellhead price: EIA, *Natural Gas Monthly*, DOE/EIA-0130(2006/04) (Washington, DC, April 2006). 2004 natural gas wellhead price: Minerals Management Service and EIA, *Natural Gas Annual 2004*, DOE/EIA-0131(2004) (Washington, DC, December 2005). 2004 and 2005 coal minemouth and delivered coal prices: EIA, *Annual Coal Report 2005*, DOE/EIA-0584(2005) (Washington, DC, October 2006). 2005 petroleum supply values and 2004 crude oil and lease condensate production: EIA, *Petroleum Supply Annual 2005*, DOE/EIA-0340(2005)/1 (Washington, DC, October 2006). Other 2004 petroleum supply values: EIA, *Petroleum Supply Annual 2004*, DOE/EIA-0340(2004)/1 (Washington, DC, June 2005). 2004 and 2005 low sulfur light crude oil price: EIA, Form EIA-856, "Monthly Foreign Crude Oil Acquisition Report." Other 2004 and 2005 coal values: *Quarterly Coal Report, October-December 2005*, DOE/EIA-0121(2005/4Q) (Washington, DC, March 2006). Other 2004 and 2005 values: EIA, *Annual Energy Review 2005*, DOE/EIA-0384(2005) (Washington, DC, July 2006). **Projections:** EIA, AEO2007 National Energy Modeling System run AEO2007.D112106A.

Table A2. Energy Consumption by Sector and Source
(Quadrillion Btu per Year, Unless Otherwise Noted)

Sector and Source	Reference Case							Annual Growth 2005-2030 (percent)
	2004	2005	2010	2015	2020	2025	2030	
Energy Consumption								
Residential								
Liquefied Petroleum Gases	0.54	0.51	0.53	0.56	0.58	0.60	0.62	0.8%
Kerosene	0.09	0.10	0.10	0.10	0.10	0.09	0.09	-0.3%
Distillate Fuel Oil	0.93	0.93	0.90	0.89	0.85	0.80	0.76	-0.8%
Liquid Fuels and Other Petroleum Subtotal	1.55	1.54	1.53	1.55	1.53	1.49	1.46	-0.2%
Natural Gas	5.02	4.98	5.18	5.35	5.43	5.45	5.47	0.4%
Coal	0.01	0.01	0.01	0.01	0.01	0.01	0.01	-1.2%
Renewable Energy[1]	0.40	0.41	0.43	0.41	0.40	0.40	0.39	-0.2%
Electricity	4.41	4.66	5.06	5.43	5.80	6.13	6.47	1.3%
Delivered Energy	**11.39**	**11.60**	**12.21**	**12.74**	**13.17**	**13.48**	**13.80**	**0.7%**
Electricity Related Losses	9.75	10.15	10.90	11.44	12.08	12.50	12.89	1.0%
Total	**21.15**	**21.75**	**23.11**	**24.18**	**25.26**	**25.98**	**26.70**	**0.8%**
Commercial								
Liquefied Petroleum Gases	0.10	0.09	0.09	0.09	0.10	0.10	0.10	0.4%
Motor Gasoline[2]	0.05	0.05	0.05	0.05	0.05	0.05	0.05	0.6%
Kerosene	0.02	0.02	0.03	0.03	0.03	0.03	0.03	0.4%
Distillate Fuel Oil	0.47	0.48	0.45	0.48	0.48	0.48	0.49	0.1%
Residual Fuel Oil	0.12	0.14	0.14	0.14	0.14	0.14	0.14	0.2%
Liquid Fuels and Other Petroleum Subtotal	0.76	0.77	0.75	0.79	0.80	0.81	0.81	0.2%
Natural Gas	3.23	3.15	3.31	3.64	3.86	4.10	4.36	1.3%
Coal	0.10	0.10	0.10	0.10	0.10	0.10	0.10	-0.1%
Renewable Energy[3]	0.12	0.12	0.12	0.12	0.12	0.12	0.12	0.0%
Electricity	4.19	4.32	4.77	5.28	5.78	6.36	7.03	2.0%
Delivered Energy	**8.40**	**8.46**	**9.05**	**9.93**	**10.66**	**11.48**	**12.43**	**1.6%**
Electricity Related Losses	9.27	9.42	10.27	11.13	12.03	12.97	14.01	1.6%
Total	**17.67**	**17.88**	**19.33**	**21.06**	**22.69**	**24.45**	**26.44**	**1.6%**
Industrial[4]								
Liquefied Petroleum Gases	2.27	2.13	2.26	2.24	2.26	2.31	2.40	0.5%
Motor Gasoline[2]	0.32	0.32	0.32	0.32	0.33	0.35	0.36	0.4%
Distillate Fuel Oil	1.21	1.23	1.18	1.19	1.22	1.23	1.26	0.1%
Residual Fuel Oil	0.22	0.23	0.18	0.18	0.17	0.18	0.18	-0.9%
Petrochemical Feedstocks	1.54	1.38	1.48	1.49	1.50	1.52	1.57	0.5%
Other Petroleum[5]	4.53	4.45	4.05	4.26	4.34	4.48	4.78	0.3%
Liquid Fuels and Other Petroleum Subtotal	10.09	9.73	9.47	9.68	9.82	10.07	10.55	0.3%
Natural Gas	7.45	6.84	7.86	7.90	8.26	8.68	8.90	1.1%
Natural-Gas-to-Liquids Heat and Power	0.00	0.00	0.00	0.00	0.00	0.00	0.00	N/A
Lease and Plant Fuel[6]	1.13	1.10	1.10	1.10	1.21	1.17	1.15	0.2%
Natural Gas Subtotal	8.58	7.94	8.95	9.00	9.46	9.85	10.05	0.9%
Metallurgical Coal	0.65	0.62	0.60	0.59	0.57	0.57	0.57	-0.3%
Other Industrial Coal	1.40	1.35	1.37	1.35	1.34	1.35	1.36	0.0%
Coal-to-Liquids Heat and Power	0.00	0.00	0.00	0.12	0.21	0.67	0.93	N/A
Net Coal Coke Imports	0.14	0.04	0.02	0.02	0.02	0.02	0.02	-3.4%
Coal Subtotal	2.18	2.01	2.00	2.07	2.14	2.61	2.89	1.5%
Biofuels Heat and Coproducts	0.21	0.24	0.69	0.74	0.78	0.83	0.88	5.2%
Renewable Energy[7]	1.70	1.44	1.60	1.71	1.81	1.93	2.05	1.4%
Electricity	3.48	3.48	3.63	3.76	3.83	3.94	4.09	0.6%
Delivered Energy	**26.24**	**24.85**	**26.33**	**26.97**	**27.84**	**29.23**	**30.51**	**0.8%**
Electricity Related Losses	7.68	7.60	7.81	7.93	7.98	8.03	8.15	0.3%
Total	**33.92**	**32.45**	**34.14**	**34.89**	**35.82**	**37.26**	**38.66**	**0.7%**

Table A2. Energy Consumption by Sector and Source (Continued)
(Quadrillion Btu per Year, Unless Otherwise Noted)

Sector and Source	Reference Case							Annual Growth 2005-2030 (percent)
	2004	2005	2010	2015	2020	2025	2030	
Transportation								
Liquefied Petroleum Gases	0.03	0.04	0.05	0.05	0.06	0.07	0.08	2.8%
E85[8]	0.00	0.00	0.00	0.00	0.01	0.02	0.02	11.8%
Motor Gasoline[2]	17.01	17.00	17.37	18.57	19.95	21.38	22.89	1.2%
Jet Fuel[9]	3.38	3.37	4.04	4.34	4.54	4.59	4.70	1.3%
Distillate Fuel Oil[10]	5.93	6.02	6.64	7.28	7.81	8.59	9.58	1.9%
Residual Fuel Oil	0.74	0.81	0.82	0.84	0.85	0.86	0.87	0.3%
Liquid Hydrogen	0.00	0.00	0.00	0.00	0.00	0.00	0.00	43.1%
Other Petroleum[11]	0.15	0.18	0.18	0.19	0.19	0.19	0.19	0.2%
Liquid Fuels and Other Petroleum Subtotal	27.25	27.42	29.11	31.26	33.41	35.69	38.34	1.3%
Pipeline Fuel Natural Gas	0.59	0.58	0.66	0.70	0.79	0.79	0.79	1.3%
Compressed Natural Gas	0.03	0.03	0.06	0.08	0.09	0.11	0.12	5.5%
Electricity	0.02	0.02	0.03	0.03	0.03	0.04	0.04	1.8%
Delivered Energy	**27.89**	**28.05**	**29.86**	**32.07**	**34.33**	**36.63**	**39.29**	**1.4%**
Electricity Related Losses	0.05	0.05	0.06	0.07	0.07	0.07	0.08	1.5%
Total	**27.94**	**28.11**	**29.92**	**32.14**	**34.40**	**36.71**	**39.37**	**1.4%**
Delivered Energy Consumption for All Sectors								
Liquefied Petroleum Gases	2.93	2.77	2.93	2.95	2.99	3.07	3.19	0.6%
E85[8]	0.00	0.00	0.00	0.00	0.01	0.02	0.02	11.8%
Motor Gasoline[2]	17.38	17.37	17.74	18.94	20.34	21.78	23.30	1.2%
Jet Fuel[9]	3.38	3.37	4.04	4.34	4.54	4.59	4.70	1.3%
Kerosene	0.13	0.14	0.14	0.15	0.14	0.14	0.14	-0.2%
Distillate Fuel Oil	8.54	8.65	9.17	9.84	10.36	11.11	12.09	1.3%
Residual Fuel Oil	1.08	1.17	1.13	1.15	1.16	1.18	1.19	0.1%
Petrochemical Feedstocks	1.54	1.38	1.48	1.49	1.50	1.52	1.57	0.5%
Liquid Hydrogen	0.00	0.00	0.00	0.00	0.00	0.00	0.00	43.1%
Other Petroleum[12]	4.66	4.61	4.22	4.42	4.51	4.65	4.96	0.3%
Liquid Fuels and Other Petroleum Subtotal	39.65	39.46	40.86	43.29	45.55	48.06	51.17	1.0%
Natural Gas	15.71	15.01	16.41	16.96	17.65	18.33	18.86	0.9%
Natural-Gas-to-Liquids Heat and Power	0.00	0.00	0.00	0.00	0.00	0.00	0.00	N/A
Lease and Plant Fuel[6]	1.13	1.10	1.10	1.10	1.21	1.17	1.15	0.2%
Pipeline Natural Gas	0.59	0.58	0.66	0.70	0.79	0.79	0.79	1.3%
Natural Gas Subtotal	17.44	16.68	18.17	18.76	19.64	20.30	20.80	0.9%
Metallurgical Coal	0.65	0.62	0.60	0.59	0.57	0.57	0.57	-0.3%
Other Coal	1.51	1.46	1.48	1.46	1.45	1.46	1.47	0.0%
Coal-to-Liquids Heat and Power	0.00	0.00	0.00	0.12	0.21	0.67	0.93	N/A
Net Coal Coke Imports	0.14	0.04	0.02	0.02	0.02	0.02	0.02	-3.4%
Coal Subtotal	2.30	2.12	2.11	2.18	2.24	2.71	2.99	1.4%
Biofuels Heat and Coproducts	0.21	0.24	0.69	0.74	0.78	0.83	0.88	5.2%
Renewable Energy[13]	2.21	1.97	2.14	2.24	2.34	2.44	2.56	1.1%
Electricity	12.11	12.49	13.49	14.51	15.45	16.47	17.63	1.4%
Delivered Energy	**73.92**	**72.97**	**77.46**	**81.71**	**86.00**	**90.82**	**96.03**	**1.1%**
Electricity Related Losses	26.75	27.23	29.04	30.56	32.17	33.57	35.13	1.0%
Total	**100.67**	**100.19**	**106.50**	**112.28**	**118.16**	**124.39**	**131.16**	**1.1%**
Electric Power[14]								
Distillate Fuel Oil	0.15	0.19	0.24	0.24	0.25	0.28	0.28	1.5%
Residual Fuel Oil	0.99	0.96	0.67	0.74	0.72	0.71	0.72	-1.1%
Liquid Fuels and Other Petroleum Subtotal	1.14	1.16	0.90	0.97	0.97	0.99	1.01	-0.6%
Natural Gas	5.61	5.95	6.56	7.31	7.40	6.78	6.09	0.1%
Steam Coal	20.30	20.75	22.13	23.45	25.05	27.90	31.14	1.6%
Nuclear Power	8.22	8.13	8.23	8.47	9.23	9.23	9.33	0.6%
Renewable Energy[15]	3.55	3.64	4.67	4.83	4.93	5.09	5.15	1.4%
Electricity Imports	0.04	0.08	0.04	0.03	0.04	0.04	0.04	-2.6%
Total	**38.86**	**39.71**	**42.53**	**45.07**	**47.62**	**50.04**	**52.77**	**1.1%**

Table A2. Energy Consumption by Sector and Source (Continued)
(Quadrillion Btu per Year, Unless Otherwise Noted)

Sector and Source	Reference Case							Annual Growth 2005-2030 (percent)
	2004	2005	2010	2015	2020	2025	2030	
Total Energy Consumption								
Liquefied Petroleum Gases	2.93	2.77	2.93	2.95	2.99	3.07	3.19	0.6%
E85[8]	0.00	0.00	0.00	0.00	0.01	0.02	0.02	11.8%
Motor Gasoline[2]	17.38	17.37	17.74	18.94	20.34	21.78	23.30	1.2%
Jet Fuel[9]	3.38	3.37	4.04	4.34	4.54	4.59	4.70	1.3%
Kerosene	0.13	0.14	0.14	0.15	0.14	0.14	0.14	-0.2%
Distillate Fuel Oil	8.69	8.84	9.40	10.08	10.61	11.38	12.37	1.4%
Residual Fuel Oil	2.07	2.14	1.80	1.89	1.88	1.89	1.91	-0.4%
Petrochemical Feedstocks	1.54	1.38	1.48	1.49	1.50	1.52	1.57	0.5%
Liquid Hydrogen	0.00	0.00	0.00	0.00	0.00	0.00	0.00	43.1%
Other Petroleum[12]	4.66	4.61	4.22	4.42	4.51	4.65	4.96	0.3%
Liquid Fuels and Other Petroleum Subtotal	40.79	40.61	41.76	44.26	46.52	49.05	52.17	1.0%
Natural Gas	21.33	20.96	22.97	24.27	25.05	25.11	24.95	0.7%
Natural-Gas-to-Liquids Heat and Power	0.00	0.00	0.00	0.00	0.00	0.00	0.00	N/A
Lease and Plant Fuel[6]	1.13	1.10	1.10	1.10	1.21	1.17	1.15	0.2%
Pipeline Natural Gas	0.59	0.58	0.66	0.70	0.79	0.79	0.79	1.3%
Natural Gas Subtotal	23.05	22.63	24.73	26.07	27.04	27.08	26.89	0.7%
Metallurgical Coal	0.65	0.62	0.60	0.59	0.57	0.57	0.57	-0.3%
Other Coal	21.81	22.21	23.61	24.91	26.50	29.36	32.61	1.5%
Coal-to-Liquids Heat and Power	0.00	0.00	0.00	0.12	0.21	0.67	0.93	N/A
Net Coal Coke Imports	0.14	0.04	0.02	0.02	0.02	0.02	0.02	-3.4%
Coal Subtotal	22.60	22.87	24.24	25.64	27.29	30.62	34.14	1.6%
Nuclear Power	8.22	8.13	8.23	8.47	9.23	9.23	9.33	0.6%
Biofuels Heat and Coproducts	0.21	0.24	0.69	0.74	0.78	0.83	0.88	5.2%
Renewable Energy[16]	5.76	5.61	6.81	7.07	7.27	7.54	7.71	1.3%
Electricity Imports	0.04	0.08	0.04	0.03	0.04	0.04	0.04	-2.6%
Total	**100.67**	**100.19**	**106.50**	**112.28**	**118.16**	**124.39**	**131.16**	**1.1%**
Energy Use and Related Statistics								
Delivered Energy Use	73.92	72.97	77.46	81.71	86.00	90.82	96.03	1.1%
Total Energy Use	100.67	100.19	106.50	112.28	118.16	124.39	131.16	1.1%
Ethanol Consumed in Motor Gasoline and E85	0.29	0.33	0.91	0.98	1.06	1.15	1.22	5.3%
Population (millions)	294.23	296.94	310.26	323.70	337.13	350.78	364.94	0.8%
Gross Domestic Product (billion 2000 dollars)	10704	11049	12790	14698	17077	19666	22494	2.9%
Carbon Dioxide Emissions (million metric tons)	5923.1	5945.3	6214.0	6588.9	6944.5	7424.6	7950.2	1.2%

[1]Includes wood used for residential heating. See Table A4 and/or Table A17 for estimates of nonmarketed renewable energy consumption for geothermal heat pumps, solar thermal hot water heating, and solar photovoltaic electricity generation.
[2]Includes ethanol (blends of 10 percent or less) and ethers blended into gasoline.
[3]Includes commercial sector consumption of wood and wood waste, landfill gas, municipal solid waste, and other biomass for combined heat and power. See Table A5 and/or Table A17 for estimates of nonmarketed renewable energy consumption for solar thermal hot water heating and solar photovoltaic electricity generation.
[4]Includes energy for combined heat and power plants, except those whose primary business is to sell electricity, or electricity and heat, to the public.
[5]Includes petroleum coke, asphalt, road oil, lubricants, still gas, tire-derived fuel, and miscellaneous petroleum products.
[6]Represents natural gas used in well, field, and lease operations, and in natural gas processing plant machinery.
[7]Includes consumption of energy produced from hydroelectric, wood and wood waste, municipal solid waste, and other biomass sources.
[8]E85 refers to a blend of 85 percent ethanol (renewable) and 15 percent motor gasoline (nonrenewable). To address cold starting issues, the percentage of ethanol actually varies seasonally. The annual average ethanol content of 74 percent is used for this forecast.
[9]Includes only kerosene type.
[10]Diesel fuel for on- and off- road use.
[11]Includes aviation gasoline and lubricants.
[12]Includes unfinished oils, natural gasoline, motor gasoline blending components, aviation gasoline, lubricants, still gas, asphalt, road oil, petroleum coke, tire-derived fuel, and miscellaneous petroleum products.
[13]Includes electricity generated for sale to the grid and for own use from renewable sources, and non-electric energy from renewable sources. Excludes nonmarketed renewable energy consumption for geothermal heat pumps, buildings photovoltaic systems, and solar thermal hot water heaters.
[14]Includes consumption of energy by electricity-only and combined heat and power plants whose primary business is to sell electricity, or electricity and heat, to the public. Includes small power producers and exempt wholesale generators.
[15]Includes conventional hydroelectric, geothermal, wood and wood waste, municipal solid waste, other biomass, petroleum coke, wind, photovoltaic and solar thermal sources. Excludes net electricity imports.
[16]Includes hydroelectric, geothermal, wood and wood waste, municipal solid waste, other biomass, wind, photovoltaic and solar thermal sources. Includes ethanol components of E85; excludes ethanol blends (10 percent or less) in motor gasoline. Excludes net electricity imports and nonmarketed renewable energy consumption for geothermal heat pumps, buildings photovoltaic systems, and solar thermal hot water heaters.
Btu = British thermal unit.
N/A = Not applicable.
Note: Totals may not equal sum of components due to independent rounding. Data for 2004 and 2005 are model results and may differ slightly from official EIA data reports. Consumption values of 0.00 are values that round to 0.00, because they are less than 0.005.
Sources: 2004 and 2005 consumption based on: Energy Information Administration (EIA), *Annual Energy Review 2005*, DOE/EIA-0384(2005) (Washington, DC, July 2006). 2004 and 2005 population and gross domestic product: Global Insight macroeconomic model CTL0806. 2004 and 2005 carbon dioxide emissions: EIA, *Emissions of Greenhouse Gases in the United States 2005,* DOE/EIA-0573(2005) (Washington, DC, November 2006). **Projections:** EIA, AEO2007 National Energy Modeling System run AEO2007.D112106A.

Table A3. Energy Prices by Sector and Source
(2005 Dollars per Million Btu, Unless Otherwise Noted)

Sector and Source	Reference Case							Annual Growth 2005-2030 (percent)
	2004	2005	2010	2015	2020	2025	2030	
Residential								
Liquefied Petroleum Gases	18.21	19.29	23.67	22.88	23.18	23.53	23.91	0.9%
Distillate Fuel Oil	12.89	14.73	14.87	12.60	13.15	13.56	14.13	-0.2%
Natural Gas	10.72	12.43	10.98	10.24	10.54	10.97	11.43	-0.3%
Electricity	27.10	27.59	26.91	25.99	26.37	26.61	26.76	-0.1%
Commercial								
Distillate Fuel Oil	10.48	12.68	12.72	10.73	11.35	11.85	12.45	-0.1%
Residual Fuel Oil	6.25	8.41	7.54	6.49	7.07	7.17	7.31	-0.6%
Natural Gas	9.40	11.20	9.34	8.48	8.67	8.96	9.30	-0.7%
Electricity	24.59	25.25	24.50	23.33	23.95	24.23	24.27	-0.2%
Industrial[1]								
Liquefied Petroleum Gases	11.18	16.96	16.42	15.57	15.91	16.22	16.55	-0.1%
Distillate Fuel Oil	10.99	13.08	12.95	11.40	12.04	12.62	13.25	0.1%
Residual Fuel Oil	5.77	7.77	9.50	8.21	8.91	9.26	9.58	0.8%
Natural Gas[2]	6.47	8.16	6.43	5.65	5.90	6.21	6.56	-0.9%
Metallurgical Coal	2.31	3.06	3.09	2.72	2.71	2.70	2.75	-0.4%
Other Industrial Coal	1.80	2.15	2.26	2.20	2.18	2.23	2.29	0.2%
Coal to Liquids	0.00	0.00	0.00	0.89	0.97	1.23	1.33	N/A
Electricity	15.88	16.69	18.01	16.46	17.07	17.35	17.43	0.2%
Transportation								
Liquefied Petroleum Gases[3]	19.68	23.92	24.34	23.49	23.66	23.95	24.29	0.1%
E85[4]	20.91	23.10	21.29	20.09	20.61	21.26	21.50	-0.3%
Motor Gasoline[5]	15.72	18.64	17.90	16.06	16.63	17.32	17.76	-0.2%
Jet Fuel[6]	9.22	13.14	10.91	9.89	10.51	11.10	11.75	-0.4%
Distillate Fuel Oil[7]	13.58	17.52	16.81	14.86	15.42	15.91	16.47	-0.2%
Residual Fuel Oil	4.85	5.51	8.05	7.04	7.36	7.90	8.27	1.6%
Natural Gas[8]	11.91	14.76	13.97	12.86	12.98	13.22	13.45	-0.4%
Electricity	26.10	25.22	24.86	23.81	24.22	24.47	24.46	-0.1%
Electric Power[9]								
Distillate Fuel Oil	9.52	11.38	11.71	9.26	9.84	10.25	10.79	-0.2%
Residual Fuel Oil	4.99	6.96	6.58	5.60	6.08	6.58	6.85	-0.1%
Natural Gas	6.11	8.18	6.22	5.50	5.76	6.05	6.33	-1.0%
Steam Coal	1.40	1.53	1.71	1.60	1.58	1.63	1.69	0.4%
Average Price to All Users[10]								
Liquefied Petroleum Gases	12.69	17.48	18.02	17.24	17.62	17.96	18.30	0.2%
E85[4]	20.91	23.10	21.29	20.09	20.61	21.26	21.50	-0.3%
Motor Gasoline[5]	15.71	18.60	17.90	16.06	16.63	17.32	17.75	-0.2%
Jet Fuel	9.22	13.14	10.91	9.89	10.51	11.10	11.75	-0.4%
Distillate Fuel Oil	12.91	16.22	15.70	13.92	14.53	15.08	15.70	-0.1%
Residual Fuel Oil	5.10	6.59	7.61	6.55	7.00	7.47	7.79	0.7%
Natural Gas	7.83	9.65	7.83	7.06	7.32	7.68	8.09	-0.7%
Metallurgical Coal	2.31	3.06	3.09	2.72	2.71	2.70	2.75	-0.4%
Other Coal	1.43	1.57	1.74	1.63	1.61	1.66	1.72	0.4%
Coal to Liquids	N/A	N/A	N/A	0.89	0.97	1.23	1.33	N/A
Electricity	23.01	23.73	23.66	22.55	23.15	23.47	23.60	-0.0%

Table A3. Energy Prices by Sector and Source (Continued)
(2005 Dollars per Million Btu, Unless Otherwise Noted)

Sector and Source	Reference Case							Annual Growth 2005-2030 (percent)
	2004	2005	2010	2015	2020	2025	2030	
Non-Renewable Energy Expenditures by Sector (billion 2005 dollars)								
Residential	195.99	215.13	220.44	221.00	236.03	248.99	262.21	0.8%
Commercial	141.78	154.38	157.97	163.37	181.74	201.11	222.08	1.5%
Industrial	184.23	196.07	200.48	184.11	194.88	206.98	222.08	0.5%
Transportation	385.00	474.66	476.38	459.42	511.07	570.28	632.79	1.2%
Total Non-Renewable Expenditures	907.00	1040.25	1055.27	1027.91	1123.73	1227.35	1339.16	1.0%
Transportation Renewable Expenditures	0.02	0.03	0.06	0.09	0.15	0.32	0.51	11.5%
Total Expenditures	**907.02**	**1040.29**	**1055.33**	**1028.00**	**1123.89**	**1227.67**	**1339.68**	**1.0%**

[1]Includes energy for combined heat and power plants, except those whose primary business is to sell electricity, or electricity and heat, to the public.
[2]Excludes use for lease and plant fuel.
[3]Includes Federal and State taxes while excluding county and local taxes.
[4]E85 refers to a blend of 85 percent ethanol (renewable) and 15 percent motor gasoline (nonrenewable). To address cold starting issues, the percentage of ethanol actually varies seasonally. The annual average ethanol content of 74 percent is used for this forecast.
[5]Sales weighted-average price for all grades. Includes Federal, State and local taxes.
[6]Kerosene-type jet fuel. Includes Federal and State taxes while excluding county and local taxes.
[7]Diesel fuel for on-road use. Includes Federal and State taxes while excluding county and local taxes.
[8]Compressed natural gas used as a vehicle fuel. Includes estimated motor vehicle fuel taxes and estimated dispensing costs or charges.
[9]Includes electricity-only and combined heat and power plants whose primary business is to sell electricity, or electricity and heat, to the public.
[10]Weighted averages of end-use fuel prices are derived from the prices shown in each sector and the corresponding sectoral consumption.
Btu = British thermal unit.
N/A = Not applicable.
Note: Data for 2004 and 2005 are model results and may differ slightly from official EIA data reports.
Sources: 2004 and 2005 prices for motor gasoline, distillate fuel oil, and jet fuel are based on prices in the Energy Information Administration (EIA), *Petroleum Marketing Annual 2005*, DOE/EIA-0487(2005) (Washington, DC, August 2006). 2004 residential and commercial natural gas delivered prices: EIA, *Natural Gas Annual 2004*, DOE/EIA-0131(2004) (Washington, DC, December 2005). 2005 residential and commercial natural gas delivered prices: EIA, *Natural Gas Monthly*, DOE/EIA-0130(2006/04) (Washington, DC, April 2006). 2004 and 2005 industrial natural gas delivered prices are estimated based on: EIA, *Manufacturing Energy Consumption Survey 1994* and industrial and wellhead prices from the *Natural Gas Annual 2004*, DOE/EIA-0131(2004) (Washington, DC, December 2005) and the *Natural Gas Monthly*, DOE/EIA-0130(2006/04) (Washington, DC, April 2006). 2004 transportation sector natural gas delivered prices are based on: EIA, *Natural Gas Annual 2004*, DOE/EIA-0131(2004) (Washington, DC, December 2005) and estimated state taxes, federal taxes, and dispensing costs or charges. 2005 transportation sector natural gas delivered prices are model results. 2004 and 2005 electric power sector natural gas prices: EIA, *Electric Power Monthly*, DOE/EIA-0226, May 2003 through April 2004, Table 4.11.A. 2004 and 2005 coal prices based on: EIA, *Quarterly Coal Report, October-December 2005*, DOE/EIA-0121(2005/4Q) (Washington, DC, March 2006) and EIA, AEO2007 National Energy Modeling System run AEO2007.D112106A. 2004 and 2005 electricity prices: EIA, *Annual Energy Review 2005*, DOE/EIA-0384(2005) (Washington, DC, July 2006). 2004 and 2005 ethanol prices derived from weekly spot prices in the Oxy Fuel News. **Projections:** EIA, AEO2007 National Energy Modeling System run AEO2007.D112106A.

Table A4. Residential Sector Key Indicators and Consumption
(Quadrillion Btu per Year, Unless Otherwise Noted)

Key Indicators and Consumption	Reference Case							Annual Growth 2005-2030 (percent)
	2004	2005	2010	2015	2020	2025	2030	
Key Indicators								
Households (millions)								
Single-Family	77.46	78.95	84.84	90.34	95.57	100.32	104.76	1.1%
Multifamily	27.37	27.67	29.10	30.39	31.81	33.31	34.88	0.9%
Mobile Homes	6.72	6.70	6.77	7.03	7.32	7.59	7.85	0.6%
Total	**111.56**	**113.32**	**120.71**	**127.75**	**134.70**	**141.21**	**147.49**	**1.1%**
Average House Square Footage	**1755**	**1770**	**1830**	**1883**	**1929**	**1969**	**2004**	**0.5%**
Energy Intensity								
(million Btu per household)								
Delivered Energy Consumption	102.1	102.3	101.1	99.7	97.8	95.4	93.6	-0.4%
Total Energy Consumption	189.6	191.9	191.4	189.3	187.5	184.0	181.0	-0.2%
(thousand Btu per square foot)								
Delivered Energy Consumption	58.2	57.8	55.3	53.0	50.7	48.5	46.7	-0.9%
Total Energy Consumption	108.0	108.4	104.6	100.5	97.2	93.4	90.3	-0.7%
Delivered Energy Consumption by Fuel								
Electricity								
Space Heating	0.39	0.40	0.44	0.47	0.49	0.50	0.51	1.0%
Space Cooling	0.65	0.74	0.73	0.76	0.80	0.85	0.90	0.8%
Water Heating	0.37	0.37	0.39	0.40	0.41	0.41	0.42	0.4%
Refrigeration	0.40	0.39	0.36	0.35	0.35	0.36	0.38	-0.1%
Cooking	0.10	0.11	0.12	0.12	0.13	0.14	0.15	1.3%
Clothes Dryers	0.24	0.25	0.26	0.27	0.29	0.30	0.32	1.0%
Freezers	0.13	0.12	0.12	0.12	0.12	0.13	0.13	0.2%
Lighting	0.72	0.73	0.79	0.87	0.94	0.99	1.03	1.3%
Clothes Washers[1]	0.03	0.03	0.03	0.03	0.03	0.03	0.03	-0.8%
Dishwashers[1]	0.02	0.02	0.03	0.03	0.03	0.03	0.03	1.2%
Color Televisions and Set-Top Boxes	0.30	0.30	0.38	0.39	0.40	0.45	0.50	2.0%
Personal Computers	0.07	0.08	0.12	0.14	0.17	0.20	0.21	4.1%
Furnace Fans	0.08	0.08	0.10	0.11	0.11	0.12	0.13	1.6%
Other Uses[2]	0.89	1.01	1.21	1.36	1.53	1.63	1.74	2.2%
Delivered Energy	**4.41**	**4.66**	**5.06**	**5.43**	**5.80**	**6.13**	**6.47**	**1.3%**
Natural Gas								
Space Heating	3.53	3.52	3.69	3.82	3.88	3.89	3.89	0.4%
Space Cooling	0.00	0.00	0.00	0.00	0.00	0.00	0.00	31.8%
Water Heating	1.16	1.14	1.15	1.17	1.18	1.18	1.20	0.2%
Cooking	0.21	0.22	0.23	0.23	0.24	0.25	0.26	0.7%
Clothes Dryers	0.07	0.07	0.07	0.08	0.08	0.09	0.09	1.0%
Other Uses[3]	0.04	0.04	0.04	0.04	0.04	0.04	0.04	-0.3%
Delivered Energy	**5.02**	**4.98**	**5.18**	**5.35**	**5.43**	**5.45**	**5.47**	**0.4%**
Distillate Fuel Oil								
Space Heating	0.81	0.82	0.79	0.79	0.76	0.72	0.68	-0.7%
Water Heating	0.12	0.12	0.11	0.10	0.09	0.08	0.08	-1.6%
Other Uses[4]	0.00	0.00	0.00	0.00	0.00	0.00	0.00	-0.5%
Delivered Energy	**0.93**	**0.93**	**0.90**	**0.89**	**0.85**	**0.80**	**0.76**	**-0.8%**
Liquefied Petroleum Gases								
Space Heating	0.29	0.26	0.27	0.27	0.27	0.27	0.27	0.1%
Water Heating	0.05	0.05	0.05	0.05	0.05	0.05	0.05	0.2%
Cooking	0.03	0.03	0.03	0.03	0.03	0.03	0.03	0.7%
Other Uses[3]	0.17	0.17	0.18	0.21	0.23	0.25	0.27	1.8%
Delivered Energy	**0.54**	**0.51**	**0.53**	**0.56**	**0.58**	**0.60**	**0.62**	**0.8%**
Marketed Renewables (wood)[5]	0.40	0.41	0.43	0.41	0.40	0.40	0.39	-0.2%
Other Fuels[6]	0.10	0.11	0.11	0.11	0.11	0.10	0.10	-0.4%

Table A4. Residential Sector Key Indicators and Consumption (Continued)
(Quadrillion Btu per Year, Unless Otherwise Noted)

Key Indicators and Consumption	Reference Case							Annual Growth 2005-2030 (percent)
	2004	2005	2010	2015	2020	2025	2030	
Delivered Energy Consumption by End Use								
Space Heating	5.52	5.51	5.73	5.87	5.91	5.87	5.83	0.2%
Space Cooling	0.65	0.74	0.73	0.76	0.80	0.85	0.90	0.8%
Water Heating	1.70	1.68	1.69	1.72	1.73	1.73	1.74	0.2%
Refrigeration	0.40	0.39	0.36	0.35	0.35	0.36	0.38	-0.1%
Cooking	0.35	0.35	0.37	0.39	0.41	0.42	0.44	0.9%
Clothes Dryers	0.31	0.32	0.34	0.35	0.37	0.39	0.41	1.0%
Freezers	0.13	0.12	0.12	0.12	0.12	0.13	0.13	0.2%
Lighting	0.72	0.73	0.79	0.87	0.94	0.99	1.03	1.3%
Clothes Washers	0.03	0.03	0.03	0.03	0.03	0.03	0.03	-0.8%
Dishwashers	0.02	0.02	0.03	0.03	0.03	0.03	0.03	1.2%
Color Televisions and Set-Top Boxes	0.30	0.30	0.38	0.39	0.40	0.45	0.50	2.0%
Personal Computers	0.07	0.08	0.12	0.14	0.17	0.20	0.21	4.1%
Furnace Fans	0.08	0.08	0.10	0.11	0.11	0.12	0.13	1.6%
Other Uses[7]	1.11	1.22	1.43	1.61	1.80	1.92	2.05	2.1%
Delivered Energy	**11.39**	**11.60**	**12.21**	**12.74**	**13.17**	**13.48**	**13.80**	**0.7%**
Electricity Related Losses	**9.75**	**10.15**	**10.90**	**11.44**	**12.08**	**12.50**	**12.89**	**1.0%**
Total Energy Consumption by End Use								
Space Heating	6.39	6.40	6.67	6.85	6.92	6.89	6.86	0.3%
Space Cooling	2.08	2.36	2.30	2.37	2.47	2.57	2.70	0.5%
Water Heating	2.52	2.49	2.53	2.57	2.59	2.57	2.57	0.1%
Refrigeration	1.28	1.24	1.15	1.09	1.09	1.10	1.13	-0.4%
Cooking	0.58	0.58	0.62	0.65	0.68	0.71	0.73	0.9%
Clothes Dryers	0.86	0.86	0.90	0.93	0.97	1.01	1.05	0.8%
Freezers	0.41	0.40	0.37	0.37	0.38	0.39	0.39	-0.1%
Lighting	2.33	2.34	2.48	2.69	2.89	3.00	3.07	1.1%
Clothes Washers	0.11	0.11	0.10	0.09	0.08	0.08	0.08	-1.0%
Dishwashers	0.08	0.08	0.08	0.09	0.09	0.10	0.10	1.0%
Color Televisions and Set-Top Boxes	0.95	0.96	1.20	1.21	1.24	1.35	1.49	1.8%
Personal Computers	0.24	0.25	0.37	0.45	0.52	0.60	0.63	3.9%
Furnace Fans	0.27	0.27	0.30	0.33	0.35	0.36	0.38	1.3%
Other Uses[7]	3.07	3.42	4.03	4.49	4.98	5.24	5.51	1.9%
Total	**21.15**	**21.75**	**23.11**	**24.18**	**25.26**	**25.98**	**26.70**	**0.8%**
Nonmarketed Renewables[8]								
Geothermal Heat Pumps	0.00	0.00	0.01	0.01	0.01	0.01	0.02	6.8%
Solar Hot Water Heating	0.03	0.03	0.03	0.04	0.05	0.06	0.06	3.5%
Solar Photovoltaic	0.00	0.00	0.00	0.00	0.00	0.00	0.00	11.1%
Total	**0.03**	**0.03**	**0.04**	**0.05**	**0.06**	**0.07**	**0.08**	**4.0%**

[1]Does not include water heating portion of load.
[2]Includes small electric devices, heating elements, and motors not listed above.
[3]Includes such appliances as swimming pool heaters, outdoor grills, and outdoor lighting (natural gas).
[4]Includes such appliances as swimming pool and spa heaters.
[5]Includes wood used for primary and secondary heating in wood stoves or fireplaces as reported in the *Residential Energy Consumption Survey 2001*.
[6]Includes kerosene and coal.
[7]Includes all other uses listed above.
[8]Represents primary energy displaced.
Btu = British thermal unit.
Note: Totals may not equal sum of components due to independent rounding. Data for 2004 and 2005 are model results and may differ slightly from official EIA data reports.
Sources: 2004 and 2005 based on: Energy Information Administration (EIA), *Annual Energy Review 2005*, DOE/EIA-0384(2005) (Washington, DC, July 2006). **Projections:** EIA, AEO2007 National Energy Modeling System run AEO2007.D112106A.

Table A5. Commercial Sector Key Indicators and Consumption
(Quadrillion Btu per Year, Unless Otherwise Noted)

Key Indicators and Consumption	Reference Case							Annual Growth 2005-2030 (percent)
	2004	2005	2010	2015	2020	2025	2030	
Key Indicators								
Total Floorspace (billion square feet)								
Surviving	71.1	72.4	78.5	84.6	90.8	97.8	105.5	1.5%
New Additions	1.9	1.9	1.8	1.9	2.1	2.3	2.5	1.1%
Total	**73.0**	**74.3**	**80.4**	**86.5**	**92.9**	**100.1**	**108.0**	**1.5%**
Energy Consumption Intensity (thousand Btu per square foot)								
Delivered Energy Consumption	115.1	113.9	112.6	114.8	114.7	114.7	115.1	0.0%
Electricity Related Losses	127.0	126.8	127.8	128.6	129.6	129.5	129.7	0.1%
Total Energy Consumption	242.1	240.8	240.4	243.4	244.3	244.2	244.8	0.1%
Delivered Energy Consumption by Fuel								
Purchased Electricity								
Space Heating[1]	0.17	0.17	0.17	0.18	0.18	0.18	0.19	0.4%
Space Cooling[1]	0.46	0.55	0.50	0.52	0.54	0.57	0.61	0.4%
Water Heating[1]	0.18	0.18	0.18	0.18	0.19	0.19	0.20	0.5%
Ventilation	0.19	0.19	0.19	0.21	0.22	0.23	0.25	1.1%
Cooking	0.04	0.04	0.04	0.04	0.04	0.04	0.04	-0.2%
Lighting	1.19	1.18	1.19	1.26	1.33	1.41	1.51	1.0%
Refrigeration	0.23	0.23	0.24	0.25	0.26	0.28	0.30	1.0%
Office Equipment (PC)	0.14	0.18	0.30	0.35	0.38	0.39	0.40	3.2%
Office Equipment (non-PC)	0.32	0.35	0.48	0.57	0.67	0.78	0.92	3.9%
Other Uses[2]	1.28	1.25	1.48	1.72	1.98	2.28	2.62	3.0%
Delivered Energy	**4.19**	**4.32**	**4.77**	**5.28**	**5.78**	**6.36**	**7.03**	**2.0%**
Natural Gas								
Space Heating[1]	1.38	1.35	1.45	1.57	1.64	1.71	1.78	1.1%
Space Cooling[1]	0.02	0.02	0.02	0.03	0.03	0.04	0.04	2.3%
Water Heating[1]	0.59	0.57	0.56	0.64	0.69	0.75	0.82	1.5%
Cooking	0.24	0.23	0.26	0.29	0.31	0.34	0.36	1.8%
Other Uses[3]	1.00	0.97	1.02	1.12	1.19	1.26	1.36	1.3%
Delivered Energy	**3.23**	**3.15**	**3.31**	**3.64**	**3.86**	**4.10**	**4.36**	**1.3%**
Distillate Fuel Oil								
Space Heating[1]	0.20	0.20	0.20	0.22	0.22	0.22	0.23	0.6%
Water Heating[1]	0.07	0.07	0.06	0.07	0.07	0.07	0.07	-0.0%
Other Uses[4]	0.20	0.21	0.18	0.19	0.19	0.19	0.19	-0.4%
Delivered Energy	**0.47**	**0.48**	**0.45**	**0.48**	**0.48**	**0.48**	**0.49**	**0.1%**
Marketed Renewables (biomass)	0.12	0.12	0.12	0.12	0.12	0.12	0.12	0.0%
Other Fuels[5]	0.39	0.40	0.40	0.41	0.41	0.42	0.42	0.2%
Delivered Energy Consumption by End Use								
Space Heating[1]	1.75	1.71	1.83	1.96	2.04	2.12	2.20	1.0%
Space Cooling[1]	0.48	0.57	0.53	0.55	0.57	0.61	0.65	0.5%
Water Heating[1]	0.84	0.82	0.80	0.89	0.95	1.02	1.09	1.2%
Ventilation	0.19	0.19	0.19	0.21	0.22	0.23	0.25	1.1%
Cooking	0.27	0.27	0.29	0.33	0.35	0.37	0.40	1.6%
Lighting	1.19	1.18	1.19	1.26	1.33	1.41	1.51	1.0%
Refrigeration	0.23	0.23	0.24	0.25	0.26	0.28	0.30	1.0%
Office Equipment (PC)	0.14	0.18	0.30	0.35	0.38	0.39	0.40	3.2%
Office Equipment (non-PC)	0.32	0.35	0.48	0.57	0.67	0.78	0.92	3.9%
Other Uses[6]	2.98	2.95	3.21	3.57	3.90	4.28	4.71	1.9%
Delivered Energy	**8.40**	**8.46**	**9.05**	**9.93**	**10.66**	**11.48**	**12.43**	**1.6%**

Table A5. Commercial Sector Key Indicators and Consumption (Continued)
(Quadrillion Btu per Year, Unless Otherwise Noted)

Key Indicators and Consumption	Reference Case							Annual Growth 2005-2030 (percent)
	2004	2005	2010	2015	2020	2025	2030	
Electricity Related Losses	**9.27**	**9.42**	**10.27**	**11.13**	**12.03**	**12.97**	**14.01**	**1.6%**
Total Energy Consumption by End Use								
Space Heating[1]	2.12	2.08	2.20	2.34	2.41	2.49	2.57	0.9%
Space Cooling[1]	1.50	1.77	1.61	1.64	1.69	1.76	1.86	0.2%
Water Heating[1]	1.22	1.20	1.18	1.27	1.34	1.41	1.49	0.9%
Ventilation	0.60	0.60	0.61	0.64	0.67	0.70	0.74	0.8%
Cooking	0.36	0.35	0.37	0.40	0.43	0.45	0.47	1.2%
Lighting	3.82	3.76	3.74	3.92	4.09	4.29	4.53	0.7%
Refrigeration	0.74	0.74	0.76	0.78	0.81	0.85	0.90	0.8%
Office Equipment (PC)	0.46	0.58	0.96	1.08	1.18	1.20	1.21	3.0%
Office Equipment (non-PC)	1.03	1.12	1.50	1.78	2.05	2.37	2.74	3.6%
Other Uses[6]	5.81	5.68	6.40	7.20	8.03	8.93	9.93	2.3%
Total	**17.67**	**17.88**	**19.33**	**21.06**	**22.69**	**24.45**	**26.44**	**1.6%**
Nonmarketed Renewable Fuels[7]								
Solar Thermal	0.02	0.02	0.03	0.03	0.03	0.03	0.03	0.6%
Solar Photovoltaic	0.00	0.00	0.00	0.00	0.00	0.01	0.01	11.3%
Total	**0.02**	**0.03**	**0.03**	**0.03**	**0.03**	**0.03**	**0.04**	**2.1%**

[1]Includes fuel consumption for district services.
[2]Includes miscellaneous uses, such as service station equipment, automated teller machines, telecommunications equipment, and medical equipment.
[3]Includes miscellaneous uses, such as pumps, emergency generators, combined heat and power in commercial buildings, and manufacturing performed in commercial buildings.
[4]Includes miscellaneous uses, such as cooking, emergency generators, and combined heat and power in commercial buildings.
[5]Includes residual fuel oil, liquefied petroleum gas, coal, motor gasoline, and kerosene.
[6]Includes miscellaneous uses, such as service station equipment, automated teller machines, telecommunications equipment, medical equipment, pumps, emergency generators, combined heat and power in commercial buildings, manufacturing performed in commercial buildings, and cooking (distillate), plus residual fuel oil, liquefied petroleum gas, coal, motor gasoline, and kerosene.
[7]Represents primary energy displaced by solar thermal space heating and water heating, and electricity generation by solar photovoltaic systems.
Btu = British thermal unit.
PC = Personal computer.
Note: Totals may not equal sum of components due to independent rounding. Data for 2004 and 2005 are model results and may differ slightly from official EIA data reports.
Sources: 2004 and 2005 based on: Energy Information Administration (EIA), *Annual Energy Review 2005*, DOE/EIA-0384(2005) (Washington, DC, July 2006).
Projections: EIA, AEO2007 National Energy Modeling System run AEO2007.D112106A.

Table A6. Industrial Sector Key Indicators and Consumption

Key Indicators and Consumption	Reference Case 2004	2005	2010	2015	2020	2025	2030	Annual Growth 2005-2030 (percent)
Key Indicators								
Value of Shipments (billion 2000 dollars)								
Manufacturing	4157	4225	4702	5332	5933	6645	7478	2.3%
Nonmanufacturing	1494	1538	1596	1701	1846	1940	2023	1.1%
Total	**5651**	**5763**	**6298**	**7033**	**7779**	**8585**	**9502**	**2.0%**
Energy Prices (2005 dollars per million Btu)								
Liquefied Petroleum Gases	11.18	16.96	16.42	15.57	15.91	16.22	16.55	-0.1%
Motor Gasoline	14.98	16.63	17.83	16.00	16.56	17.25	17.68	0.2%
Distillate Fuel Oil	10.99	13.08	12.95	11.40	12.04	12.62	13.25	0.1%
Residual Fuel Oil	5.77	7.77	9.50	8.21	8.91	9.26	9.58	0.8%
Petrochemical Feedstocks	13.75	8.30	8.67	7.67	8.44	8.40	9.13	0.4%
Asphalt and Road Oil	10.05	6.06	8.24	7.14	7.40	7.82	8.28	1.3%
Natural Gas Heat and Power	5.42	7.17	5.52	4.73	5.04	5.38	5.72	-0.9%
Natural Gas Feedstocks	7.16	8.81	7.13	6.34	6.59	6.90	7.24	-0.8%
Metallurgical Coal	2.31	3.06	3.09	2.72	2.71	2.70	2.75	-0.4%
Other Industrial Coal	1.80	2.15	2.26	2.20	2.18	2.23	2.29	0.2%
Coal to Liquids	N/A	N/A	N/A	0.89	0.97	1.23	1.33	N/A
Electricity	15.88	16.69	18.01	16.46	17.07	17.35	17.43	0.2%
Energy Consumption (quadrillion Btu)[1]								
Industrial Consumption Excluding Refining								
Liquefied Petroleum Gases Heat and Power	0.13	0.13	0.08	0.08	0.08	0.08	0.08	-1.8%
Liquefied Petroleum Gases Feedstocks	2.13	1.98	2.17	2.16	2.18	2.21	2.29	0.6%
Motor Gasoline	0.32	0.32	0.32	0.32	0.33	0.35	0.36	0.4%
Distillate Fuel Oil	1.21	1.22	1.18	1.19	1.22	1.23	1.26	0.1%
Residual Fuel Oil	0.20	0.22	0.14	0.14	0.14	0.14	0.14	-1.6%
Petrochemical Feedstocks	1.54	1.38	1.48	1.49	1.50	1.52	1.57	0.5%
Petroleum Coke	0.36	0.33	0.31	0.31	0.31	0.32	0.34	0.1%
Asphalt and Road Oil	1.27	1.31	1.24	1.24	1.29	1.33	1.37	0.2%
Miscellaneous Petroleum[2]	0.55	0.59	0.45	0.40	0.38	0.38	0.38	-1.7%
Petroleum Subtotal	7.72	7.48	7.38	7.34	7.43	7.58	7.79	0.2%
Natural Gas Heat and Power	5.80	5.30	5.83	6.00	6.22	6.57	6.97	1.1%
Natural Gas Feedstocks	0.60	0.57	0.58	0.57	0.57	0.57	0.58	0.0%
Lease and Plant Fuel[3]	1.13	1.10	1.10	1.10	1.21	1.17	1.15	0.2%
Natural Gas Subtotal	7.54	6.97	7.51	7.68	7.99	8.31	8.70	0.9%
Metallurgical Coal and Coke[4]	0.79	0.66	0.63	0.60	0.59	0.59	0.59	-0.5%
Other Industrial Coal	1.30	1.23	1.26	1.23	1.23	1.24	1.25	0.1%
Coal Subtotal	2.08	1.89	1.88	1.84	1.81	1.82	1.84	-0.1%
Renewables[5]	1.70	1.44	1.60	1.71	1.81	1.93	2.05	1.4%
Purchased Electricity	3.34	3.35	3.44	3.56	3.63	3.73	3.87	0.6%
Delivered Energy	**22.38**	**21.14**	**21.81**	**22.13**	**22.67**	**23.36**	**24.24**	**0.5%**
Electricity Related Losses	7.39	7.31	7.41	7.51	7.56	7.60	7.70	0.2%
Total	**29.77**	**28.45**	**29.22**	**29.64**	**30.23**	**30.96**	**31.94**	**0.5%**
Refining Consumption								
Liquefied Petroleum Gases Heat and Power	0.01	0.02	0.00	0.00	0.00	0.01	0.03	2.5%
Distillate Fuel Oil	0.00	0.00	0.00	0.00	0.00	0.00	0.00	N/A
Residual Fuel Oil	0.01	0.01	0.03	0.03	0.03	0.04	0.04	4.0%
Petroleum Coke	0.57	0.56	0.60	0.67	0.77	0.84	0.87	1.8%
Still Gas	1.74	1.62	1.44	1.62	1.56	1.57	1.78	0.4%
Miscellaneous Petroleum[2]	0.03	0.03	0.02	0.02	0.03	0.03	0.04	1.0%
Petroleum Subtotal	2.37	2.25	2.09	2.35	2.39	2.49	2.76	0.8%
Natural Gas Heat and Power	1.04	0.97	1.45	1.32	1.47	1.54	1.36	1.4%
Natural-Gas-to-Liquids Heat and Power	0.00	0.00	0.00	0.00	0.00	0.00	0.00	N/A
Natural Gas Subtotal	1.04	0.97	1.45	1.32	1.47	1.54	1.36	1.4%
Other Industrial Coal	0.10	0.12	0.11	0.11	0.11	0.11	0.11	-0.0%
Coal-to-Liquids Heat and Power	0.00	0.00	0.00	0.12	0.21	0.67	0.93	N/A
Coal Subtotal	0.10	0.12	0.11	0.24	0.32	0.78	1.05	9.2%
Biofuels Heat and Coproducts	0.21	0.24	0.69	0.74	0.78	0.83	0.88	5.2%
Purchased Electricity	0.13	0.13	0.19	0.20	0.20	0.22	0.22	2.1%
Delivered Energy	**3.86**	**3.72**	**4.52**	**4.83**	**5.17**	**5.86**	**6.26**	**2.1%**
Electricity Related Losses	0.29	0.29	0.40	0.42	0.42	0.44	0.45	1.8%
Total	**4.15**	**4.01**	**4.92**	**5.25**	**5.59**	**6.30**	**6.71**	**2.1%**

Table A6. Industrial Sector Key Indicators and Consumption (Continued)

Key Indicators and Consumption	Reference Case							Annual Growth 2005-2030 (percent)
	2004	2005	2010	2015	2020	2025	2030	
Total Industrial Sector Consumption								
Liquefied Petroleum Gases Heat and Power	0.14	0.15	0.08	0.08	0.08	0.10	0.11	-1.1%
Liquefied Petroleum Gases Feedstocks	2.13	1.98	2.17	2.16	2.18	2.21	2.29	0.6%
Motor Gasoline	0.32	0.32	0.32	0.32	0.33	0.35	0.36	0.4%
Distillate Fuel Oil	1.21	1.23	1.18	1.19	1.22	1.23	1.26	0.1%
Residual Fuel Oil	0.22	0.23	0.18	0.18	0.17	0.18	0.18	-0.9%
Petrochemical Feedstocks	1.54	1.38	1.48	1.49	1.50	1.52	1.57	0.5%
Petroleum Coke	0.93	0.89	0.91	0.98	1.08	1.16	1.21	1.2%
Asphalt and Road Oil	1.27	1.31	1.24	1.24	1.29	1.33	1.37	0.2%
Still Gas	1.74	1.62	1.44	1.62	1.56	1.57	1.78	0.4%
Miscellaneous Petroleum[2]	0.59	0.62	0.47	0.42	0.41	0.42	0.42	-1.5%
Petroleum Subtotal	10.09	9.73	9.47	9.68	9.82	10.07	10.55	0.3%
Natural Gas Heat and Power	6.85	6.27	7.28	7.32	7.69	8.11	8.33	1.1%
Natural-Gas-to-Liquids Heat and Power	0.00	0.00	0.00	0.00	0.00	0.00	0.00	N/A
Natural Gas Feedstocks	0.60	0.57	0.58	0.57	0.57	0.57	0.58	0.0%
Lease and Plant Fuel[3]	1.13	1.10	1.10	1.10	1.21	1.17	1.15	0.2%
Natural Gas Subtotal	8.58	7.94	8.95	9.00	9.46	9.85	10.05	0.9%
Metallurgical Coal and Coke[4]	0.79	0.66	0.63	0.60	0.59	0.59	0.59	-0.5%
Other Industrial Coal	1.40	1.35	1.37	1.35	1.34	1.35	1.36	0.0%
Coal-to-Liquids Heat and Power	0.00	0.00	0.00	0.12	0.21	0.67	0.93	N/A
Coal Subtotal	2.18	2.01	2.00	2.07	2.14	2.61	2.89	1.5%
Biofuels Heat and Coproducts	0.21	0.24	0.69	0.74	0.78	0.83	0.88	5.2%
Renewables[5]	1.70	1.44	1.60	1.71	1.81	1.93	2.05	1.4%
Purchased Electricity	3.48	3.48	3.63	3.76	3.83	3.94	4.09	0.6%
Delivered Energy	**26.24**	**24.85**	**26.33**	**26.97**	**27.84**	**29.23**	**30.51**	**0.8%**
Electricity Related Losses	7.68	7.60	7.81	7.93	7.98	8.03	8.15	0.3%
Total	**33.92**	**32.45**	**34.14**	**34.89**	**35.82**	**37.26**	**38.66**	**0.7%**
Energy Consumption per dollar of Shipment (thousand Btu per 2000 dollars)								
Liquefied Petroleum Gases Heat and Power	0.02	0.03	0.01	0.01	0.01	0.01	0.01	-3.0%
Liquefied Petroleum Gases Feedstocks	0.38	0.34	0.34	0.31	0.28	0.26	0.24	-1.4%
Motor Gasoline	0.06	0.06	0.05	0.05	0.04	0.04	0.04	-1.5%
Distillate Fuel Oil	0.21	0.21	0.19	0.17	0.16	0.14	0.13	-1.9%
Residual Fuel Oil	0.04	0.04	0.03	0.02	0.02	0.02	0.02	-2.9%
Petrochemical Feedstocks	0.27	0.24	0.24	0.21	0.19	0.18	0.17	-1.5%
Petroleum Coke	0.17	0.15	0.14	0.14	0.14	0.14	0.13	-0.8%
Asphalt and Road Oil	0.23	0.23	0.20	0.18	0.17	0.16	0.14	-1.8%
Still Gas	0.31	0.28	0.23	0.23	0.20	0.18	0.19	-1.6%
Miscellaneous Petroleum[2]	0.10	0.11	0.07	0.06	0.05	0.05	0.04	-3.5%
Petroleum Subtotal	1.79	1.69	1.50	1.38	1.26	1.17	1.11	-1.7%
Natural Gas Heat and Power	1.21	1.09	1.16	1.04	0.99	0.94	0.88	-0.9%
Natural-Gas-to-Liquids Heat and Power	0.00	0.00	0.00	0.00	0.00	0.00	0.00	N/A
Natural Gas Feedstocks	0.11	0.10	0.09	0.08	0.07	0.07	0.06	-2.0%
Lease and Plant Fuel[3]	0.20	0.19	0.17	0.16	0.16	0.14	0.12	-1.8%
Natural Gas Subtotal	1.52	1.38	1.42	1.28	1.22	1.15	1.06	-1.1%
Metallurgical Coal and Coke[4]	0.14	0.11	0.10	0.09	0.08	0.07	0.06	-2.4%
Other Industrial Coal	0.25	0.23	0.22	0.19	0.17	0.16	0.14	-1.9%
Coal-to-Liquids Heat and Power	0.00	0.00	0.00	0.02	0.03	0.08	0.10	N/A
Coal Subtotal	0.39	0.35	0.32	0.29	0.27	0.30	0.30	-0.5%
Biofuels Heat and Coproducts	0.04	0.04	0.11	0.10	0.10	0.10	0.09	3.1%
Renewables[5]	0.30	0.25	0.25	0.24	0.23	0.22	0.22	-0.6%
Purchased Electricity	0.61	0.60	0.58	0.54	0.49	0.46	0.43	-1.4%
Delivered Energy	**4.64**	**4.31**	**4.18**	**3.83**	**3.58**	**3.40**	**3.21**	**-1.2%**
Electricity Related Losses	1.36	1.32	1.24	1.13	1.03	0.94	0.86	-1.7%
Total	**6.00**	**5.63**	**5.42**	**4.96**	**4.60**	**4.34**	**4.07**	**-1.3%**

Table A6. Industrial Sector Key Indicators and Consumption (Continued)

Key Indicators and Consumption	Reference Case							Annual Growth 2005-2030 (percent)
	2004	2005	2010	2015	2020	2025	2030	
Industrial Combined Heat and Power								
Capacity (gigawatts)	26.31	25.53	29.28	34.19	39.05	48.82	56.54	3.2%
Generation (billion kilowatthours)	153.21	143.13	169.93	206.73	243.02	317.63	375.86	3.9%

[1]Includes energy for combined heat and power plants, except those whose primary business is to sell electricity, or electricity and heat, to the public.
[2]Includes lubricants and miscellaneous petroleum products.
[3]Represents natural gas used in well, field, and lease operations, and in natural gas processing plant machinery.
[4]Includes net coal coke imports.
[5]Includes consumption of energy produced from hydroelectric, wood and wood waste, municipal solid waste, and other biomass sources.
Btu = British thermal unit.
N/A = Not applicable.
Note: Totals may not equal sum of components due to independent rounding. Data for 2004 and 2005 are model results and may differ slightly from official EIA data reports.

Sources: 2004 and 2005 prices for motor gasoline and distillate fuel oil are based on: Energy Information Administration (EIA), *Petroleum Marketing Annual 2005*, DOE/EIA-0487(2005) (Washington, DC, August 2006). 2004 and 2005 coal prices are based on: EIA, *Quarterly Coal Report, October-December 2005*, DOE/EIA-0121(2005/4Q) (Washington, DC, March 2006) and EIA, AEO2007 National Energy Modeling System run AEO2007.D112106A. 2004 and 2005 electricity prices: EIA, *Annual Energy Review 2005*, DOE/EIA-0384(2005) (Washington, DC, July 2006). 2004 and 2005 natural gas prices are based on: EIA, *Manufacturing Energy Consumption Survey 1994* and industrial and wellhead prices from the *Natural Gas Annual 2004*, DOE/EIA-0131(2004) (Washington, DC, December 2005) and the *Natural Gas Monthly*, DOE/EIA-0130(2006/04) (Washington, DC, April 2006). 2004 refining consumption based on: *Petroleum Supply Annual 2004*, DOE/EIA-0340(2004)/1 (Washington, DC, June 2005). 2005 refining consumption based on: *Petroleum Supply Annual 2005*, DOE/EIA-0340(2005)/1 (Washington, DC, October 2006). Other 2004 and 2005 consumption values are based on: EIA, *Annual Energy Review 2005*, DOE/EIA-0384(2005) (Washington, DC, July 2006). 2004 and 2005 shipments: Global Insight industry model, July 2006. **Projections:** EIA, AEO2007 National Energy Modeling System run AEO2007.D112106A.

Table A7. Transportation Sector Key Indicators and Delivered Energy Consumption

Key Indicators and Consumption	Reference Case							Annual Growth 2005-2030 (percent)
	2004	2005	2010	2015	2020	2025	2030	
Key Indicators								
Travel Indicators								
(billion vehicle miles traveled)								
Light-Duty Vehicles less than 8,500 pounds	2652	2655	2799	3125	3474	3839	4226	1.9%
Commercial Light Trucks[1]	66	67	72	81	89	99	110	2.0%
Freight Trucks greater than 10,000 pounds	226	230	255	287	318	355	397	2.2%
(billion seat miles available)								
Air	992	1027	1172	1302	1410	1478	1544	1.6%
(billion ton miles traveled)								
Rail	1590	1590	1714	1864	2000	2223	2445	1.7%
Domestic Shipping	618	613	661	699	730	751	775	0.9%
Energy Efficiency Indicators								
(miles per gallon)								
New Light-Duty Vehicle[2]	24.6	25.2	27.3	27.9	28.2	28.9	29.2	0.6%
New Car[2]	29.1	30.0	31.7	32.4	32.8	33.4	33.7	0.5%
New Light Truck[2]	21.5	21.8	23.7	24.7	25.3	26.1	26.5	0.8%
Light-Duty Stock[3]	19.6	19.6	19.8	20.6	21.2	21.8	22.2	0.5%
New Commercial Light Truck[1]	14.5	14.6	15.8	16.4	16.7	17.2	17.4	0.7%
Stock Commercial Light Truck[1]	14.0	14.1	14.7	15.5	16.2	16.7	17.0	0.8%
Freight Truck	6.0	6.0	6.0	6.2	6.4	6.6	6.7	0.4%
(seat miles per gallon)								
Aircraft	55.5	55.7	58.2	61.9	66.4	71.6	75.6	1.2%
(ton miles per thousand Btu)								
Rail	2.9	2.9	2.9	2.9	3.0	3.0	3.0	0.1%
Domestic Shipping	2.4	2.4	2.4	2.4	2.4	2.5	2.5	0.1%
Energy Use by Mode (quadrillion Btu)								
Light-Duty Vehicles	16.34	16.36	16.76	17.99	19.44	20.98	22.66	1.3%
Commercial Light Trucks[1]	0.59	0.59	0.61	0.65	0.69	0.75	0.81	1.3%
Bus Transportation	0.26	0.26	0.27	0.28	0.28	0.29	0.30	0.5%
Freight Trucks	4.70	4.77	5.29	5.80	6.18	6.71	7.40	1.8%
Rail, Passenger	0.04	0.04	0.05	0.05	0.05	0.05	0.06	1.1%
Rail, Freight	0.55	0.55	0.59	0.63	0.68	0.75	0.82	1.6%
Shipping, Domestic	0.26	0.26	0.27	0.29	0.30	0.31	0.32	0.9%
Shipping, International	0.69	0.76	0.77	0.78	0.79	0.80	0.80	0.2%
Recreational Boats	0.19	0.18	0.20	0.23	0.24	0.25	0.27	1.5%
Air	2.85	2.84	3.50	3.79	3.97	4.00	4.11	1.5%
Military Use	0.71	0.71	0.73	0.75	0.77	0.79	0.80	0.5%
Lubricants	0.15	0.15	0.15	0.15	0.16	0.16	0.16	0.4%
Pipeline Fuel	0.59	0.58	0.66	0.70	0.79	0.79	0.79	1.3%
Total	**27.92**	**28.05**	**29.86**	**32.07**	**34.33**	**36.63**	**39.29**	**1.4%**

Table A7. Transportation Sector Key Indicators and Delivered Energy Consumption (Continued)

Key Indicators and Consumption	Reference Case							Annual Growth 2005-2030 (percent)
	2004	2005	2010	2015	2020	2025	2030	
Energy Use by Mode (million barrels per day oil equivalent)								
Light-Duty Vehicles	8.57	8.58	9.00	9.66	10.43	11.25	12.15	1.4%
Commercial Light Trucks[1]	0.31	0.31	0.33	0.35	0.37	0.40	0.44	1.4%
Bus Transportation	0.13	0.13	0.13	0.13	0.14	0.14	0.14	0.6%
Freight Trucks	2.24	2.27	2.53	2.78	2.96	3.21	3.54	1.8%
Rail, Passenger	0.02	0.02	0.02	0.02	0.02	0.03	0.03	1.1%
Rail, Freight	0.26	0.26	0.28	0.30	0.32	0.36	0.39	1.7%
Shipping, Domestic	0.12	0.12	0.13	0.13	0.14	0.14	0.15	0.9%
Shipping, International	0.30	0.33	0.34	0.34	0.35	0.35	0.35	0.2%
Recreational Boats	0.10	0.10	0.11	0.12	0.13	0.14	0.15	1.6%
Air	1.38	1.37	1.69	1.83	1.92	1.94	1.99	1.5%
Military Use	0.34	0.34	0.35	0.36	0.37	0.38	0.38	0.5%
Lubricants	0.07	0.07	0.07	0.07	0.07	0.08	0.08	0.4%
Pipeline Fuel	0.30	0.29	0.33	0.35	0.40	0.40	0.40	1.3%
Total	**14.14**	**14.19**	**15.31**	**16.45**	**17.62**	**18.81**	**20.18**	**1.4%**

[1]Commercial trucks 8,500 to 10,000 pounds.
[2]Environmental Protection Agency rated miles per gallon.
[3]Combined car and light truck "on-the-road" estimate.
Btu = British thermal unit.
Note: Totals may not equal sum of components due to independent rounding. Data for 2004 and 2005 are model results and may differ slightly from official EIA data reports.
Sources: 2004 and 2005: Energy Information Administration (EIA), *Natural Gas Annual 2004*, DOE/EIA-0131(2004) (Washington, DC, December 2005); EIA, *Annual Energy Review 2005*, DOE/EIA-0384(2005) (Washington, DC, July 2006); Federal Highway Administration, *Highway Statistics 2004* (Washington, DC, October 2005); Oak Ridge National Laboratory, *Transportation Energy Data Book: Edition 25 and Annual* (Oak Ridge, TN, 2005); National Highway Traffic and Safety Administration, *Summary of Fuel Economy Performance* (Washington, DC, March 2004); U.S. Department of Commerce, Bureau of the Census, "Vehicle Inventory and Use Survey," EC97TV (Washington, DC, October 1999); EIA, *State Energy Data Report 2002*, DOE/EIA-0214(2002) (Washington, DC, December 2005); EIA, *Estimated Number of Alternative-Fueled Vehicles*, http://www.eia.doe.gov/cneaf/alternate/page/datatables/aft1-13_03.html; U.S. Department of Transportation, Research and Special Programs Administration, *Air Carrier Statistics Monthly, December 2005/2004* (Washington, DC, 2005); EIA, *Fuel Oil and Kerosene Sales 2004*, DOE/EIA-0535(2004) (Washington, DC, November 2005); and United States Department of Defense, Defense Fuel Supply Center.
Projections: EIA, AEO2007 National Energy Modeling System run AEO2007.D112106A.

Table A8. Electricity Supply, Disposition, Prices, and Emissions
(Billion Kilowatthours, Unless Otherwise Noted)

Supply, Disposition, and Prices	Reference Case							Annual Growth 2005-2030 (percent)
	2004	2005	2010	2015	2020	2025	2030	
Generation by Fuel Type								
Electric Power Sector[1]								
Power Only[2]								
Coal	1921	1956	2090	2233	2418	2766	3191	2.0%
Petroleum	111	111	82	89	89	91	92	-0.7%
Natural Gas[3]	491	546	658	756	776	702	609	0.4%
Nuclear Power	789	780	789	812	885	886	896	0.6%
Pumped Storage/Other	-8	-7	-9	-9	-9	-9	-9	1.1%
Renewable Sources[4]	321	319	426	440	445	458	461	1.5%
Distributed Generation (Natural Gas)	0	0	0	0	1	2	5	N/A
Total	**3624**	**3705**	**4037**	**4322**	**4605**	**4897**	**5245**	**1.4%**
Combined Heat and Power[5]								
Coal	36	37	31	29	29	29	29	-1.0%
Petroleum	5	5	1	1	1	1	1	-5.3%
Natural Gas	137	129	137	145	142	132	123	-0.2%
Renewable Sources	4	4	4	3	3	3	4	0.1%
Total	**185**	**178**	**172**	**179**	**176**	**166**	**157**	**-0.5%**
Total Net Generation	**3808**	**3883**	**4209**	**4501**	**4781**	**5063**	**5402**	**1.3%**
Less Direct Use	35	35	34	34	34	34	34	-0.1%
Net Available to the Grid	**3773**	**3849**	**4175**	**4467**	**4747**	**5029**	**5368**	**1.3%**
End-Use Generation[6]								
Coal	22	22	22	33	41	85	110	6.6%
Petroleum	6	6	11	13	13	13	14	3.4%
Natural Gas	82	77	97	117	142	169	200	3.9%
Other Gaseous Fuels[7]	6	5	4	5	5	5	6	0.8%
Renewable Sources[4]	37	34	37	40	44	48	54	1.9%
Other[8]	14	12	11	11	11	11	11	-0.0%
Total	**167**	**155**	**183**	**220**	**256**	**332**	**395**	**3.8%**
Less Direct Use	136	126	144	168	194	236	276	3.2%
Total Sales to the Grid	**31**	**29**	**39**	**51**	**62**	**96**	**119**	**5.9%**
Total Electricity Generation	**3975**	**4038**	**4392**	**4721**	**5037**	**5395**	**5797**	**1.5%**
Total Net Generation to the Grid	**3804**	**3877**	**4214**	**4519**	**4810**	**5125**	**5487**	**1.4%**
Net Imports	**11**	**25**	**11**	**8**	**11**	**13**	**13**	**-2.6%**
Electricity Sales by Sector								
Residential	1294	1365	1483	1591	1701	1797	1896	1.3%
Commercial	1229	1267	1398	1548	1694	1864	2062	2.0%
Industrial	1018	1021	1063	1103	1123	1155	1199	0.6%
Transportation	7	7	8	9	10	11	12	1.8%
Total	**3548**	**3660**	**3953**	**4251**	**4528**	**4827**	**5168**	**1.4%**
Direct Use	171	161	178	202	228	270	310	2.6%
Total Electricity Use	**3719**	**3821**	**4132**	**4453**	**4756**	**5097**	**5478**	**1.5%**
End-Use Prices (2005 cents per kilowatthour)								
Residential	9.2	9.4	9.2	8.9	9.0	9.1	9.1	-0.1%
Commercial	8.4	8.6	8.4	8.0	8.2	8.3	8.3	-0.2%
Industrial	5.4	5.7	6.1	5.6	5.8	5.9	5.9	0.2%
Transportation	8.9	8.6	8.5	8.1	8.3	8.3	8.3	-0.1%
All Sectors Average	**7.9**	**8.1**	**8.1**	**7.7**	**7.9**	**8.0**	**8.1**	**-0.0%**
Prices by Service Category (2005 cents per kilowatthour)								
Generation	5.2	5.4	5.4	5.0	5.2	5.4	5.4	0.0%
Transmission	0.6	0.6	0.6	0.7	0.7	0.7	0.7	1.0%
Distribution	2.1	2.1	2.1	2.1	2.0	2.0	1.9	-0.3%

Table A8. Electricity Supply, Disposition, Prices, and Emissions (Continued)
(Billion Kilowatthours, Unless Otherwise Noted)

Supply, Disposition, and Prices	Reference Case							Annual Growth 2005-2030 (percent)
	2004	2005	2010	2015	2020	2025	2030	
Electric Power Sector Emissions[1]								
Sulfur Dioxide (million tons)	10.26	10.21	6.56	4.46	3.90	3.68	3.63	-4.1%
Nitrogen Oxide (million tons)	3.75	3.60	2.41	2.20	2.22	2.25	2.28	-1.8%
Mercury (tons)	47.15	51.25	37.21	24.64	19.24	16.86	15.48	-4.7%

[1]Includes electricity-only and combined heat and power plants whose primary business is to sell electricity, or electricity and heat, to the public.
[2]Includes plants that only produce electricity.
[3]Includes electricity generation from fuel cells.
[4]Includes conventional hydroelectric, geothermal, wood, wood waste, municipal solid waste, landfill gas, other biomass, solar, and wind power.
[5]Includes combined heat and power plants whose primary business is to sell electricity and heat to the public (i.e., those that report North American Industry Classification System code 22).
[6]Includes combined heat and power plants and electricity-only plants in the commercial and industrial sectors; and small on-site generating systems in the residential, commercial, and industrial sectors used primarily for own-use generation, but which may also sell some power to the grid.
[7]Includes refinery gas and still gas.
[8]Includes batteries, chemicals, hydrogen, pitch, purchased steam, sulfur, and miscellaneous technologies.
N/A = Not applicable.
Note: Totals may not equal sum of components due to independent rounding. Data for 2004 and 2005 are model results and may differ slightly from official EIA data reports.
Sources: 2004 and 2005 electric power sector generation; sales to utilities; net imports; electricity sales; and emissions: Energy Information Administration (EIA), *Annual Energy Review 2005*, DOE/EIA-0384(2005) (Washington, DC, July 2006), and supporting databases. 2004 and 2005 prices: EIA, AEO2007 National Energy Modeling System run AEO2007.D112106A. **Projections**: EIA, AEO2007 National Energy Modeling System run AEO2007.D112106A.

Table A9. Electricity Generating Capacity
(Gigawatts)

Net Summer Capacity[1]	Reference Case 2004	2005	2010	2015	2020	2025	2030	Annual Growth 2005-2030 (percent)
Electric Power Sector[2]								
Power Only[3]								
Coal Steam	306.3	306.0	316.2	318.9	343.0	389.5	445.8	1.5%
Other Fossil Steam[4]	123.3	120.8	119.0	89.3	88.8	88.4	87.0	-1.3%
Combined Cycle	133.0	144.2	160.9	163.2	171.4	178.4	179.2	0.9%
Combustion Turbine/Diesel	128.1	130.3	134.2	118.0	124.3	133.0	152.3	0.6%
Nuclear Power[5]	99.6	100.0	100.5	102.5	111.7	111.7	112.6	0.5%
Pumped Storage	20.8	20.8	20.8	20.8	20.8	20.8	20.8	0.0%
Fuel Cells	0.0	0.0	0.0	0.0	0.0	0.0	0.0	N/A
Renewable Sources[6]	94.4	97.2	105.3	106.6	107.3	108.2	109.6	0.5%
Distributed Generation[7]	0.0	0.0	0.2	0.6	2.1	5.5	11.4	N/A
Total	**905.3**	**919.2**	**956.9**	**919.7**	**969.6**	**1035.4**	**1118.6**	**0.8%**
Combined Heat and Power[8]								
Coal Steam	4.7	4.7	4.7	4.2	4.2	4.2	4.2	-0.5%
Other Fossil Steam[4]	0.5	0.5	0.5	0.5	0.5	0.5	0.5	0.0%
Combined Cycle	32.5	32.5	32.4	32.4	32.4	32.4	32.4	-0.0%
Combustion Turbine/Diesel	2.9	2.9	2.9	2.9	2.9	2.9	2.9	-0.0%
Renewable Sources[6]	0.4	0.4	0.4	0.4	0.4	0.4	0.4	0.2%
Total	**40.9**	**40.9**	**40.9**	**40.3**	**40.3**	**40.3**	**40.3**	**-0.1%**
Cumulative Planned Additions[9]								
Coal Steam	0.0	0.0	8.5	9.9	9.9	9.9	9.9	N/A
Other Fossil Steam[4]	0.0	0.0	0.0	0.0	0.0	0.0	0.0	N/A
Combined Cycle	0.0	0.0	16.7	17.6	17.6	17.6	17.6	N/A
Combustion Turbine/Diesel	0.0	0.0	3.7	3.7	3.7	3.7	3.7	N/A
Nuclear Power	0.0	0.0	0.0	0.0	0.0	0.0	0.0	N/A
Pumped Storage	0.0	0.0	0.0	0.0	0.0	0.0	0.0	N/A
Fuel Cells	0.0	0.0	0.0	0.0	0.0	0.0	0.0	N/A
Renewable Sources[6]	0.0	0.0	8.1	8.9	9.0	9.1	9.2	N/A
Distributed Generation[7]	0.0	0.0	0.0	0.0	0.0	0.0	0.0	N/A
Total	**0.0**	**0.0**	**37.0**	**40.1**	**40.2**	**40.4**	**40.5**	**N/A**
Cumulative Unplanned Additions[9]								
Coal Steam	0.0	0.0	3.0	8.1	32.4	78.8	135.1	N/A
Other Fossil Steam[4]	0.0	0.0	0.0	0.0	0.0	0.0	0.0	N/A
Combined Cycle	0.0	0.0	0.0	1.4	9.7	16.6	17.5	N/A
Combustion Turbine/Diesel	0.0	0.0	0.7	3.8	10.1	18.8	38.0	N/A
Nuclear Power	0.0	0.0	0.0	0.5	9.0	9.0	12.5	N/A
Pumped Storage	0.0	0.0	0.0	0.0	0.0	0.0	0.0	N/A
Fuel Cells	0.0	0.0	0.0	0.0	0.0	0.0	0.0	N/A
Renewable Sources[6]	0.0	0.0	0.0	0.4	1.1	1.9	3.2	N/A
Distributed Generation[7]	0.0	0.0	0.2	0.6	2.1	5.5	11.4	N/A
Total	**0.0**	**0.0**	**3.9**	**14.8**	**64.4**	**130.6**	**217.7**	**N/A**
Cumulative Electric Power Sector Additions	**0.0**	**0.0**	**40.9**	**55.0**	**104.7**	**171.0**	**258.2**	**N/A**
Cumulative Retirements[10]								
Coal Steam	0.0	0.0	1.2	5.6	5.7	5.7	5.7	N/A
Other Fossil Steam[4]	0.0	0.0	1.8	31.5	32.0	32.4	33.8	N/A
Combined Cycle	0.0	0.0	0.1	0.1	0.1	0.1	0.1	N/A
Combustion Turbine/Diesel	0.0	0.0	0.6	19.8	19.8	19.8	19.8	N/A
Nuclear Power	0.0	0.0	0.0	0.0	0.0	0.0	2.6	N/A
Pumped Storage	0.0	0.0	0.0	0.0	0.0	0.0	0.0	N/A
Fuel Cells	0.0	0.0	0.0	0.0	0.0	0.0	0.0	N/A
Renewable Sources[6]	0.0	0.0	0.0	0.0	0.0	0.0	0.0	N/A
Total	**0.0**	**0.0**	**3.7**	**57.0**	**57.6**	**58.1**	**62.0**	**N/A**
Total Electric Power Sector Capacity	**946.3**	**960.1**	**997.8**	**960.0**	**1010.0**	**1075.8**	**1159.0**	**0.8%**

Table A9. Electricity Generating Capacity (Continued)
(Gigawatts)

Net Summer Capacity[1]	Reference Case 2004	2005	2010	2015	2020	2025	2030	Annual Growth 2005-2030 (percent)
End-Use Generators[11]								
Coal	4.1	4.2	4.1	5.6	6.6	12.2	15.4	5.4%
Petroleum	1.6	1.6	1.6	1.8	1.8	1.9	1.9	0.7%
Natural Gas	13.9	13.5	16.9	19.5	22.8	26.4	30.4	3.3%
Other Gaseous Fuels	1.8	1.8	1.9	1.9	1.9	2.0	2.0	0.4%
Renewable Sources[6]	5.8	5.6	6.4	7.0	7.7	8.7	10.7	2.6%
Other	1.1	0.8	0.8	0.8	0.8	0.8	0.8	0.0%
Total	**28.2**	**27.5**	**31.7**	**36.7**	**41.7**	**52.0**	**61.3**	**3.3%**
Cumulative Capacity Additions[9]	**0.0**	**0.0**	**4.2**	**9.2**	**14.2**	**24.5**	**33.8**	**N/A**

[1]Net summer capacity is the steady hourly output that generating equipment is expected to supply to system load (exclusive of auxiliary power), as demonstrated by tests during summer peak demand.
[2]Includes electricity-only and combined heat and power plants whose primary business is to sell electricity, or electricity and heat, to the public.
[3]Includes plants that only produce electricity. Includes capacity increases (uprates) at existing units.
[4]Includes oil-, gas-, and dual-fired capacity.
[5]Nuclear capacity includes 2.7 gigawatts of uprates through 2030.
[6]Includes conventional hydroelectric, geothermal, wood, wood waste, municipal solid waste, landfill gas, other biomass, solar, and wind power. Facilities co-firing biomass and coal are classified as coal.
[7]Primarily peak load capacity fueled by natural gas.
[8]Includes combined heat and power plants whose primary business is to sell electricity and heat to the public (i.e., those that report North American Industry Classification System code 22).
[9]Cumulative additions after December 31, 2005.
[10]Cumulative retirements after December 31, 2005.
[11]Includes combined heat and power plants and electricity-only plants in the commercial and industrial sectors; and small on-site generating systems in the residential, commercial, and industrial sectors used primarily for own-use generation, but which may also sell some power to the grid.
N/A = Not applicable.
Note: Totals may not equal sum of components due to independent rounding. Data for 2004 and 2005 are model results and may differ slightly from official EIA data reports.
Sources: 2004 and 2005 capacity and projected planned additions: Energy Information Administration (EIA), Form EIA-860, "Annual Electric Generator Report" (preliminary). **Projections:** EIA, AEO2007 National Energy Modeling System run AEO2007.D112106A.

Table A10. Electricity Trade
(Billion Kilowatthours, Unless Otherwise Noted)

Electricity Trade	Reference Case 2004	2005	2010	2015	2020	2025	2030	Annual Growth 2005-2030 (percent)
Interregional Electricity Trade								
Gross Domestic Sales								
Firm Power	142.4	127.0	105.5	82.4	50.6	37.9	37.9	-4.7%
Economy	118.3	180.5	178.6	181.7	167.6	164.1	149.9	-0.7%
Total	**260.7**	**307.5**	**284.0**	**264.1**	**218.2**	**201.9**	**187.7**	**-2.0%**
Gross Domestic Sales (million 2005 dollars)								
Firm Power	7675.9	6845.3	5684.4	4441.8	2727.5	2041.8	2041.8	-4.7%
Economy	7058.7	11082.0	8729.7	7772.8	7568.7	7572.2	7117.4	-1.8%
Total	**14734.6**	**17927.2**	**14414.1**	**12214.6**	**10296.2**	**9613.9**	**9159.2**	**-2.7%**
International Electricity Trade								
Imports from Canada and Mexico								
Firm Power	12.5	13.1	2.5	1.9	0.8	0.4	0.4	-13.2%
Economy	21.7	31.4	26.7	23.9	25.9	26.0	26.1	-0.7%
Total	**34.2**	**44.5**	**29.2**	**25.7**	**26.7**	**26.4**	**26.5**	**-2.1%**
Exports to Canada and Mexico								
Firm Power	7.4	2.9	1.0	0.7	0.2	0.0	0.0	N/A
Economy	15.5	16.9	17.1	16.6	15.6	13.7	13.7	-0.8%
Total	**22.9**	**19.8**	**18.1**	**17.2**	**15.8**	**13.7**	**13.7**	**-1.5%**

N/A = Not applicable.

Note: Totals may not equal sum of components due to independent rounding. Data for 2004 and 2005 are model results and may differ slightly from official EIA data reports. Firm Power Sales are capacity sales, meaning the delivery of the power is scheduled as part of the normal operating conditions of the affected electric systems. Economy Sales are subject to curtailment or cessation of delivery by the supplier in accordance with prior agreements or under specified conditions.

Sources: 2004 and 2005 interregional firm electricity trade data: North American Electric Reliability Council (NERC), Electricity Sales and Demand Database 2004. 2004 and 2005 Mexican electricity trade data: Energy Information Administration (EIA), *Electric Power Annual 2004* DOE/EIA-0348(2004) (Washington, DC, November 2005). 2004 Canadian international electricity trade data: National Energy Board, *Annual Report 2003.* 2005 Canadian electricity trade data: National Energy Board, *Annual Report 2004.* **Projections:** EIA, AEO2007 National Energy Modeling System run AEO2007.D112106A.

Table A11. Liquid Fuels Supply and Disposition
(Million Barrels per Day, Unless Otherwise Noted)

Supply and Disposition	Reference Case							Annual Growth 2005-2030 (percent)
	2004	2005	2010	2015	2020	2025	2030	
Crude Oil								
Domestic Crude Production[1]	5.47	5.18	5.67	5.91	5.89	5.58	5.39	0.2%
Alaska	0.91	0.86	0.69	0.65	0.74	0.47	0.27	-4.6%
Lower 48 States	4.56	4.31	4.98	5.26	5.15	5.12	5.12	0.7%
Net Imports	10.06	10.09	9.99	10.49	11.29	12.20	13.09	1.0%
Gross Imports	10.09	10.12	10.03	10.52	11.33	12.24	13.12	1.0%
Exports	0.03	0.03	0.04	0.04	0.04	0.04	0.03	0.2%
Other Crude Supply[2]	-0.00	-0.06	0.00	0.00	0.00	0.00	0.00	N/A
Total Crude Supply	**15.52**	**15.22**	**15.66**	**16.40**	**17.19**	**17.78**	**18.47**	**0.8%**
Other Supply								
Natural Gas Plant Liquids	1.81	1.72	1.80	1.82	1.76	1.72	1.72	-0.0%
Net Product Imports	2.06	2.48	1.80	2.03	2.27	2.67	3.28	1.1%
Gross Refined Product Imports[3]	2.07	2.45	1.78	1.84	1.98	2.17	2.52	0.1%
Unfinished Oil Imports	0.49	0.58	0.41	0.46	0.51	0.60	0.67	0.6%
Ethanol Imports	0.01	0.01	0.02	0.03	0.04	0.05	0.05	8.4%
Blending Component Imports	0.49	0.54	0.82	0.96	1.03	1.15	1.36	3.8%
Exports	0.96	1.07	1.23	1.25	1.29	1.30	1.33	0.9%
Refinery Processing Gain[4]	1.05	0.99	1.21	1.27	1.41	1.45	1.49	1.7%
Other Inputs	0.34	0.39	1.02	1.25	1.31	1.60	1.88	6.5%
Ethanol	0.22	0.26	0.69	0.74	0.79	0.85	0.90	5.2%
Liquids from Gas	0.00	0.00	0.00	0.00	0.00	0.00	0.00	N/A
Liquids from Coal	0.00	0.00	0.00	0.06	0.10	0.32	0.44	N/A
Other[5]	0.12	0.13	0.33	0.46	0.43	0.43	0.53	5.7%
Total Primary Supply[6]	**20.79**	**20.79**	**21.49**	**22.78**	**23.94**	**25.22**	**26.84**	**1.0%**
Liquid Fuels Consumption								
by Fuel								
Liquefied Petroleum Gases	2.13	2.03	2.22	2.23	2.26	2.33	2.42	0.7%
E85[7]	0.00	0.00	0.00	0.00	0.01	0.01	0.02	11.9%
Motor Gasoline[8]	9.10	9.16	9.53	10.18	10.93	11.71	12.53	1.3%
Jet Fuel[9]	1.63	1.68	1.95	2.10	2.19	2.22	2.27	1.2%
Distillate Fuel Oil[10]	4.06	4.12	4.53	4.86	5.11	5.48	5.95	1.5%
Residual Fuel Oil	0.87	0.92	0.79	0.82	0.82	0.82	0.83	-0.4%
Other[11]	2.97	2.84	2.57	2.66	2.70	2.78	2.93	0.1%
by Sector								
Residential and Commercial	1.27	1.26	1.25	1.29	1.29	1.28	1.28	0.1%
Industrial[12]	5.28	5.07	5.01	5.10	5.16	5.29	5.53	0.3%
Transportation	13.80	13.87	14.93	16.04	17.15	18.33	19.69	1.4%
Electric Power[13]	0.50	0.51	0.40	0.43	0.43	0.44	0.45	-0.5%
Total	**20.76**	**20.75**	**21.59**	**22.86**	**24.03**	**25.34**	**26.95**	**1.1%**
Discrepancy[14]	**0.03**	**0.04**	**-0.10**	**-0.08**	**-0.09**	**-0.12**	**-0.11**	**N/A**

Table A11. Liquid Fuels Supply and Disposition (Continued)
(Million Barrels per Day, Unless Otherwise Noted)

Supply and Disposition	Reference Case							Annual Growth 2005-2030 (percent)
	2004	2005	2010	2015	2020	2025	2030	
Domestic Refinery Distillation Capacity[15]	16.9	17.1	17.8	18.0	18.7	19.4	20.0	0.6%
Capacity Utilization Rate (percent)[16]	93.0	91.0	89.1	92.2	93.4	93.1	93.5	0.1%
Net Import Share of Product Supplied (percent)	58.3	60.5	54.9	55.0	56.6	59.0	61.0	0.0%
Net Expenditures for Imported Crude Oil and Petroleum Products (billion 2005 dollars)	179.47	236.65	222.76	203.97	229.80	264.31	300.51	1.0%

[1]Includes lease condensate.
[2]Strategic petroleum reserve stock additions plus unaccounted for crude oil and crude stock withdrawals minus crude product supplied.
[3]Includes other hydrocarbons and alcohols.
[4]Represents volumetric gain in refinery distillation and cracking processes.
[5]Includes petroleum product stock withdrawals, domestic sources of blending components, other hydrocarbons, ethers, and renewable fuels such as biodiesel.
[6]Total crude supply plus natural gas plant liquids, other inputs, refinery processing gain, and net product imports.
[7]E85 refers to a blend of 85 percent ethanol (renewable) and 15 percent motor gasoline (nonrenewable). To address cold starting issues, the percentage of ethanol actually varies seasonally. The annual average ethanol content of 74 percent is used for this forecast.
[8]Includes ethanol and ethers blended into gasoline.
[9]Includes only kerosene type.
[10]Includes distillate fuel oil and kerosene from petroleum and biomass feedstocks.
[11]Includes aviation gasoline, petrochemical feedstocks, lubricants, waxes, asphalt, road oil, still gas, special naphthas, petroleum coke, crude oil product supplied, tire-derived fuel, methanol, liquid hydrogen,and miscellaneous petroleum products.
[12]Includes consumption for combined heat and power, which produces electricity and other useful thermal energy.
[13]Includes consumption of energy by electricity-only and combined heat and power plants whose primary business is to sell electricity, or electricity and heat, to the public. Includes small power producers and exempt wholesale generators.
[14]Balancing item. Includes unaccounted for supply, losses, and gains.
[15]End-of-year operable capacity.
[16]Rate is calculated by dividing the gross annual input to atmospheric crude oil distillation units by their operable refining capacity in barrels per calendar day.
N/A = Not applicable.
Note: Totals may not equal sum of components due to independent rounding. Data for 2004 and 2005 are model results and may differ slightly from official EIA data reports.
Sources: 2004 and 2005 imported crude oil price and petroleum product supplied based on: Energy Information Administration (EIA), *Annual Energy Review 2005*, DOE/EIA-0384(2005) (Washington, DC, July 2006). 2004 and 2005 imported low sulfur light crude oil price: EIA, Form EIA-856, "Monthly Foreign Crude Oil Acquisition Report." Other 2004 data: EIA, *Petroleum Supply Annual 2004,* DOE/EIA-0340(2004)/1 (Washington, DC, June 2005). Other 2005 data: EIA, *Petroleum Supply Annual 2005,* DOE/EIA-0340(2005)/1 (Washington, DC, October 2006). **Projections:** EIA, AEO2007 National Energy Modeling System run AEO2007.D112106A.

Table A12. Petroleum Product Prices
(2005 Cents per Gallon, Unless Otherwise Noted)

Sector and Fuel	Reference Case							Annual Growth 2005-2030 (percent)
	2004	2005	2010	2015	2020	2025	2030	
Crude Oil Prices (2005 dollars per barrel)								
Imported Low Sulfur Light Crude Oil[1]	42.87	56.76	57.47	49.87	52.04	56.37	59.12	0.2%
Imported Crude Oil[1]	37.09	49.19	51.20	44.61	46.47	49.57	51.63	0.2%
Delivered Sector Product Prices								
Residential								
Liquefied Petroleum Gases	156.9	166.3	204.0	197.2	199.8	202.8	206.1	0.9%
Distillate Fuel Oil	178.8	204.3	206.3	174.7	182.3	188.1	195.9	-0.2%
Commercial								
Distillate Fuel Oil	145.0	175.4	175.5	147.9	156.5	163.4	171.7	-0.1%
Residual Fuel Oil	93.5	126.0	112.8	97.2	105.8	107.3	109.4	-0.6%
Residual Fuel Oil (2005 dollars per barrel)	39.29	52.90	47.39	40.83	44.44	45.08	45.97	-0.6%
Industrial[2]								
Liquefied Petroleum Gases	96.3	146.2	141.5	134.1	137.1	139.8	142.6	-0.1%
Distillate Fuel Oil	152.2	181.1	178.1	156.5	165.3	173.2	181.8	0.0%
Residual Fuel Oil	86.4	116.3	142.2	122.9	133.4	138.6	143.5	0.8%
Residual Fuel Oil (2005 dollars per barrel)	36.30	48.86	59.74	51.64	56.04	58.21	60.25	0.8%
Transportation								
Liquefied Petroleum Gases	169.5	206.1	209.8	202.4	203.9	206.4	209.3	0.1%
Ethanol (E85)[3]	196.5	217.1	198.5	187.3	192.1	198.2	200.4	-0.3%
Ethanol Wholesale Price	177.2	180.4	181.4	166.0	168.2	171.1	170.2	-0.2%
Motor Gasoline[4]	195.2	231.6	217.3	194.9	201.9	210.2	215.4	-0.3%
Jet Fuel[5]	124.4	177.4	147.2	133.5	141.8	149.8	158.6	-0.4%
Diesel Fuel (distillate fuel oil)[6]	187.1	241.3	230.4	203.6	211.2	218.1	225.7	-0.3%
Residual Fuel Oil	72.7	82.4	120.5	105.4	110.2	118.2	123.8	1.6%
Residual Fuel Oil (2005 dollars per barrel)	30.51	34.62	50.60	44.27	46.27	49.65	52.02	1.6%
Electric Power[7]								
Distillate Fuel Oil	132.0	157.9	162.3	128.4	136.5	142.2	149.6	-0.2%
Residual Fuel Oil	74.7	104.2	98.5	83.8	91.0	98.5	102.5	-0.1%
Residual Fuel Oil (2005 dollars per barrel)	31.38	43.76	41.37	35.20	38.24	41.37	43.05	-0.1%
Refined Petroleum Product Prices[8]								
Liquefied Petroleum Gases	109.3	150.7	155.3	148.6	151.9	154.8	157.7	0.2%
Motor Gasoline[4]	195.0	231.1	217.3	194.9	201.8	210.2	215.4	-0.3%
Jet Fuel[5]	124.4	177.4	147.2	133.5	141.8	149.8	158.6	-0.4%
Distillate Fuel Oil	178.3	223.9	215.9	191.0	199.4	207.0	215.5	-0.2%
Residual Fuel Oil	76.3	98.6	113.9	98.0	104.7	111.9	116.6	0.7%
Residual Fuel Oil (2005 dollars per barrel)	32.05	41.42	47.84	41.16	43.98	46.99	48.96	0.7%
Average	**168.6**	**204.5**	**195.0**	**175.7**	**183.4**	**191.3**	**198.1**	**-0.1%**

[1]Weighted average price delivered to U.S. refiners.
[2]Includes energy for combined heat and power plants, except those whose primary business is to sell electricity, or electricity and heat, to the public.
[3]E85 refers to a blend of 85 percent ethanol (renewable) and 15 percent motor gasoline (nonrenewable). To address cold starting issues, the percentage of ethanol actually varies seasonally. The annual average ethanol content of 74 percent is used for this forecast.
[4]Sales weighted-average price for all grades. Includes Federal, State and local taxes.
[5]Includes only kerosene type.
[6]Diesel fuel for on-road use. Includes Federal and State taxes while excluding county and local taxes.
[7]Includes electricity-only and combined heat and power plants whose primary business is to sell electricity, or electricity and heat, to the public. Includes small power producers and exempt wholesale generators.
[8]Weighted averages of end-use fuel prices are derived from the prices in each sector and the corresponding sectoral consumption.
Note: Data for 2004 and 2005 are model results and may differ slightly from official EIA data reports.
Sources: 2004 and 2005 imported low sulfur light crude oil price: Energy Information Administration (EIA), Form EIA-856, "Monthly Foreign Crude Oil Acquisition Report." 2004 and 2005 imported crude oil price: EIA, *Annual Energy Review 2005*, DOE/EIA-0384(2005) (Washington, DC, July 2006). 2004 and 2005 prices for motor gasoline, distillate fuel oil, and jet fuel are based on: EIA, *Petroleum Marketing Annual 2005*, DOE/EIA-0487(2005) (Washington, DC, August 2006). 2004 and 2005 residential, commercial, industrial, and transportation sector petroleum product prices are derived from: EIA, Form EIA-782A, "Refiners'/Gas Plant Operators' Monthly Petroleum Product Sales Report." 2004 and 2005 electric power prices based on: Federal Energy Regulatory Commission, FERC Form 423, "Monthly Report of Cost and Quality of Fuels for Electric Plants." 2004 and 2005 ethanol prices derived from weekly spot prices in the Oxy Fuel News. 2004 and 2005 wholesale ethanol prices derived from Bloomburg U.S. average rack price. **Projections:** EIA, AEO2007 National Energy Modeling System run AEO2007.D112106A.

Table A13. Natural Gas Supply, Disposition, and Prices
(Trillion Cubic Feet per Year, Unless Otherwise Noted)

Supply, Disposition, and Prices	Reference Case							Annual Growth 2005-2030 (percent)
	2004	2005	2010	2015	2020	2025	2030	
Production								
Dry Gas Production[1]	18.76	18.23	19.35	19.60	20.79	20.59	20.53	0.5%
Supplemental Natural Gas[2]	0.07	0.07	0.07	0.07	0.07	0.07	0.07	0.1%
Net Imports	**3.40**	**3.57**	**4.55**	**5.62**	**5.35**	**5.58**	**5.45**	**1.7%**
Pipeline[3]	2.81	3.01	2.74	2.63	1.65	1.20	0.92	-4.6%
Liquefied Natural Gas	0.59	0.57	1.81	2.99	3.69	4.38	4.53	8.7%
Total Supply	**22.22**	**21.87**	**23.97**	**25.29**	**26.21**	**26.24**	**26.06**	**0.7%**
Consumption by Sector								
Residential	4.87	4.84	5.03	5.19	5.27	5.29	5.31	0.4%
Commercial	3.13	3.05	3.22	3.53	3.75	3.98	4.24	1.3%
Industrial[4]	7.22	6.64	7.63	7.67	8.02	8.42	8.65	1.1%
Natural-Gas-to-Liquids Heat and Power[5]	0.00	0.00	0.00	0.00	0.00	0.00	0.00	N/A
Natural Gas to Liquids Production[6]	0.00	0.00	0.00	0.00	0.00	0.00	0.00	N/A
Electric Power[7]	5.48	5.78	6.38	7.11	7.19	6.59	5.92	0.1%
Transportation[8]	0.03	0.03	0.06	0.08	0.09	0.11	0.12	5.7%
Pipeline Fuel	0.57	0.56	0.64	0.68	0.76	0.77	0.77	1.3%
Lease and Plant Fuel[9]	1.10	1.07	1.07	1.07	1.17	1.14	1.12	0.2%
Total	**22.39**	**21.98**	**24.02**	**25.32**	**26.26**	**26.30**	**26.12**	**0.7%**
Discrepancy[10]	**-0.17**	**-0.11**	**-0.05**	**-0.03**	**-0.05**	**-0.06**	**-0.06**	**N/A**
Natural Gas Prices								
(2005 dollars per million Btu)								
Henry Hub Spot Price	6.08	8.60	6.28	5.46	5.71	6.14	6.52	-1.1%
Average Lower 48 Wellhead Price[11]	5.63	7.29	5.59	4.84	5.07	5.46	5.80	-0.9%
(2005 dollars per thousand cubic feet)								
Average Lower 48 Wellhead Price[11]	5.80	7.51	5.76	4.99	5.22	5.62	5.98	-0.9%
Delivered Prices								
Residential	11.05	12.80	11.31	10.55	10.86	11.30	11.77	-0.3%
Commercial	9.69	11.54	9.62	8.73	8.93	9.23	9.58	-0.7%
Industrial[4]	6.67	8.41	6.62	5.82	6.08	6.40	6.76	-0.9%
Electric Power[7]	6.27	8.42	6.40	5.66	5.93	6.22	6.51	-1.0%
Transportation[12]	12.28	15.20	14.38	13.25	13.36	13.62	13.86	-0.4%
Average[13]	**8.06**	**9.94**	**8.07**	**7.28**	**7.54**	**7.91**	**8.33**	**-0.7%**

[1]Marketed production (wet) minus extraction losses.
[2]Synthetic natural gas, propane air, coke oven gas, refinery gas, biomass gas, air injected for Btu stabilization, and manufactured gas commingled and distributed with natural gas.
[3]Includes any natural gas regasified in the Bahamas and transported via pipeline to Florida.
[4]Includes energy for combined heat and power plants, except those whose primary business is to sell electricity, or electricity and heat, to the public.
[5]Includes any natural gas used in the process of converting natural gas to liquid fuel that is not actually converted.
[6]Includes any natural gas that is converted into liquid fuel.
[7]Includes consumption of energy by electricity-only and combined heat and power plants whose primary business is to sell electricity, or electricity and heat, to the public. Includes small power producers and exempt wholesale generators.
[8]Compressed natural gas used as vehicle fuel.
[9]Represents natural gas used in well, field, and lease operations, and in natural gas processing plant machinery.
[10]Balancing item. Natural gas lost as a result of converting flow data measured at varying temperatures and pressures to a standard temperature and pressure and the merger of different data reporting systems which vary in scope, format, definition, and respondent type. In addition, 2004 and 2005 values include net storage injections.
[11]Represents lower 48 onshore and offshore supplies.
[12]Compressed natural gas used as a vehicle fuel. Price includes estimated motor vehicle fuel taxes and estimated dispensing costs or charges.
[13]Weighted average prices. Weights used are the sectoral consumption values excluding lease, plant, and pipeline fuel.
N/A = Not applicable.
Note: Totals may not equal sum of components due to independent rounding. Data for 2004 and 2005 are model results and may differ slightly from official EIA data reports.
Sources: 2004 supply values; and lease, plant, and pipeline fuel consumption: Energy Information Administration (EIA), *Natural Gas Annual 2004*, DOE/EIA-0131(2004) (Washington, DC, December 2005). 2005 supply values; and lease, plant, and pipeline fuel consumption; and wellhead price: EIA, *Natural Gas Monthly*, DOE/EIA-0130(2006/04) (Washington, DC, April 2006). Other 2004 and 2005 consumption based on: EIA, *Annual Energy Review 2005*, DOE/EIA-0384(2005) (Washington, DC, July 2006). 2004 wellhead price: Minerals Management Service and EIA, *Natural Gas Annual 2004*, DOE/EIA-0131(2004) (Washington, DC, December 2005). 2004 residential and commercial delivered prices: EIA, *Natural Gas Annual 2004*, DOE/EIA-0131(2004) (Washington, DC, December 2005). 2005 residential and commercial delivered prices: EIA, *Natural Gas Monthly*, DOE/EIA-0130(2006/04) (Washington, DC, April 2006). 2004 and 2005 electric power prices: EIA, *Electric Power Monthly*, DOE/EIA-0226, May 2005 through April 2006, Table 4.11.A. 2004 and 2005 industrial delivered prices are estimated based on: EIA, *Manufacturing Energy Consumption Survey 1994* and industrial and wellhead prices from the *Natural Gas Annual 2004*, DOE/EIA-0131(2004) (Washington, DC, December 2005) and the *Natural Gas Monthly*, DOE/EIA-0130(2006/04) (Washington, DC, April 2006). 2004 transportation sector delivered prices are based on: EIA, *Natural Gas Annual 2004*, DOE/EIA-0131(2004) (Washington, DC, December 2005) and estimated state taxes, federal taxes, and dispensing costs or charges. 2005 transportation sector delivered prices are model results. **Projections:** EIA, AEO2007 National Energy Modeling System run AEO2007.D112106A.

Table A14. Oil and Gas Supply

Production and Supply	Reference Case							Annual Growth 2005-2030 (percent)
	2004	2005	2010	2015	2020	2025	2030	
Crude Oil								
Lower 48 Average Wellhead Price[1] (2005 dollars per barrel)	**38.69**	**50.76**	**48.54**	**41.71**	**44.88**	**48.37**	**51.25**	**0.0%**
Production (million barrels per day)[2]								
United States Total	5.45	5.18	5.67	5.91	5.89	5.58	5.39	0.2%
Lower 48 Onshore	2.94	2.89	2.93	2.91	2.94	2.95	2.92	0.0%
Lower 48 Offshore	1.61	1.42	2.05	2.35	2.21	2.16	2.20	1.8%
Alaska	0.91	0.86	0.69	0.65	0.74	0.47	0.27	-4.6%
Lower 48 End of Year Reserves[2] (billion barrels)	**18.27**	**16.98**	**19.53**	**20.46**	**19.98**	**19.47**	**17.94**	**0.2%**
Natural Gas								
Lower 48 Average Wellhead Price[1] (2005 dollars per million Btu)								
Henry Hub Spot Price	6.08	8.60	6.28	5.46	5.71	6.14	6.52	-1.1%
Average Lower 48 Wellhead Price[1]	5.63	7.29	5.59	4.84	5.07	5.46	5.80	-0.9%
(2005 dollars per thousand cubic feet)								
Average Lower 48 Wellhead Price[1]	5.80	7.51	5.76	4.99	5.22	5.62	5.98	-0.9%
Dry Production (trillion cubic feet)[3]								
United States Total	18.76	18.23	19.35	19.60	20.79	20.59	20.53	0.5%
Lower 48 Onshore	14.10	14.36	15.22	14.79	14.66	14.84	15.13	0.2%
Associated-Dissolved[4]	1.40	1.43	1.39	1.32	1.28	1.23	1.19	-0.8%
Non-Associated	12.69	12.93	13.83	13.46	13.38	13.61	13.94	0.3%
Conventional	5.19	4.94	5.27	4.71	4.30	3.98	3.75	-1.1%
Unconventional	7.50	7.99	8.56	8.75	9.09	9.63	10.19	1.0%
Lower 48 Offshore	4.23	3.41	3.88	4.56	4.09	3.55	3.25	-0.2%
Associated-Dissolved[4]	0.88	0.71	0.92	1.13	1.05	0.94	0.85	0.7%
Non-Associated	3.35	2.69	2.96	3.43	3.04	2.61	2.40	-0.5%
Alaska	0.44	0.45	0.25	0.25	2.05	2.20	2.16	6.4%
Lower 48 End of Year Dry Reserves[3] (trillion cubic feet)	**184.11**	**189.91**	**205.23**	**210.31**	**208.32**	**208.61**	**210.60**	**0.4%**
Supplemental Gas Supplies (trillion cubic feet)[5]	**0.07**	**0.07**	**0.07**	**0.07**	**0.07**	**0.07**	**0.07**	**0.1%**
Total Lower 48 Wells Drilled (thousands)	**32.67**	**41.66**	**37.17**	**32.01**	**31.84**	**32.78**	**30.65**	**-1.2%**

[1]Represents lower 48 onshore and offshore supplies.
[2]Includes lease condensate.
[3]Marketed production (wet) minus extraction losses.
[4]Gas which occurs in crude oil reservoirs either as free gas (associated) or as gas in solution with crude oil (dissolved).
[5]Synthetic natural gas, propane air, coke oven gas, refinery gas, biomass gas, air injected for Btu stabilization, and manufactured gas commingled and distributed with natural gas.

Note: Totals may not equal sum of components due to independent rounding. Data for 2004 and 2005 are model results and may differ slightly from official EIA data reports.

Sources: 2004 and 2005 crude oil lower 48 average wellhead price: Energy Information Administration (EIA), *Petroleum Marketing Annual 2005*, DOE/EIA-0487(2005) (Washington, DC, August 2006). 2004 and 2005 lower 48 onshore, lower 48 offshore, and Alaska crude oil production: EIA, *Petroleum Supply Annual 2005*, DOE/EIA-0340(2005)/1 (Washington, DC, October 2006). 2004 U.S. crude oil and natural gas reserves: EIA, *U.S. Crude Oil, Natural Gas, and Natural Gas Liquids Reserves*, DOE/EIA-0216(2004) (Washington, DC, November 2005). 2004 Alaska and total natural gas production, and supplemental gas supplies: EIA, *Natural Gas Annual 2004*, DOE/EIA-0131(2004) (Washington, DC, December 2005). 2004 natural gas lower 48 average wellhead price: Minerals Management Service and EIA, *Natural Gas Annual 2004*, DOE/EIA-0131(2004) (Washington, DC, December 2005). 2005 natural gas lower 48 average wellhead price, Alaska and total natural gas production, and supplemental gas supplies: EIA, *Natural Gas Monthly*, DOE/EIA-0130(2006/04) (Washington, DC, April 2006). Other 2004 and 2005 values: EIA, Office of Integrated Analysis and Forecasting. **Projections:** EIA, AEO2007 National Energy Modeling System run AEO2007.D112106A.

Table A15. Coal Supply, Disposition, and Prices
(Million Short Tons per Year, Unless Otherwise Noted)

Supply, Disposition, and Prices	Reference Case							Annual Growth 2005-2030 (percent)
	2004	2005	2010	2015	2020	2025	2030	
Production[1]								
Appalachia	391	397	381	371	348	351	373	-0.3%
Interior	146	149	171	199	203	215	247	2.0%
West	575	585	637	697	772	951	1072	2.5%
East of the Mississippi	485	494	498	507	487	499	545	0.4%
West of the Mississippi	627	638	691	759	837	1018	1146	2.4%
Total	**1112**	**1131**	**1189**	**1266**	**1323**	**1517**	**1691**	**1.6%**
Waste Coal Supplied[2]	**11**	**13**	**13**	**13**	**13**	**13**	**13**	**-0.0%**
Net Imports								
Imports[3]	26	29	37	42	72	79	95	4.9%
Exports	48	50	44	37	31	27	27	-2.4%
Total	**-22**	**-21**	**-7**	**5**	**41**	**52**	**68**	**N/A**
Total Supply[4]	**1101**	**1124**	**1195**	**1284**	**1377**	**1582**	**1772**	**1.8%**
Consumption by Sector								
Residential and Commercial	5	5	5	5	5	5	5	-0.3%
Coke Plants	24	23	22	21	21	21	21	-0.5%
Other Industrial[5]	62	61	64	62	63	63	64	0.2%
Coal-to-Liquids Heat and Power	0	0	0	8	13	42	57	N/A
Coal to Liquids Production	0	0	0	8	13	40	55	N/A
Electric Power[6]	1016	1039	1104	1178	1262	1411	1570	1.7%
Total	**1107**	**1128**	**1195**	**1282**	**1377**	**1582**	**1772**	**1.8%**
Discrepancy and Stock Change[7]	**-6**	**-5**	**0**	**1**	**0**	**0**	**0**	**N/A**
Average Minemouth Price[8]								
(2005 dollars per short ton)	20.68	23.34	24.20	22.41	21.58	21.55	22.60	-0.1%
(2005 dollars per million Btu)	1.01	1.15	1.18	1.11	1.08	1.09	1.15	-0.0%
Delivered Prices (2005 dollars per short ton)[9]								
Coke Plants	63.36	83.79	84.86	74.51	74.25	73.93	75.55	-0.4%
Other Industrial[5]	40.49	47.63	48.86	47.45	46.55	47.60	48.54	0.1%
Coal to Liquids	N/A	N/A	N/A	13.79	15.05	19.82	21.89	N/A
Electric Power								
(2005 dollars per short ton)	28.12	30.83	34.17	31.84	31.39	32.20	33.52	0.3%
(2005 dollars per million Btu)	1.40	1.53	1.71	1.60	1.58	1.63	1.69	0.4%
Average	**29.58**	**32.82**	**35.89**	**33.10**	**32.42**	**32.72**	**33.82**	**0.1%**
Exports[10]	55.75	67.10	69.35	64.51	64.49	61.66	63.81	-0.2%

[1]Includes anthracite, bituminous coal, and lignite.
[2]Includes waste coal consumed by the electric power and industrial sectors. Waste coal supplied is counted as a supply-side item to balance the same amount of waste coal included in the consumption data.
[3]Excludes imports to Puerto Rico and the U.S. Virgin Islands.
[4]Production plus waste coal supplied plus net imports.
[5]Includes consumption for combined heat and power plants, except those plants whose primary business is to sell electricity, or electricity and heat, to the public. Excludes all coal use in the coal-to-liquids process.
[6]Includes all electricity-only and combined heat and power plants whose primary business is to sell electricity, or electricity and heat, to the public.
[7]Balancing item: the sum of production, net imports, and waste coal supplied minus total consumption.
[8]Includes reported prices for both open market and captive mines.
[9]Prices weighted by consumption; weighted average excludes residential and commercial prices, and export free-alongside-ship (f.a.s.) prices.
[10]F.a.s. price at U.S. port of exit.
N/A = Not applicable.
Btu = British thermal unit.
Note: Totals may not equal sum of components due to independent rounding. Data for 2004 and 2005 are model results and may differ slightly from official EIA data reports.
Sources: 2004 and 2005 data based on: Energy Information Administration (EIA), *Annual Coal Report 2005*, DOE/EIA-0584(2005) (Washington, DC, October 2006); EIA, *Quarterly Coal Report, October-December 2005*, DOE/EIA-0121(2005/4Q) (Washington, DC, March 2006); and EIA, AEO2007 National Energy Modeling System run AEO2007.D112106A. **Projections:** EIA, AEO2007 National Energy Modeling System run AEO2007.D112106A.

Table A16. Renewable Energy Generating Capacity and Generation
(Gigawatts, Unless Otherwise Noted)

Capacity and Generation	Reference Case							Annual Growth 2005-2030 (percent)
	2004	2005	2010	2015	2020	2025	2030	
Electric Power Sector[1]								
Net Summer Capacity								
Conventional Hydropower	79.96	79.97	79.99	79.99	80.12	80.18	80.18	0.0%
Geothermal[2]	2.17	2.28	2.46	2.54	2.79	2.95	3.15	1.3%
Municipal Waste[3]	3.19	3.23	3.43	3.79	3.80	3.80	3.87	0.7%
Wood and Other Biomass[4,5]	2.04	2.06	2.22	2.22	2.37	2.89	3.80	2.5%
Solar Thermal	0.40	0.40	0.54	0.56	0.58	0.60	0.63	1.8%
Solar Photovoltaic[6]	0.03	0.03	0.07	0.14	0.22	0.31	0.39	10.6%
Wind	6.97	9.62	16.97	17.70	17.85	17.89	17.98	2.5%
Total	**94.75**	**97.59**	**105.69**	**106.94**	**107.72**	**108.62**	**110.00**	**0.5%**
Generation (billion kilowatthours)								
Conventional Hydropower	265.06	261.89	297.50	302.83	303.85	304.36	304.51	0.6%
Geothermal[2]	14.81	15.12	17.34	17.73	19.79	21.05	22.66	1.6%
Municipal Waste[3]	19.86	20.56	21.56	24.38	24.42	24.43	24.95	0.8%
Wood and Other Biomass[5]	9.73	9.92	43.29	46.22	47.47	58.01	58.21	7.3%
Dedicated Plants	8.54	5.38	11.11	10.49	11.61	16.07	23.80	6.1%
Cofiring	1.19	4.53	32.18	35.74	35.86	41.93	34.41	8.4%
Solar Thermal	0.57	0.54	1.16	1.22	1.28	1.36	1.43	4.0%
Solar Photovoltaic[6]	0.01	0.01	0.18	0.34	0.54	0.76	0.98	22.6%
Wind	14.14	14.60	48.26	50.85	51.35	51.52	51.85	5.2%
Total	**324.19**	**322.64**	**429.28**	**443.57**	**448.71**	**461.47**	**464.59**	**1.5%**
End-Use Generators[7]								
Net Summer Capacity								
Conventional Hydropower[8]	0.65	0.63	0.63	0.63	0.63	0.63	0.63	0.0%
Geothermal	0.00	0.00	0.00	0.00	0.00	0.00	0.00	N/A
Municipal Waste[9]	0.33	0.33	0.33	0.33	0.33	0.33	0.33	0.0%
Biomass	4.66	4.49	4.79	5.38	5.90	6.50	7.19	1.9%
Solar Photovoltaic[6]	0.12	0.18	0.63	0.69	0.80	1.22	2.52	11.2%
Total	**5.76**	**5.63**	**6.39**	**7.04**	**7.66**	**8.69**	**10.68**	**2.6%**
Generation (billion kilowatthours)								
Conventional Hydropower[8]	4.99	3.18	3.18	3.18	3.18	3.18	3.18	0.0%
Geothermal	0.00	0.00	0.00	0.00	0.00	0.00	0.00	N/A
Municipal Waste[9]	2.64	2.75	2.75	2.75	2.75	2.75	2.75	0.0%
Biomass	28.90	27.91	29.69	33.15	36.17	39.67	43.70	1.8%
Solar Photovoltaic[6]	0.23	0.33	1.21	1.33	1.53	2.32	4.78	11.2%
Total	**36.77**	**34.18**	**36.84**	**40.42**	**43.63**	**47.92**	**54.41**	**1.9%**

[1]Includes electricity-only and combined heat and power plants whose primary business is to sell electricity, or electricity and heat, to the public.
[2]Includes hydrothermal resources only (hot water and steam).
[3]Includes municipal solid waste, landfill gas, and municipal sewage sludge. Incremental growth is assumed to be for landfill gas facilities. All municipal solid waste is included, although a portion of the municipal solid waste stream contains petroleum-derived plastics and other non-renewable sources.
[4]Facilities co-firing biomass and coal are classified as coal.
[5]Includes projections for energy crops after 2010.
[6]Does not include off-grid photovoltaics (PV). Based on annual PV shipments from 1989 through 2004, EIA estimates that as much as 167 megawatts of remote electricity generation PV applications (i.e., off-grid power systems) were in service in 2004, plus an additional 447 megawatts in communications, transportation, and assorted other non-grid-connected, specialized applications. See Energy Information Administration, *Annual Energy Review 2005*, DOE/EIA-0384(2005) (Washington, DC, July 2006), Table 10.6 (annual PV shipments, 1989-2004). The approach used to develop the estimate, based on shipment data, provides an upper estimate of the size of the PV stock, including both grid-based and off-grid PV. It will overestimate the size of the stock, because shipments include a substantial number of units that are exported, and each year some of the PV units installed earlier will be retired from service or abandoned.
[7]Includes combined heat and power plants and electricity-only plants in the commercial and industrial sectors; and small on-site generating systems in the residential, commercial, and industrial sectors used primarily for own-use generation, but which may also sell some power to the grid.
[8]Represents own-use industrial hydroelectric power.
[9]Includes municipal solid waste, landfill gas, and municipal sewage sludge. All municipal solid waste is included, although a portion of the municipal solid waste stream contains petroleum-derived plastics and other non-renewable sources.
N/A = Not applicable.
Note: Totals may not equal sum of components due to independent rounding. Data for 2004 and 2005 are model results and may differ slightly from official EIA data reports.
Sources: 2004 and 2005 capacity: Energy Information Administration (EIA), Form EIA-860, "Annual Electric Generator Report" (preliminary). 2004 and 2005 generation: EIA, *Annual Energy Review 2005*, DOE/EIA-0384(2005) (Washington, DC, July 2006). **Projections:** EIA, AEO2007 National Energy Modeling System run AEO2007.D112106A.

Table A17. Renewable Energy, Consumption by Sector and Source[1]
(Quadrillion Btu per Year)

Sector and Source	Reference Case 2004	2005	2010	2015	2020	2025	2030	Annual Growth 2005-2030 (percent)
Marketed Renewable Energy[2]								
Residential (wood)	**0.40**	**0.41**	**0.43**	**0.41**	**0.40**	**0.40**	**0.39**	**-0.2%**
Commercial (biomass)	**0.12**	**0.12**	**0.12**	**0.12**	**0.12**	**0.12**	**0.12**	**0.0%**
Industrial[3]	**1.91**	**1.69**	**2.28**	**2.45**	**2.59**	**2.76**	**2.93**	**2.2%**
Conventional Hydroelectric	0.05	0.03	0.03	0.03	0.03	0.03	0.03	N/A
Municipal Waste[4]	0.01	0.01	0.01	0.01	0.01	0.01	0.01	N/A
Biomass	1.64	1.40	1.55	1.67	1.77	1.88	2.01	1.4%
Biofuels Heat and Coproducts	0.21	0.24	0.69	0.74	0.78	0.83	0.88	5.2%
Transportation	**0.30**	**0.34**	**0.95**	**1.01**	**1.10**	**1.19**	**1.27**	**5.5%**
Ethanol used in E85[5]	0.00	0.00	0.00	0.00	0.00	0.01	0.02	11.8%
Ethanol used in Gasoline Blending	0.29	0.33	0.91	0.98	1.05	1.14	1.20	5.3%
Biodiesel used in Distillate Blending	0.00	0.00	0.04	0.03	0.04	0.05	0.05	N/A
Electric Power[6]	**3.55**	**3.64**	**4.67**	**4.83**	**4.93**	**5.09**	**5.15**	**1.4%**
Conventional Hydroelectric	2.66	2.68	2.99	3.04	3.05	3.06	3.06	0.5%
Geothermal	0.31	0.32	0.36	0.37	0.44	0.48	0.53	2.1%
Municipal Waste[7]	0.27	0.28	0.29	0.33	0.33	0.33	0.34	0.8%
Biomass	0.16	0.21	0.51	0.55	0.56	0.68	0.67	4.8%
Dedicated Plants	0.14	0.09	0.11	0.11	0.12	0.17	0.26	4.1%
Cofiring	0.02	0.11	0.40	0.44	0.44	0.50	0.41	5.2%
Solar Thermal	0.01	0.01	0.01	0.02	0.02	0.02	0.02	6.2%
Solar Photovoltaic	0.00	0.00	0.00	0.00	0.00	0.00	0.00	N/A
Wind	0.14	0.15	0.50	0.52	0.53	0.53	0.53	5.2%
Total Marketed Renewable Energy	**6.27**	**6.19**	**8.45**	**8.82**	**9.15**	**9.56**	**9.86**	**1.9%**
Sources of Ethanol								
From Corn	0.28	0.33	0.87	0.93	0.99	1.07	1.13	5.1%
From Cellulose	0.00	0.00	0.01	0.02	0.02	0.02	0.02	N/A
Imports	0.01	0.01	0.02	0.03	0.05	0.06	0.07	8.4%
Total	**0.29**	**0.33**	**0.91**	**0.98**	**1.06**	**1.15**	**1.22**	**5.3%**
Nonmarketed Renewable Energy[8] Selected Consumption								
Residential	**0.03**	**0.03**	**0.04**	**0.05**	**0.06**	**0.07**	**0.08**	**4.0%**
Solar Hot Water Heating	0.03	0.03	0.03	0.04	0.05	0.06	0.06	3.5%
Geothermal Heat Pumps	0.00	0.00	0.01	0.01	0.01	0.01	0.02	6.8%
Solar Photovoltaic	0.00	0.00	0.00	0.00	0.00	0.00	0.00	11.1%
Commercial	**0.02**	**0.03**	**0.03**	**0.03**	**0.03**	**0.03**	**0.04**	**2.1%**
Solar Thermal	0.02	0.02	0.03	0.03	0.03	0.03	0.03	0.6%
Solar Photovoltaic	0.00	0.00	0.00	0.00	0.00	0.01	0.01	11.3%

[1]Actual heat rates used to determine fuel consumption for all renewable fuels except hydropower, solar, and wind. Consumption at hydroelectric, solar, and wind facilities determined by using the fossil fuel equivalent of 10,280 Btu per kilowatthour.
[2]Includes nonelectric renewable energy groups for which the energy source is bought and sold in the marketplace, although all transactions may not necessarily be marketed, and marketed renewable energy inputs for electricity entering the marketplace on the electric power grid. Excludes electricity imports; see Table A2.
[3]Includes all electricity production by industrial and other combined heat and power for the grid and for own use.
[4]Includes municipal solid waste, landfill gas, and municipal sewage sludge. All municipal solid waste is included, although a portion of the municipal solid waste stream contains petroleum-derived plastics and other non-renewable sources.
[5]Excludes motor gasoline component of E85.
[6]Includes consumption of energy by electricity-only and combined heat and power plants whose primary business is to sell electricity, or electricity and heat, to the public. Includes small power producers and exempt wholesale generators.
[7]Includes municipal solid waste, landfill gas, and municipal sewage sludge. Incremental growth is assumed to be for landfill gas facilities. All municipal solid waste is included, although a portion of the municipal solid waste stream contains petroleum-derived plastics and other non-renewable sources.
[8]Includes selected renewable energy consumption data for which the energy is not bought or sold, either directly or indirectly as an input to marketed energy. The Energy Information Administration does not estimate or project total consumption of nonmarketed renewable energy.
N/A = Not applicable.
Btu = British thermal unit.
Note: Totals may not equal sum of components due to independent rounding. Data for 2004 and 2005 are model results and may differ slightly from official EIA data reports.
Sources: 2004 and 2005 ethanol: Energy Information Administration (EIA), *Annual Energy Review 2005*, DOE/EIA-0384(2005) (Washington, DC, July 2006). 2004 and 2005 electric power sector: EIA, Form EIA-860, "Annual Electric Generator Report" (preliminary). Other 2004 and 2005 values: EIA, Office of Integrated Analysis and Forecasting. **Projections:** EIA, AEO2007 National Energy Modeling System run AEO2007.D112106A.

Table A18. Carbon Dioxide Emissions by Sector and Source
(Million Metric Tons, Unless Otherwise Noted)

Sector and Source	Reference Case							Annual Growth 2005-2030 (percent)
	2004	2005	2010	2015	2020	2025	2030	
Residential								
Petroleum	106	105	105	107	105	102	100	-0.2%
Natural Gas	265	262	274	282	287	288	289	0.4%
Coal	1	1	1	1	1	1	1	-0.2%
Electricity[1]	842	886	940	1002	1064	1143	1225	1.3%
Total	**1214**	**1254**	**1320**	**1392**	**1456**	**1534**	**1614**	**1.0%**
Commercial								
Petroleum	54	55	54	57	57	58	59	0.2%
Natural Gas	170	166	175	192	204	216	230	1.3%
Coal	10	8	9	9	9	9	9	0.6%
Electricity[1]	800	822	886	975	1059	1185	1332	1.9%
Total	**1034**	**1051**	**1124**	**1233**	**1330**	**1469**	**1630**	**1.8%**
Industrial[2]								
Petroleum	437	431	406	419	423	433	457	0.2%
Natural Gas[3]	435	400	466	468	493	513	524	1.1%
Coal	201	189	186	194	199	243	269	1.4%
Electricity[1]	663	663	674	694	702	734	774	0.6%
Total	**1736**	**1682**	**1732**	**1775**	**1817**	**1924**	**2024**	**0.7%**
Transportation								
Petroleum[4]	1902	1922	1994	2142	2288	2443	2626	1.3%
Natural Gas[5]	32	32	38	41	47	48	48	1.7%
Electricity[1]	5	5	5	6	6	7	7	1.8%
Total	**1939**	**1958**	**2037**	**2189**	**2341**	**2498**	**2682**	**1.3%**
Electric Power[6]								
Petroleum	98	100	69	75	74	76	77	-1.1%
Natural Gas	296	319	346	385	390	357	321	0.0%
Coal	1904	1944	2078	2203	2354	2623	2927	1.6%
Other[7]	11	12	12	14	14	14	14	0.8%
Total	**2309**	**2375**	**2505**	**2677**	**2832**	**3070**	**3338**	**1.4%**
Total by Fuel								
Petroleum[3]	2598	2614	2629	2799	2947	3112	3318	1.0%
Natural Gas	1198	1178	1298	1369	1420	1422	1412	0.7%
Coal	2115	2142	2275	2407	2563	2877	3206	1.6%
Other[7]	11	12	12	14	14	14	14	0.8%
Total	**5923**	**5945**	**6214**	**6589**	**6944**	**7425**	**7950**	**1.2%**
Carbon Dioxide Emissions (tons per person)	**20.1**	**20.0**	**20.0**	**20.4**	**20.6**	**21.2**	**21.8**	**0.3%**

[1]Emissions from the electric power sector are distributed to the end-use sectors.
[2]Fuel consumption includes energy for combined heat and power plants (CHP), except those plants whose primary business is to sell electricity, or electricity and heat, to the public.
[3]Includes lease and plant fuel.
[4]This includes carbon dioxide from international bunker fuels, both civilian and military, which are excluded from the accounting of carbon dioxide emissions under the United Nations convention. From 1990 through 2004, international bunker fuels accounted for 83 to 115 million metric tons annually.
[5]Includes pipeline fuel natural gas and compressed natural gas used as vehicle fuel.
[6]Includes electricity-only and combined heat and power plants whose primary business is to sell electricity, or electricity and heat, to the public.
[7]Includes emissions from geothermal power and nonbiogenic emissions from municipal solid waste.
Note: Totals may not equal sum of components due to independent rounding. Data for 2004 and 2005 are model results and may differ slightly from official EIA data reports.
Sources: 2004 and 2005 emissions and emission factors: Energy Information Administration (EIA), *Emissions of Greenhouse Gases in the United States 2005,* DOE/EIA-0573(2005) (Washington, DC, November 2006). **Projections:** EIA, AEO2007 National Energy Modeling System run AEO2007.D112106A.

Table A19. Macroeconomic Indicators
(Billion 2000 Chain-Weighted Dollars, Unless Otherwise Noted)

Indicators	Reference Case							Annual Growth 2005-2030 (percent)
	2004	2005	2010	2015	2020	2025	2030	
Real Gross Domestic Product	**10704**	**11049**	**12790**	**14698**	**17077**	**19666**	**22494**	**2.9%**
Components of Real Gross Domestic Product								
Real Consumption	7577	7841	9111	10423	12006	13731	15590	2.8%
Real Investment	1771	1866	2139	2478	3030	3773	4735	3.8%
Real Government Spending	1941	1958	2117	2242	2396	2541	2709	1.3%
Real Exports	1120	1196	1767	2543	3584	4894	6581	7.1%
Real Imports	1711	1815	2321	2911	3761	4963	6649	5.3%
Energy Intensity (thousand Btu per 2000 dollar of GDP)								
Delivered Energy	6.91	6.60	6.06	5.56	5.04	4.62	4.27	-1.7%
Total Energy	9.41	9.07	8.33	7.64	6.92	6.33	5.83	-1.8%
Price Indices								
GDP Chain-type Price Index (2000=1.000)	1.094	1.127	1.253	1.366	1.495	1.648	1.815	1.9%
Consumer Price Index (1982-4=1.00)								
All-urban	1.89	1.95	2.16	2.36	2.61	2.90	3.23	2.0%
Energy Commodities and Services	1.51	1.77	1.93	1.94	2.19	2.48	2.80	1.8%
Wholesale Price Index (1982=1.00)								
All Commodities	1.47	1.57	1.68	1.72	1.82	1.94	2.06	1.1%
Fuel and Power	1.27	1.57	1.64	1.62	1.84	2.11	2.39	1.7%
Interest Rates (percent, nominal)								
Federal Funds Rate	1.35	3.21	4.71	4.93	5.11	5.07	5.14	N/A
10-Year Treasury Note	4.27	4.29	5.52	5.66	5.75	5.78	5.80	N/A
AA Utility Bond Rate	6.04	5.44	7.36	7.64	7.72	7.78	7.77	N/A
Value of Shipments (billion 2000 dollars)								
Total Industrial	5651	5763	6298	7033	7779	8585	9502	2.0%
Nonmanufacturing	1494	1538	1596	1701	1846	1940	2023	1.1%
Manufacturing	4157	4225	4702	5332	5933	6645	7478	2.3%
Energy-Intensive	1161	1160	1262	1347	1426	1522	1631	1.4%
Non-energy Intensive	2996	3065	3440	3985	4507	5123	5848	2.6%
Population and Employment (millions)								
Population, with Armed Forces Overseas	294.2	296.9	310.3	323.7	337.1	350.8	364.9	0.8%
Population, aged 16 and over	229.2	231.8	244.2	254.7	265.4	276.7	288.6	0.9%
Population, over age 65	36.4	36.8	40.4	47.0	54.9	63.8	71.6	2.7%
Employment, Nonfarm	131.4	133.4	141.9	147.0	154.6	162.3	169.2	1.0%
Employment, Manufacturing	14.3	14.2	13.8	13.7	13.4	13.0	12.5	-0.5%
Key Labor Indicators								
Labor Force (millions)	147.4	149.3	157.5	162.2	167.0	172.7	180.4	0.8%
Nonfarm Labor Productivity (1992=1.00)	1.32	1.36	1.50	1.69	1.90	2.15	2.42	2.3%
Unemployment Rate (percent)	5.52	5.06	4.83	4.98	4.46	4.55	4.71	N/A
Key Indicators for Energy Demand								
Real Disposable Personal Income	8011	8105	9568	11077	13000	15172	17535	3.1%
Housing Starts (millions)	2.08	2.22	1.94	1.91	1.90	1.85	1.80	-0.8%
Commercial Floorspace (billion square feet)	73.0	74.3	80.4	86.5	92.9	100.1	108.0	1.5%
Unit Sales of Light-Duty Vehicles (millions)	16.87	16.95	17.14	18.05	19.04	20.01	21.10	0.9%

GDP = Gross domestic product.
Btu = British thermal unit.
N/A = Not applicable.
Sources: 2004 and 2005: Global Insight macroeconomic model CTL0806 and Global Insight industry model, July 2005. **Projections:** Energy Information Administration, AEO2007 National Energy Modeling System run AEO2007.D112106A.

Table A20. International Petroleum Supply and Disposition Summary
(Million Barrels per Day, Unless Otherwise Noted)

Supply and Disposition	Reference Case							Annual Growth 2005-2030 (percent)
	2004	2005	2010	2015	2020	2025	2030	
Crude Oil Prices (2005 dollars per barrel)[1]								
Imported Low Sulfur Light Crude Oil	42.87	56.76	57.47	49.87	52.04	56.37	59.12	0.2%
Imported Crude Oil	37.09	49.19	51.20	44.61	46.47	49.57	51.63	0.2%
Conventional Production (Conventional)[2]								
OPEC[3]								
Asia	1.18	1.17	1.11	1.10	1.09	1.08	1.10	-0.2%
Middle East	22.60	22.96	22.23	24.03	26.60	29.72	33.20	1.5%
North Africa	3.55	3.78	4.29	4.51	4.24	4.07	3.93	0.2%
West Africa	2.47	2.78	3.07	3.81	4.10	4.32	4.48	1.9%
South America	2.75	2.71	2.59	2.42	2.30	2.24	2.24	-0.8%
Total OPEC	**32.55**	**33.41**	**33.30**	**35.87**	**38.33**	**41.44**	**44.95**	**1.2%**
Non-OPEC								
OECD								
United States (50 states)	8.46	8.03	8.98	9.45	9.48	9.18	9.12	0.5%
Canada	2.12	2.12	1.93	2.01	1.89	1.76	1.62	-1.1%
Mexico	3.85	3.78	3.15	3.01	3.18	3.35	3.52	-0.3%
OECD Europe[4]	6.39	5.96	5.73	4.91	4.22	3.64	3.16	-2.5%
Japan	0.12	0.10	0.10	0.10	0.10	0.10	0.10	0.0%
Australia and New Zealand	0.58	0.60	0.56	0.51	0.51	0.55	0.60	-0.0%
Total OECD	**21.53**	**20.59**	**20.45**	**20.00**	**19.39**	**18.57**	**18.12**	**-0.5%**
Non-OECD								
Russia	9.27	9.51	9.98	10.30	10.79	11.23	11.54	0.8%
Other Eurasia[5]	2.21	2.48	3.98	4.91	5.41	5.99	6.55	4.0%
China	3.64	3.74	3.53	3.20	3.30	3.30	3.20	-0.6%
Other Asia[6]	2.80	2.53	2.29	2.50	2.60	2.60	2.50	-0.1%
Middle East[7]	1.68	1.67	2.00	2.20	2.40	2.70	2.90	2.2%
Africa	3.40	3.59	5.19	6.45	7.38	8.51	9.83	4.1%
Brazil	1.58	1.76	2.39	2.90	3.20	3.50	3.90	3.2%
Other Central and South America	2.35	2.31	2.32	2.54	2.66	2.75	2.90	0.9%
Total Non-OECD	**26.94**	**27.59**	**31.67**	**35.00**	**37.75**	**40.59**	**43.32**	**1.8%**
Total Conventional Production	**81.01**	**81.59**	**85.42**	**90.86**	**95.47**	**100.59**	**106.40**	**1.1%**
Unconventional Production[8]								
United States (50 states)	0.22	0.25	0.71	0.81	0.91	1.20	1.37	7.0%
Other North America	1.09	1.09	1.91	2.32	2.74	3.25	3.66	5.0%
OECD Europe[3]	0.04	0.08	0.15	0.18	0.19	0.23	0.27	5.1%
Middle East[7]	0.08	0.02	0.57	0.64	0.75	0.89	1.11	16.8%
Africa.	0.16	0.16	0.32	0.42	0.52	0.62	0.73	6.3%
Central and South America	0.83	0.93	1.35	1.59	1.81	2.18	2.40	3.9%
Other	0.02	0.28	0.62	0.81	0.90	1.05	1.41	6.7%
Total Unconventional Production	**2.44**	**2.80**	**5.63**	**6.78**	**7.83**	**9.42**	**10.93**	**5.6%**
Total Production	**83.45**	**84.39**	**91.05**	**97.64**	**103.29**	**110.01**	**117.33**	**1.3%**

Table A20. International Petroleum Supply and Disposition Summary (Continued)
(Million Barrels per Day, Unless Otherwise Noted)

Supply and Disposition	Reference Case							Annual Growth 2005-2030 (percent)
	2004	2005	2010	2015	2020	2025	2030	
Consumption[9]								
OECD								
United States (50 states)	20.76	20.75	21.59	22.85	24.02	25.33	26.93	1.0%
United States Territories	0.37	0.38	0.43	0.47	0.51	0.54	0.59	1.8%
Canada	2.32	2.28	2.42	2.54	2.49	2.56	2.59	0.5%
Mexico	2.00	2.09	2.22	2.47	2.68	2.93	3.19	1.7%
OECD Europe[3]	15.86	15.73	15.82	15.89	15.76	16.00	16.26	0.1%
Japan	5.43	5.58	5.42	5.48	5.43	5.46	5.45	-0.1%
South Korea	2.18	2.30	2.58	2.85	3.04	3.24	3.45	1.6%
Australia and New Zealand	1.04	1.05	1.08	1.10	1.13	1.17	1.22	0.6%
Total OECD	**49.95**	**50.16**	**51.54**	**53.65**	**55.05**	**57.25**	**59.69**	**0.7%**
Non-OECD								
Russia	2.81	2.75	2.85	3.05	3.11	3.28	3.39	0.8%
Other Non-OECD Eurasia[5]	2.03	2.33	2.63	2.95	3.18	3.46	3.75	1.9%
China	6.49	6.86	8.70	9.99	11.66	13.24	15.05	3.2%
India	2.48	2.52	2.94	3.32	3.66	4.07	4.45	2.3%
Other Non-OECD Asia	6.03	6.02	6.89	7.70	8.51	9.36	10.29	2.2%
Middle East[7]	5.74	5.56	6.06	6.60	7.00	7.43	7.81	1.4%
Africa	2.83	3.01	3.70	4.05	4.30	4.54	4.93	2.0%
Brazil	2.17	2.20	2.39	2.63	2.82	3.09	3.29	1.6%
Other Central and South America	2.94	2.99	3.36	3.71	4.00	4.29	4.68	1.8%
Total Non-OECD	**33.50**	**34.23**	**39.52**	**44.00**	**48.23**	**52.76**	**57.64**	**2.1%**
Total Consumption	**83.45**	**84.39**	**91.05**	**97.64**	**103.29**	**110.01**	**117.33**	**1.3%**
OPEC Production[10]	33.20	34.04	34.72	37.47	40.19	43.71	47.65	1.4%
Non-OPEC Production[10]	50.25	50.35	56.34	60.18	63.10	66.30	69.68	1.3%
Net Eurasia Exports	8.24	8.67	10.87	12.10	13.12	13.98	14.85	2.2%
OPEC Market Share	39.8	40.3	38.1	38.4	38.9	39.7	40.6	0.0%

[1]Weighted average price delivered to U.S. refiners.
[2]Includes production of crude oil (including lease condensates), natural gas plant liquids, other hydrogen and hydrocarbons for refinery feedstocks, alcohol and other sources, and refinery gains.
[3]OPEC = Organization of Petroleum Exporting Countries - Algeria, Indonesia, Iran, Iraq, Kuwait, Libya, Nigeria, Qatar, Saudi Arabia, the United Arab Emirates, and Venezuela. Does not include Angola, which was admitted as a full member to OPEC on December 14, 2006.
[4]OECD Europe = Organization for Economic Cooperation and Development - Austria, Belgium, Czech Republic, Denmark, Finland, France, Germany, Greece, Hungary, Iceland, Ireland, Italy, Luxembourg, the Netherlands, Norway, Poland, Portugal, Slovakia, Spain, Sweden, Switzerland, Turkey, and the United Kingdom.
[5]Eurasia consists of Armenia, Azerbaijan, Belarus, Estonia, Georgia, Kazakhstan, Kyrgyzstan, Latvia, Lithuania, Moldova, Russia, Tajikistan, Turkmenistan, Ukraine, and Uzbekistan.
[6]Other Asia = Afghanistan, Bangladesh, Bhutan, Brunei, Cambodia (Kampuchea), Fiji, French Polynesia, Guam, Hong Kong, Indonesia, Kiribati, Laos, Malaysia, Macau, Maldives, Mongolia, Myanmar (Burma), Nauru, Nepal, New Caledonia, Niue, North Korea, Pakistan, Papua New Guinea, Philippines, Samoa, Singapore, Solomon Islands, Sri Lanka, Taiwan, Thailand, Tonga, Vanuatu, and Vietnam.
[7]Non-OPEC Middle East includes Turkey.
[8]Includes liquids produced from energy crops, natural gas, coal, oil sands, and shale. Includes both OPEC and non-OPEC producers in the regional breakdown.
[9]Includes both OPEC and non-OPEC consumers in the regional breakdown.
[10]Includes both conventional and nonconventional liquids production.
Note: Totals may not equal sum of components due to independent rounding. Data for 2004 and 2005 are model results and may differ slightly from official EIA data reports.
Sources: 2004 and 2005 low sulfur light crude oil price: Energy Information Administration (EIA), Form EIA-856, "Monthly Foreign Crude Oil Acquisition Report." 2004 and 2005 imported crude oil price: EIA, *Annual Energy Review 2005*, DOE/EIA-0384(2005) (Washington, DC, July 2006). 2004 quantities derived from: EIA, *International Energy Annual 2004*, DOE/EIA-0219(2004) (Washington, DC, May-July 2006). **2005 quantities and projections:** EIA, AEO2007 National Energy Modeling System run AEO2007.D112106A.

Economic Growth Case Comparisons

Table B1. Total Energy Supply and Disposition Summary
(Quadrillion Btu per Year, Unless Otherwise Noted)

Supply, Disposition, and Prices	2005	Projections 2010 Low Economic Growth	2010 Reference	2010 High Economic Growth	2020 Low Economic Growth	2020 Reference	2020 High Economic Growth	2030 Low Economic Growth	2030 Reference	2030 High Economic Growth
Production										
Crude Oil and Lease Condensate	10.96	11.98	11.99	12.01	12.25	12.48	12.65	11.25	11.40	11.53
Natural Gas Plant Liquids	2.33	2.39	2.43	2.46	2.30	2.38	2.42	2.22	2.31	2.39
Dry Natural Gas	18.77	19.61	19.93	20.19	20.81	21.41	21.80	20.36	21.15	21.88
Coal[1]	23.20	24.36	24.47	24.58	25.19	26.61	28.22	29.64	33.52	36.90
Nuclear Power	8.13	8.23	8.23	8.23	8.91	9.23	9.28	8.80	9.33	10.53
Hydropower	2.71	3.02	3.02	3.02	3.09	3.08	3.08	3.09	3.09	3.10
Biomass[2]	2.71	4.15	4.22	4.32	4.47	4.69	5.04	4.66	5.26	6.06
Other Renewable Energy[3]	0.76	1.20	1.18	1.18	1.28	1.33	1.32	1.38	1.44	1.51
Other[4]	0.22	0.66	0.67	0.73	0.90	0.89	1.09	0.99	1.12	1.21
Total	**69.80**	**75.60**	**76.13**	**76.71**	**79.20**	**82.09**	**84.91**	**82.37**	**88.63**	**95.10**
Imports										
Crude Oil[5]	22.09	21.39	21.88	22.55	23.16	24.72	26.54	25.19	28.63	33.17
Liquid Fuels and Other Petroleum[6]	7.16	5.64	6.02	6.27	6.30	7.05	7.80	7.27	9.02	10.13
Natural Gas	4.42	5.12	5.36	5.59	5.63	6.17	6.80	5.41	6.47	8.03
Other Imports[7]	0.85	0.92	0.92	0.94	1.66	1.73	1.81	2.03	2.26	2.16
Total	**34.52**	**33.07**	**34.18**	**35.34**	**36.75**	**39.66**	**42.96**	**39.89**	**46.37**	**53.49**
Exports										
Petroleum[8]	2.31	2.69	2.71	2.75	2.78	2.84	2.92	2.80	2.90	3.03
Natural Gas	0.75	0.70	0.69	0.68	0.72	0.69	0.66	0.95	0.87	0.78
Coal	1.27	1.12	1.12	1.12	0.88	0.80	0.74	0.66	0.69	0.67
Total	**4.33**	**4.50**	**4.52**	**4.55**	**4.38**	**4.33**	**4.31**	**4.41**	**4.47**	**4.49**
Discrepancy[9]	**-0.20**	**-0.66**	**-0.70**	**-0.71**	**-0.60**	**-0.74**	**-0.60**	**-0.64**	**-0.63**	**-0.30**
Consumption										
Liquid Fuels and Other Petroleum[10]	40.61	40.80	41.76	42.78	43.72	46.52	49.44	46.52	52.17	57.99
Natural Gas	22.63	24.17	24.73	25.23	25.86	27.04	28.10	24.94	26.89	29.28
Coal	22.87	24.13	24.24	24.35	25.80	27.29	28.97	30.16	34.14	37.29
Nuclear Power	8.13	8.23	8.23	8.23	8.91	9.23	9.28	8.80	9.33	10.53
Hydropower	2.71	3.02	3.02	3.02	3.09	3.08	3.08	3.09	3.09	3.10
Biomass[11]	2.38	3.24	3.30	3.38	3.47	3.64	3.91	3.56	4.06	4.66
Other Renewable Energy[3]	0.76	1.20	1.18	1.18	1.28	1.33	1.32	1.38	1.44	1.51
Other[12]	0.08	0.04	0.04	0.05	0.04	0.04	0.04	0.04	0.04	0.05
Total	**100.19**	**104.83**	**106.50**	**108.21**	**112.16**	**118.16**	**124.14**	**118.50**	**131.16**	**144.40**

Table B1. Total Energy Supply and Disposition Summary (Continued)
(Quadrillion Btu per Year, Unless Otherwise Noted)

Supply, Disposition, and Prices	2005	2010 Low Economic Growth	2010 Reference	2010 High Economic Growth	2020 Low Economic Growth	2020 Reference	2020 High Economic Growth	2030 Low Economic Growth	2030 Reference	2030 High Economic Growth
Prices (2005 dollars per unit)										
Petroleum (dollars per barrel)										
Imported Low Sulfur Light Crude Oil Price[13]	56.76	57.47	57.47	57.13	52.51	52.04	52.04	59.12	59.12	59.12
Imported Crude Oil Price[13]	49.19	51.20	51.20	51.20	46.47	46.47	46.47	51.63	51.63	51.63
Natural Gas (dollars per million Btu)										
Price at Henry Hub	8.60	6.05	6.28	6.50	5.41	5.71	5.73	6.16	6.52	6.87
Wellhead Price[14]	7.29	5.38	5.59	5.79	4.80	5.07	5.09	5.48	5.80	6.13
Natural Gas (dollars per thousand cubic feet)										
Wellhead Price[14]	7.51	5.54	5.76	5.96	4.95	5.22	5.24	5.64	5.98	6.31
Coal (dollars per ton)										
Minemouth Price[15]	23.34	23.98	24.20	24.38	20.95	21.58	22.17	20.99	22.60	23.64
Coal (dollars per million Btu)										
Minemouth Price[15]	1.15	1.17	1.18	1.19	1.05	1.08	1.11	1.07	1.15	1.20
Average Delivered Price[16]	1.61	1.76	1.77	1.78	1.59	1.62	1.66	1.62	1.71	1.77
Average Electricity Price (cents per kilowatthour)	8.1	7.9	8.1	8.2	7.6	7.9	8.1	7.8	8.1	8.4

[1]Includes waste coal.
[2]Includes grid-connected electricity from wood and waste; biomass, such as corn, used for liquid fuels production; and non-electric energy demand from wood. Refer to Table A17 for details.
[3]Includes grid-connected electricity from landfill gas; municipal solid waste; wind; photovoltaic and solar thermal sources; and non-electric energy from renewable sources, such as active and passive solar systems. Excludes electricity imports using renewable sources and nonmarketed renewable energy. See Table A17 for selected nonmarketed residential and commercial renewable energy.
[4]Includes liquid hydrogen, methanol, and some domestic inputs to refineries.
[5]Includes imports of crude oil for the Strategic Petroleum Reserve.
[6]Includes imports of finished petroleum products, unfinished oils, alcohols, ethers, blending components, and renewable fuels such as ethanol.
[7]Includes coal, coal coke (net), and electricity (net).
[8]Includes crude oil and petroleum products.
[9]Balancing item. Includes unaccounted for supply, losses, gains, and net storage withdrawals.
[10]Includes petroleum-derived fuels and non-petroleum derived fuels, such as ethanol and biodiesel. Petroleum coke, which is a solid, is included. Also included are natural gas plant liquids, crude oil consumed as a fuel, and liquid hydrogen. Refer to Table A17 for detailed renewable liquid fuels consumption.
[11]Includes grid-connected electricity from wood and wood waste, non-electric energy from wood, and biofuels heat and coproducts used in the production of liquid fuels, but excludes the energy content of the liquid fuels.
[12]Includes net electricity imports.
[13]Weighted average price delivered to U.S. refiners.
[14]Represents lower 48 onshore and offshore supplies.
[15]Includes reported prices for both open market and captive mines.
[16]Prices weighted by consumption; weighted average excludes residential and commercial prices, and export free-alongside-ship (f.a.s.) prices.
Btu = British thermal unit.
N/A = Not applicable.
Note: Totals may not equal sum of components due to independent rounding. Data for 2005 are model results and may differ slightly from official EIA data reports.
Sources: 2005 natural gas supply values and natural gas wellhead price: EIA, *Natural Gas Monthly*, DOE/EIA-0130(2006/04) (Washington, DC, April 2006). 2005 coal minemouth and delivered coal prices: EIA, *Annual Coal Report 2005*, DOE/EIA-0584(2005) (Washington, DC, October 2006). 2005 petroleum supply values: EIA, *Petroleum Supply Annual 2005*, DOE/EIA-0340(2005)/1 (Washington, DC, October 2006). 2005 low sulfur light crude oil price: EIA, Form EIA-856, "Monthly Foreign Crude Oil Acquisition Report." Other 2005 coal values: *Quarterly Coal Report, October-December 2005*, DOE/EIA-0121(2005/4Q) (Washington, DC, March 2006). Other 2005 values: EIA, *Annual Energy Review 2005*, DOE/EIA-0384(2005) (Washington, DC, July 2006). **Projections:** EIA, AEO2007 National Energy Modeling System runs LM2007.D112106A, AEO2007.D112106A, and HM2007.D112106A.

Table B2. Energy Consumption by Sector and Source
(Quadrillion Btu per Year, Unless Otherwise Noted)

Sector and Source	2005	Projections								
		2010			2020			2030		
		Low Economic Growth	Reference	High Economic Growth	Low Economic Growth	Reference	High Economic Growth	Low Economic Growth	Reference	High Economic Growth
Energy Consumption										
Residential										
Liquefied Petroleum Gases	0.51	0.53	0.53	0.54	0.56	0.58	0.60	0.58	0.62	0.66
Kerosene	0.10	0.10	0.10	0.10	0.09	0.10	0.10	0.09	0.09	0.09
Distillate Fuel Oil	0.93	0.90	0.90	0.90	0.85	0.85	0.85	0.75	0.76	0.76
Liquid Fuels and Other Petroleum Subtotal	1.54	1.52	1.53	1.53	1.50	1.53	1.55	1.41	1.46	1.51
Natural Gas	4.98	5.16	5.18	5.20	5.30	5.43	5.58	5.19	5.47	5.74
Coal	0.01	0.01	0.01	0.01	0.01	0.01	0.01	0.01	0.01	0.01
Renewable Energy[1]	0.41	0.42	0.43	0.43	0.40	0.40	0.41	0.38	0.39	0.41
Electricity	4.66	5.02	5.06	5.10	5.62	5.80	6.00	6.05	6.47	6.88
Delivered Energy	**11.60**	**12.14**	**12.21**	**12.27**	**12.82**	**13.17**	**13.55**	**13.04**	**13.80**	**14.55**
Electricity Related Losses	10.15	10.87	10.90	10.93	11.79	12.08	12.32	12.32	12.89	13.52
Total	**21.75**	**23.01**	**23.11**	**23.20**	**24.61**	**25.26**	**25.87**	**25.36**	**26.70**	**28.07**
Commercial										
Liquefied Petroleum Gases	0.09	0.09	0.09	0.09	0.09	0.10	0.10	0.10	0.10	0.10
Motor Gasoline[2]	0.05	0.05	0.05	0.05	0.05	0.05	0.05	0.05	0.05	0.06
Kerosene	0.02	0.03	0.03	0.03	0.03	0.03	0.03	0.03	0.03	0.03
Distillate Fuel Oil	0.48	0.45	0.45	0.45	0.47	0.48	0.49	0.47	0.49	0.51
Residual Fuel Oil	0.14	0.14	0.14	0.14	0.14	0.14	0.14	0.14	0.14	0.14
Liquid Fuels and Other Petroleum Subtotal	0.77	0.75	0.75	0.75	0.78	0.80	0.81	0.78	0.81	0.85
Natural Gas	3.15	3.31	3.31	3.31	3.73	3.86	4.02	4.01	4.36	4.71
Coal	0.10	0.10	0.10	0.10	0.10	0.10	0.10	0.10	0.10	0.10
Renewable Energy[3]	0.12	0.12	0.12	0.12	0.12	0.12	0.12	0.12	0.12	0.12
Electricity	4.32	4.76	4.77	4.78	5.59	5.78	5.98	6.47	7.03	7.58
Delivered Energy	**8.46**	**9.04**	**9.05**	**9.07**	**10.32**	**10.66**	**11.03**	**11.48**	**12.43**	**13.36**
Electricity Related Losses	9.42	10.30	10.27	10.26	11.74	12.03	12.30	13.18	14.01	14.90
Total	**17.88**	**19.34**	**19.33**	**19.32**	**22.06**	**22.69**	**23.32**	**24.66**	**26.44**	**28.26**
Industrial[4]										
Liquefied Petroleum Gases	2.13	2.15	2.26	2.37	1.97	2.26	2.56	1.88	2.40	3.00
Motor Gasoline[2]	0.32	0.30	0.32	0.34	0.30	0.33	0.37	0.31	0.36	0.41
Distillate Fuel Oil	1.23	1.11	1.18	1.24	1.11	1.22	1.33	1.09	1.26	1.44
Residual Fuel Oil	0.23	0.17	0.18	0.18	0.17	0.17	0.18	0.17	0.18	0.20
Petrochemical Feedstocks	1.38	1.40	1.48	1.57	1.30	1.50	1.72	1.19	1.57	1.99
Other Petroleum[5]	4.45	3.89	4.05	4.26	4.00	4.34	4.75	4.17	4.78	5.30
Liquid Fuels and Other Petroleum Subtotal	9.73	9.02	9.47	9.95	8.84	9.82	10.91	8.81	10.55	12.33
Natural Gas	6.84	7.65	7.86	8.01	7.58	8.26	8.76	7.58	8.90	10.42
Natural-Gas-to-Liquids Heat and Power	0.00	0.00	0.00	0.00	0.00	0.00	0.00	0.00	0.00	0.00
Lease and Plant Fuel[6]	1.10	1.08	1.10	1.11	1.17	1.21	1.22	1.11	1.15	1.18
Natural Gas Subtotal	7.94	8.74	8.95	9.12	8.75	9.46	9.98	8.69	10.05	11.60
Metallurgical Coal	0.62	0.59	0.60	0.62	0.51	0.57	0.62	0.44	0.57	0.69
Other Industrial Coal	1.35	1.35	1.37	1.39	1.29	1.34	1.39	1.27	1.36	1.45
Coal-to-Liquids Heat and Power	0.00	0.00	0.00	0.00	0.13	0.21	0.28	0.82	0.93	1.07
Net Coal Coke Imports	0.04	0.02	0.02	0.02	0.01	0.02	0.02	0.00	0.02	0.03
Coal Subtotal	2.01	1.96	2.00	2.04	1.95	2.14	2.32	2.54	2.89	3.25
Biofuels Heat and Coproducts	0.24	0.68	0.69	0.71	0.74	0.78	0.83	0.80	0.88	0.98
Renewable Energy[7]	1.44	1.54	1.60	1.65	1.67	1.81	1.97	1.74	2.05	2.38
Electricity	3.48	3.51	3.63	3.75	3.53	3.83	4.16	3.42	4.09	4.79
Delivered Energy	**24.85**	**25.46**	**26.33**	**27.21**	**25.48**	**27.84**	**30.17**	**25.99**	**30.51**	**35.34**
Electricity Related Losses	7.60	7.60	7.81	8.03	7.40	7.98	8.55	6.97	8.15	9.41
Total	**32.45**	**33.06**	**34.14**	**35.24**	**32.88**	**35.82**	**38.72**	**32.96**	**38.66**	**44.74**

Table B2. Energy Consumption by Sector and Source (Continued)
(Quadrillion Btu per Year, Unless Otherwise Noted)

Sector and Source	2005	Projections								
		2010			2020			2030		
		Low Economic Growth	Reference	High Economic Growth	Low Economic Growth	Reference	High Economic Growth	Low Economic Growth	Reference	High Economic Growth
Transportation										
Liquefied Petroleum Gases	0.04	0.05	0.05	0.05	0.06	0.06	0.07	0.07	0.08	0.09
E85[8]	0.00	0.00	0.00	0.00	0.01	0.01	0.01	0.02	0.02	0.04
Motor Gasoline[2]	17.00	17.17	17.37	17.60	19.03	19.95	20.86	21.02	22.89	24.72
Jet Fuel[9]	3.37	3.98	4.04	4.09	4.41	4.54	4.70	4.32	4.70	5.20
Distillate Fuel Oil[10]	6.02	6.42	6.64	6.88	7.14	7.81	8.53	8.07	9.58	11.11
Residual Fuel Oil	0.81	0.82	0.82	0.83	0.84	0.85	0.85	0.85	0.87	0.88
Liquid Hydrogen	0.00	0.00	0.00	0.00	0.00	0.00	0.00	0.00	0.00	0.00
Other Petroleum[11]	0.18	0.18	0.18	0.18	0.19	0.19	0.19	0.19	0.19	0.20
Liquid Fuels and Other Petroleum Subtotal	27.42	28.62	29.11	29.63	31.66	33.41	35.21	34.55	38.34	42.24
Pipeline Fuel Natural Gas	0.58	0.65	0.66	0.67	0.77	0.79	0.81	0.75	0.79	0.83
Compressed Natural Gas	0.03	0.06	0.06	0.06	0.09	0.09	0.10	0.10	0.12	0.14
Electricity	0.02	0.03	0.03	0.03	0.03	0.03	0.03	0.04	0.04	0.04
Delivered Energy	**28.05**	**29.35**	**29.86**	**30.39**	**32.55**	**34.33**	**36.16**	**35.44**	**39.29**	**43.25**
Electricity Related Losses	0.05	0.06	0.06	0.06	0.07	0.07	0.07	0.08	0.08	0.08
Total	**28.11**	**29.42**	**29.92**	**30.45**	**32.62**	**34.40**	**36.23**	**35.52**	**39.37**	**43.33**
Delivered Energy Consumption for All Sectors										
Liquefied Petroleum Gases	2.77	2.82	2.93	3.05	2.68	2.99	3.32	2.62	3.19	3.84
E85[8]	0.00	0.00	0.00	0.00	0.01	0.01	0.01	0.02	0.02	0.04
Motor Gasoline[2]	17.37	17.51	17.74	17.98	19.38	20.34	21.28	21.38	23.30	25.18
Jet Fuel[9]	3.37	3.98	4.04	4.09	4.41	4.54	4.70	4.32	4.70	5.20
Kerosene	0.14	0.14	0.14	0.14	0.14	0.14	0.14	0.14	0.14	0.14
Distillate Fuel Oil	8.65	8.88	9.17	9.46	9.56	10.36	11.20	10.38	12.09	13.82
Residual Fuel Oil	1.17	1.13	1.13	1.14	1.14	1.16	1.18	1.16	1.19	1.22
Petrochemical Feedstocks	1.38	1.40	1.48	1.57	1.30	1.50	1.72	1.19	1.57	1.99
Liquid Hydrogen	0.00	0.00	0.00	0.00	0.00	0.00	0.00	0.00	0.00	0.00
Other Petroleum[12]	4.61	4.05	4.22	4.43	4.17	4.51	4.92	4.34	4.96	5.48
Liquid Fuels and Other Petroleum Subtotal	39.46	39.92	40.86	41.86	42.78	45.55	48.47	45.55	51.17	56.93
Natural Gas	15.01	16.19	16.41	16.59	16.69	17.65	18.46	16.89	18.86	21.02
Natural-Gas-to-Liquids Heat and Power	0.00	0.00	0.00	0.00	0.00	0.00	0.00	0.00	0.00	0.00
Lease and Plant Fuel[6]	1.10	1.08	1.10	1.11	1.17	1.21	1.22	1.11	1.15	1.18
Pipeline Natural Gas	0.58	0.65	0.66	0.67	0.77	0.79	0.81	0.75	0.79	0.83
Natural Gas Subtotal	16.68	17.92	18.17	18.37	18.63	19.64	20.49	18.75	20.80	23.03
Metallurgical Coal	0.62	0.59	0.60	0.62	0.51	0.57	0.62	0.44	0.57	0.69
Other Coal	1.46	1.46	1.48	1.50	1.40	1.45	1.50	1.37	1.47	1.56
Coal-to-Liquids Heat and Power	0.00	0.00	0.00	0.00	0.13	0.21	0.28	0.82	0.93	1.07
Net Coal Coke Imports	0.04	0.02	0.02	0.02	0.01	0.02	0.02	0.00	0.02	0.03
Coal Subtotal	2.12	2.07	2.11	2.14	2.06	2.24	2.43	2.64	2.99	3.36
Biofuels Heat and Coproducts	0.24	0.68	0.69	0.71	0.74	0.78	0.83	0.80	0.88	0.98
Renewable Energy[13]	1.97	2.09	2.14	2.20	2.19	2.34	2.50	2.23	2.56	2.91
Electricity	12.49	13.33	13.49	13.66	14.76	15.45	16.18	15.98	17.63	19.29
Delivered Energy	**72.97**	**76.00**	**77.46**	**78.93**	**81.17**	**86.00**	**90.90**	**85.96**	**96.03**	**106.50**
Electricity Related Losses	27.23	28.83	29.04	29.28	31.00	32.17	33.24	32.54	35.13	37.90
Total	**100.19**	**104.83**	**106.50**	**108.21**	**112.16**	**118.16**	**124.14**	**118.50**	**131.16**	**144.40**
Electric Power[14]										
Distillate Fuel Oil	0.19	0.23	0.24	0.23	0.24	0.25	0.27	0.26	0.28	0.31
Residual Fuel Oil	0.96	0.66	0.67	0.68	0.70	0.72	0.71	0.71	0.72	0.74
Liquid Fuels and Other Petroleum Subtotal	1.16	0.89	0.90	0.91	0.94	0.97	0.97	0.97	1.01	1.06
Natural Gas	5.95	6.25	6.56	6.86	7.23	7.40	7.60	6.19	6.09	6.25
Steam Coal	20.75	22.06	22.13	22.21	23.74	25.05	26.54	27.52	31.14	33.93
Nuclear Power	8.13	8.23	8.23	8.23	8.91	9.23	9.28	8.80	9.33	10.53
Renewable Energy[15]	3.64	4.69	4.67	4.67	4.91	4.93	4.99	4.99	5.15	5.37
Electricity Imports	0.08	0.04	0.04	0.05	0.04	0.04	0.04	0.04	0.04	0.05
Total	**39.71**	**42.15**	**42.53**	**42.93**	**45.76**	**47.62**	**49.42**	**48.52**	**52.77**	**57.19**

Table B2. Energy Consumption by Sector and Source (Continued)
(Quadrillion Btu per Year, Unless Otherwise Noted)

Sector and Source	2005	Projections								
		2010			2020			2030		
		Low Economic Growth	Reference	High Economic Growth	Low Economic Growth	Reference	High Economic Growth	Low Economic Growth	Reference	High Economic Growth
Total Energy Consumption										
Liquefied Petroleum Gases	2.77	2.82	2.93	3.05	2.68	2.99	3.32	2.62	3.19	3.84
E85[8]	0.00	0.00	0.00	0.00	0.01	0.01	0.01	0.02	0.02	0.04
Motor Gasoline[2]	17.37	17.51	17.74	17.98	19.38	20.34	21.28	21.38	23.30	25.18
Jet Fuel[9]	3.37	3.98	4.04	4.09	4.41	4.54	4.70	4.32	4.70	5.20
Kerosene	0.14	0.14	0.14	0.14	0.14	0.14	0.14	0.14	0.14	0.14
Distillate Fuel Oil	8.84	9.11	9.40	9.70	9.79	10.61	11.47	10.64	12.37	14.13
Residual Fuel Oil	2.14	1.78	1.80	1.82	1.85	1.88	1.88	1.87	1.91	1.97
Petrochemical Feedstocks	1.38	1.40	1.48	1.57	1.30	1.50	1.72	1.19	1.57	1.99
Liquid Hydrogen	0.00	0.00	0.00	0.00	0.00	0.00	0.00	0.00	0.00	0.00
Other Petroleum[12]	4.61	4.05	4.22	4.43	4.17	4.51	4.92	4.34	4.96	5.48
Liquid Fuels and Other Petroleum Subtotal	40.61	40.80	41.76	42.78	43.72	46.52	49.44	46.52	52.17	57.99
Natural Gas	20.96	22.44	22.97	23.45	23.92	25.05	26.06	23.08	24.95	27.27
Natural-Gas-to-Liquids Heat and Power	0.00	0.00	0.00	0.00	0.00	0.00	0.00	0.00	0.00	0.00
Lease and Plant Fuel[6]	1.10	1.08	1.10	1.11	1.17	1.21	1.22	1.11	1.15	1.18
Pipeline Natural Gas	0.58	0.65	0.66	0.67	0.77	0.79	0.81	0.75	0.79	0.83
Natural Gas Subtotal	22.63	24.17	24.73	25.23	25.86	27.04	28.10	24.94	26.89	29.28
Metallurgical Coal	0.62	0.59	0.60	0.62	0.51	0.57	0.62	0.44	0.57	0.69
Other Coal	22.21	23.52	23.61	23.71	25.14	26.50	28.05	28.89	32.61	35.49
Coal-to-Liquids Heat and Power	0.00	0.00	0.00	0.00	0.13	0.21	0.28	0.82	0.93	1.07
Net Coal Coke Imports	0.04	0.02	0.02	0.02	0.01	0.02	0.02	0.00	0.02	0.03
Coal Subtotal	22.87	24.13	24.24	24.35	25.80	27.29	28.97	30.16	34.14	37.29
Nuclear Power	8.13	8.23	8.23	8.23	8.91	9.23	9.28	8.80	9.33	10.53
Biofuels Heat and Coproducts	0.24	0.68	0.69	0.71	0.74	0.78	0.83	0.80	0.88	0.98
Renewable Energy[16]	5.61	6.78	6.81	6.87	7.10	7.27	7.49	7.22	7.71	8.28
Electricity Imports	0.08	0.04	0.04	0.05	0.04	0.04	0.04	0.04	0.04	0.05
Total	**100.19**	**104.83**	**106.50**	**108.21**	**112.16**	**118.16**	**124.14**	**118.50**	**131.16**	**144.40**
Energy Use and Related Statistics										
Delivered Energy Use	72.97	76.00	77.46	78.93	81.17	86.00	90.90	85.96	96.03	106.50
Total Energy Use	100.19	104.83	106.50	108.21	112.16	118.16	124.14	118.50	131.16	144.40
Ethanol Consumed in Motor Gasoline and E85	0.33	0.90	0.91	0.94	1.00	1.06	1.13	1.12	1.22	1.37
Population (millions)	296.94	308.47	310.26	312.77	323.58	337.13	351.39	334.24	364.94	395.64
Gross Domestic Product (billion 2000 dollars)	11049	12359	12790	13219	15686	17077	18490	19249	22494	25757
Carbon Dioxide Emissions (million metric tons)	5945.3	6124.7	6214.0	6304.2	6582.8	6944.5	7322.2	7141.4	7950.2	8711.2

[1]Includes wood used for residential heating. See Table A4 and/or Table A17 for estimates of nonmarketed renewable energy consumption for geothermal heat pumps, solar thermal hot water heating, and solar photovoltaic electricity generation.
[2]Includes ethanol (blends of 10 percent or less) and ethers blended into gasoline.
[3]Includes commercial sector consumption of wood and wood waste, landfill gas, municipal solid waste, and other biomass for combined heat and power. See Table A5 and/or Table A17 for estimates of nonmarketed renewable energy consumption for solar thermal hot water heating and solar photovoltaic electricity generation.
[4]Includes energy for combined heat and power plants, except those whose primary business is to sell electricity, or electricity and heat, to the public.
[5]Includes petroleum coke, asphalt, road oil, lubricants, still gas, tire-derived fuel, and miscellaneous petroleum products.
[6]Represents natural gas used in well, field, and lease operations, and in natural gas processing plant machinery.
[7]Includes consumption of energy produced from hydroelectric, wood and wood waste, municipal solid waste, and other biomass sources.
[8]E85 refers to a blend of 85 percent ethanol (renewable) and 15 percent motor gasoline (nonrenewable). To address cold starting issues, the percentage of ethanol actually varies seasonally. The annual average ethanol content of 74 percent is used for this forecast.
[9]Includes only kerosene type.
[10]Diesel fuel for on- and off- road use.
[11]Includes aviation gasoline and lubricants.
[12]Includes unfinished oils, natural gasoline, motor gasoline blending components, aviation gasoline, lubricants, still gas, asphalt, road oil, petroleum coke, tire-derived fuel, and miscellaneous petroleum products.
[13]Includes electricity generated for sale to the grid and for own use from renewable sources, and non-electric energy from renewable sources. Excludes nonmarketed renewable energy consumption for geothermal heat pumps, buildings photovoltaic systems, and solar thermal hot water heaters.
[14]Includes consumption of energy by electricity-only and combined heat and power plants whose primary business is to sell electricity, or electricity and heat, to the public. Includes small power producers and exempt wholesale generators.
[15]Includes conventional hydroelectric, geothermal, wood and wood waste, municipal solid waste, other biomass, petroleum coke, wind, photovoltaic and solar thermal sources. Excludes net electricity imports.
[16]Includes hydroelectric, geothermal, wood and wood waste, municipal solid waste, other biomass, wind, photovoltaic and solar thermal sources. Includes ethanol components of E85; excludes ethanol blends (10 percent or less) in motor gasoline. Excludes net electricity imports and nonmarketed renewable energy consumption for geothermal heat pumps, buildings photovoltaic systems, and solar thermal hot water heaters.
Btu = British thermal unit.
N/A = Not applicable.
Note: Totals may not equal sum of components due to independent rounding. Data for 2005 are model results and may differ slightly from official EIA data reports. Consumption values of 0.00 are values that round to 0.00, because they are less than 0.005.
Sources: 2005 consumption based on: Energy Information Administration (EIA), *Annual Energy Review 2005*, DOE/EIA-0384(2005) (Washington, DC, July 2006). 2005 population and gross domestic product: Global Insight macroeconomic model CTL0806. 2005 carbon dioxide emissions: EIA, *Emissions of Greenhouse Gases in the United States 2005*, DOE/EIA-0573(2005) (Washington, DC, November 2006). **Projections:** EIA, AEO2007 National Energy Modeling System runs LM2007.D112106A, AEO2007.D112106A, and HM2007.D112106A.

Table B3. Energy Prices by Sector and Source
(2005 Dollars per Million Btu, Unless Otherwise Noted)

Sector and Source	2005	Projections								
		2010			2020			2030		
		Low Economic Growth	Reference	High Economic Growth	Low Economic Growth	Reference	High Economic Growth	Low Economic Growth	Reference	High Economic Growth
Residential										
Liquefied Petroleum Gases	19.29	23.46	23.67	23.88	22.88	23.18	23.24	23.54	23.91	24.26
Distillate Fuel Oil	14.73	14.82	14.87	14.98	12.95	13.15	13.54	13.59	14.13	14.60
Natural Gas	12.43	10.77	10.98	11.18	10.23	10.54	10.62	11.00	11.43	11.83
Electricity	27.59	26.50	26.91	27.31	25.50	26.37	27.03	25.70	26.76	28.05
Commercial										
Distillate Fuel Oil	12.68	12.65	12.72	12.82	11.01	11.35	11.68	11.87	12.45	13.01
Residual Fuel Oil	8.41	7.53	7.54	7.55	7.07	7.07	7.09	7.32	7.31	7.35
Natural Gas	11.20	9.13	9.34	9.53	8.40	8.67	8.70	8.98	9.30	9.62
Electricity	25.25	24.03	24.50	24.96	23.00	23.95	24.61	23.24	24.27	25.59
Industrial[1]										
Liquefied Petroleum Gases	16.96	16.22	16.42	16.63	15.66	15.91	15.90	16.36	16.55	16.80
Distillate Fuel Oil	13.08	12.87	12.95	13.04	11.64	12.04	12.37	12.64	13.25	13.88
Residual Fuel Oil	7.77	9.44	9.50	9.60	8.90	8.91	9.05	9.61	9.58	10.28
Natural Gas[2]	8.16	6.22	6.43	6.62	5.64	5.90	5.95	6.24	6.56	6.86
Metallurgical Coal	3.06	3.08	3.09	3.11	2.72	2.71	2.71	2.69	2.75	2.83
Other Industrial Coal	2.15	2.25	2.26	2.27	2.14	2.18	2.23	2.19	2.29	2.36
Coal to Liquids	0.00	0.00	0.00	0.00	0.90	0.97	1.02	1.20	1.33	1.40
Electricity	16.69	17.59	18.01	18.40	16.28	17.07	17.57	16.55	17.43	18.53
Transportation										
Liquefied Petroleum Gases[3]	23.92	24.13	24.34	24.55	23.37	23.66	23.72	23.90	24.29	24.65
E85[4]	23.10	21.03	21.29	21.35	20.38	20.61	20.74	21.26	21.50	21.56
Motor Gasoline[5]	18.64	17.79	17.90	18.01	16.30	16.63	16.99	17.16	17.76	18.20
Jet Fuel[6]	13.14	10.82	10.91	11.01	10.15	10.51	10.93	11.14	11.75	12.80
Distillate Fuel Oil[7]	17.52	16.72	16.81	16.90	14.87	15.42	15.88	15.70	16.47	17.38
Residual Fuel Oil	5.51	7.91	8.05	8.14	7.37	7.36	7.42	8.26	8.27	9.36
Natural Gas[8]	14.76	13.75	13.97	14.18	12.59	12.98	13.22	12.93	13.45	13.98
Electricity	25.22	24.38	24.86	25.35	23.50	24.22	24.64	23.80	24.46	25.53
Electric Power[9]										
Distillate Fuel Oil	11.38	11.63	11.71	11.83	9.68	9.84	10.17	10.33	10.79	11.39
Residual Fuel Oil	6.96	6.54	6.58	6.63	6.04	6.08	6.20	6.74	6.85	7.43
Natural Gas	8.18	5.98	6.22	6.45	5.47	5.76	5.79	6.02	6.33	6.63
Steam Coal	1.53	1.70	1.71	1.72	1.55	1.58	1.62	1.60	1.69	1.74
Average Price to All Users[10]										
Liquefied Petroleum Gases	17.48	17.87	18.02	18.17	17.49	17.62	17.52	18.30	18.30	18.38
E85[4]	23.10	21.03	21.29	21.35	20.38	20.61	20.74	21.26	21.50	21.56
Motor Gasoline[5]	18.60	17.79	17.90	18.01	16.30	16.63	16.99	17.16	17.75	18.20
Jet Fuel	13.14	10.82	10.91	11.01	10.15	10.51	10.93	11.14	11.75	12.80
Distillate Fuel Oil	16.22	15.62	15.70	15.81	14.03	14.53	14.98	14.95	15.70	16.54
Residual Fuel Oil	6.59	7.53	7.61	7.68	6.98	7.00	7.09	7.73	7.79	8.57
Natural Gas	9.65	7.65	7.83	8.01	7.06	7.32	7.36	7.76	8.09	8.37
Metallurgical Coal	3.06	3.08	3.09	3.11	2.72	2.71	2.71	2.69	2.75	2.83
Other Coal	1.57	1.73	1.74	1.75	1.58	1.61	1.65	1.63	1.72	1.77
Coal to Liquids	0.00	0.00	0.00	0.00	0.90	0.97	1.02	1.20	1.33	1.40
Electricity	23.73	23.27	23.66	24.04	22.35	23.15	23.70	22.74	23.60	24.71

Table B3. Energy Prices by Sector and Source (Continued)
(2005 Dollars per Million Btu, Unless Otherwise Noted)

Sector and Source	2005	Projections								
		2010			2020			2030		
		Low Economic Growth	Reference	High Economic Growth	Low Economic Growth	Reference	High Economic Growth	Low Economic Growth	Reference	High Economic Growth
Non-Renewable Energy Expenditures by Sector (billion 2005 dollars)										
Residential	215.13	215.77	220.44	225.03	222.24	236.03	247.96	237.41	262.21	289.21
Commercial	154.38	154.68	157.97	161.22	169.35	181.74	192.39	196.37	222.08	250.92
Industrial	196.07	188.55	200.48	212.61	169.48	194.88	220.13	174.11	222.08	276.29
Transportation	474.66	465.42	476.38	488.17	471.76	511.07	552.61	548.13	632.79	724.25
Total Non-Renewable Expenditures	1040.25	1024.42	1055.27	1087.03	1032.83	1123.73	1213.09	1156.01	1339.16	1540.67
Transportation Renewable Expenditures	0.03	0.06	0.06	0.07	0.13	0.15	0.19	0.39	0.51	0.79
Total Expenditures	**1040.29**	**1024.49**	**1055.33**	**1087.10**	**1032.96**	**1123.89**	**1213.28**	**1156.40**	**1339.68**	**1541.47**

[1]Includes energy for combined heat and power plants, except those whose primary business is to sell electricity, or electricity and heat, to the public.
[2]Excludes use for lease and plant fuel.
[3]Includes Federal and State taxes while excluding county and local taxes.
[4]E85 refers to a blend of 85 percent ethanol (renewable) and 15 percent motor gasoline (nonrenewable). To address cold starting issues, the percentage of ethanol actually varies seasonally. The annual average ethanol content of 74 percent is used for this forecast.
[5]Sales weighted-average price for all grades. Includes Federal, State and local taxes.
[6]Kerosene-type jet fuel. Includes Federal and State taxes while excluding county and local taxes.
[7]Diesel fuel for on-road use. Includes Federal and State taxes while excluding county and local taxes.
[8]Compressed natural gas used as a vehicle fuel. Includes estimated motor vehicle fuel taxes and estimated dispensing costs or charges.
[9]Includes electricity-only and combined heat and power plants whose primary business is to sell electricity, or electricity and heat, to the public.
[10]Weighted averages of end-use fuel prices are derived from the prices shown in each sector and the corresponding sectoral consumption.
Btu = British thermal unit.
N/A = Not applicable.
Note: Data for 2005 are model results and may differ slightly from official EIA data reports.
Sources: 2005 prices for motor gasoline, distillate fuel oil, and jet fuel are based on prices in the Energy Information Administration (EIA), *Petroleum Marketing Annual 2005*, DOE/EIA-0487(2005) (Washington, DC, August 2006). 2005 residential and commercial natural gas delivered prices: EIA, *Natural Gas Monthly*, DOE/EIA-0130(2006/04) (Washington, DC, April 2006). 2005 industrial natural gas delivered prices are estimated based on: EIA, *Manufacturing Energy Consumption Survey 1994* and industrial and wellhead prices from the *Natural Gas Annual 2004*, DOE/EIA-0131(2004) (Washington, DC, December 2005) and the *Natural Gas Monthly*, DOE/EIA-0130(2006/04) (Washington, DC, April 2006). 2005 transportation sector natural gas delivered prices are model results. 2005 electric power sector natural gas prices: EIA, *Electric Power Monthly*, DOE/EIA-0226, May 2003 through April 2004, Table 4.11.A. 2005 coal prices based on: EIA, *Quarterly Coal Report, October-December 2005*, DOE/EIA-0121(2005/4Q) (Washington, DC, March 2006) and EIA, AEO2007 National Energy Modeling System run AEO2007.D112106A. 2005 electricity prices: EIA, *Annual Energy Review 2005*, DOE/EIA-0384(2005) (Washington, DC, July 2006). 2005 ethanol prices derived from weekly spot prices in the Oxy Fuel News. **Projections:** EIA, AEO2007 National Energy Modeling System runs LM2007.D112106A, AEO2007.D112106A, and HM2007.D112106A.

Economic Growth Case Comparisons

Table B4. Macroeconomic Indicators
(Billion 2000 Chain-Weighted Dollars, Unless Otherwise Noted)

Indicators	2005	Projections								
		2010			2020			2030		
		Low Economic Growth	Reference	High Economic Growth	Low Economic Growth	Reference	High Economic Growth	Low Economic Growth	Reference	High Economic Growth
Real Gross Domestic Product	**11049**	**12359**	**12790**	**13219**	**15686**	**17077**	**18490**	**19249**	**22494**	**25757**
Components of Real Gross Domestic Product										
Real Consumption	7841	8867	9111	9353	11140	12006	12888	13629	15590	17564
Real Investment	1866	1936	2139	2342	2657	3030	3407	3760	4735	5711
Real Government Spending	1958	2074	2117	2160	2236	2396	2555	2358	2709	3060
Real Exports	1196	1741	1767	1792	3191	3584	3984	5530	6581	7654
Real Imports	1815	2254	2321	2386	3584	3761	3898	6231	6649	7022
Energy Intensity (thousand Btu per 2000 dollar of GDP)										
Delivered Energy	6.60	6.15	6.06	5.97	5.17	5.04	4.92	4.47	4.27	4.13
Total Energy	9.07	8.48	8.33	8.19	7.15	6.92	6.71	6.16	5.83	5.61
Price Indices										
GDP Chain-Type Price Index (2000=1.000)	1.127	1.276	1.253	1.231	1.620	1.495	1.370	2.059	1.815	1.576
Consumer Price Index (1982-4=1)										
All-Urban	1.95	2.21	2.16	2.13	2.83	2.61	2.39	3.65	3.23	2.82
Energy Commodities and Services	1.77	1.95	1.93	1.92	2.31	2.19	2.05	3.06	2.80	2.53
Wholesale Price Index (1982=1.00)										
All Commodities	1.57	1.72	1.68	1.65	2.00	1.82	1.63	2.41	2.06	1.73
Fuel and Power	1.57	1.65	1.64	1.64	1.94	1.84	1.72	2.62	2.39	2.17
Interest Rates (percent, nominal)										
Federal Funds Rate	3.21	5.03	4.71	4.43	5.59	5.11	4.60	5.67	5.14	4.61
10-Year Treasury Note	4.29	5.96	5.52	5.11	6.29	5.75	5.18	6.39	5.80	5.21
AA Utility Bond Rate	5.44	7.65	7.36	7.09	8.30	7.72	7.13	8.40	7.77	7.14
Value of Shipments (billion 2000 dollars)										
Total Industrial	5763	6001	6298	6587	6962	7779	8614	7712	9502	11357
Non-manufacturing	1538	1469	1596	1723	1614	1846	2082	1698	2023	2353
Manufacturing	4225	4532	4702	4865	5347	5933	6533	6014	7478	9004
Energy-Intensive	1160	1229	1262	1295	1316	1426	1541	1396	1631	1874
Non-Energy Intensive	3065	3304	3440	3569	4031	4507	4991	4618	5848	7130
Population and Employment (millions)										
Population with Armed Forces Overseas	296.9	308.5	310.3	312.8	323.6	337.1	351.4	334.2	364.9	395.6
Population (aged 16 and over)	231.8	242.5	244.2	246.6	256.2	265.4	274.7	268.6	288.6	308.7
Population, over age 65	36.8	40.3	40.4	40.6	53.9	54.9	55.8	69.1	71.6	74.1
Employment, Nonfarm	133.4	135.8	141.9	148.0	143.7	154.6	165.6	153.4	169.2	185.0
Employment, Manufacturing	14.2	13.5	13.8	14.1	12.8	13.4	13.8	11.4	12.5	13.4
Key Labor Indicators										
Labor Force (millions)	149.3	155.5	157.5	159.6	161.2	167.0	173.3	170.9	180.4	190.2
Non-farm Labor Productivity (1992=1.00)	1.36	1.48	1.50	1.53	1.79	1.90	2.03	2.16	2.42	2.69
Unemployment Rate (percent)	5.06	4.94	4.83	4.72	4.65	4.46	4.25	4.87	4.71	4.56
Key Indicators for Energy Demand										
Real Disposable Personal Income	8105	9317	9568	9814	12184	13000	13834	15691	17535	19397
Housing Starts (millions)	2.22	1.62	1.94	2.26	1.48	1.90	2.33	1.20	1.80	2.41
Commercial Floorspace (billion square feet)	74.3	80.0	80.4	80.8	88.9	92.9	97.0	98.1	108.0	118.1
Unit Sales of Light-Duty Vehicles (millions)	16.95	16.60	17.14	17.90	17.63	19.04	20.65	18.62	21.10	23.97

GDP = Gross domestic product.
Btu = British thermal unit.
Sources: 2005: Global Insight macroeconomic model CTL0806, and Global Insight industry model, July 2005. **Projections:** Energy Information Administration, AEO2007 National Energy Modeling System runs LM2007.D112106A, AEO2007.D112106A, and HM2007.D112106A.

Price Case Comparisons

Table C1. Total Energy Supply and Disposition Summary
(Quadrillion Btu per Year, Unless Otherwise Noted)

Supply, Disposition, and Prices	2005	Projections								
		2010			2020			2030		
		Low Price	Reference	High Price	Low Price	Reference	High Price	Low Price	Reference	High Price
Production										
Crude Oil and Lease Condensate	10.96	12.31	11.99	11.66	12.59	12.48	12.67	11.11	11.40	12.79
Natural Gas Plant Liquids	2.33	2.45	2.43	2.37	2.39	2.38	2.28	2.33	2.31	2.31
Dry Natural Gas	18.77	20.23	19.93	19.38	21.34	21.41	21.01	21.26	21.15	21.53
Coal[1]	23.20	24.09	24.47	24.68	24.56	26.61	29.80	27.22	33.52	38.32
Nuclear Power	8.13	8.23	8.23	8.23	8.49	9.23	9.39	8.32	9.33	10.31
Hydropower	2.71	3.02	3.02	3.02	3.07	3.08	3.09	3.09	3.09	3.09
Biomass[2]	2.71	4.00	4.22	4.30	4.18	4.69	4.96	4.51	5.26	5.62
Other Renewable Energy[3]	0.76	1.18	1.18	1.22	1.24	1.33	1.41	1.34	1.44	1.51
Other[4]	0.22	0.77	0.67	0.85	1.03	0.89	0.98	1.24	1.12	1.12
Total	**69.80**	**76.27**	**76.13**	**75.70**	**78.89**	**82.09**	**85.58**	**80.42**	**88.63**	**96.61**
Imports										
Crude Oil[5]	22.09	21.77	21.88	21.92	25.59	24.72	21.94	31.38	28.63	23.18
Liquid Fuels and Other Petroleum[6]	7.16	6.54	6.02	5.78	9.66	7.05	5.35	12.83	9.02	5.24
Natural Gas	4.42	5.45	5.36	4.99	8.12	6.17	4.34	10.32	6.47	3.74
Other Imports[7]	0.85	0.92	0.92	0.93	1.51	1.73	1.89	1.90	2.26	2.23
Total	**34.52**	**34.68**	**34.18**	**33.62**	**44.88**	**39.66**	**33.53**	**56.43**	**46.37**	**34.39**
Exports										
Petroleum[8]	2.31	2.76	2.71	2.69	2.92	2.84	2.75	3.15	2.90	2.81
Natural Gas	0.75	0.70	0.69	0.68	0.84	0.69	0.56	1.24	0.87	0.49
Coal	1.27	1.12	1.12	1.11	0.81	0.80	0.83	0.65	0.69	0.69
Total	**4.33**	**4.59**	**4.52**	**4.49**	**4.58**	**4.33**	**4.15**	**5.04**	**4.47**	**3.98**
Discrepancy[9]	**-0.20**	**-0.54**	**-0.70**	**-0.63**	**-0.11**	**-0.74**	**-0.79**	**0.07**	**-0.63**	**-0.74**
Consumption										
Liquid Fuels and Other Petroleum[10]	40.61	42.25	41.76	41.32	49.08	46.52	43.68	56.22	52.17	47.52
Natural Gas	22.63	25.11	24.73	23.83	28.78	27.04	24.70	30.62	26.89	24.60
Coal	22.87	23.86	24.24	24.45	25.20	27.29	29.60	28.43	34.14	36.39
Nuclear Power	8.13	8.23	8.23	8.23	8.49	9.23	9.39	8.32	9.33	10.31
Hydropower	2.71	3.02	3.02	3.02	3.07	3.08	3.09	3.09	3.09	3.09
Biomass[11]	2.38	3.22	3.30	3.35	3.40	3.64	3.85	3.69	4.06	4.24
Other Renewable Energy[3]	0.76	1.18	1.18	1.22	1.24	1.33	1.41	1.34	1.44	1.51
Other[12]	0.08	0.04	0.04	0.05	0.04	0.04	0.04	0.04	0.04	0.08
Total	**100.19**	**106.90**	**106.50**	**105.46**	**119.29**	**118.16**	**115.75**	**131.75**	**131.16**	**127.74**

Price Case Comparisons

Table C1. Total Energy Supply and Disposition Summary (Continued)
(Quadrillion Btu per Year, Unless Otherwise Noted)

Supply, Disposition, and Prices	2005	Projections								
		2010			2020			2030		
		Low Price	Reference	High Price	Low Price	Reference	High Price	Low Price	Reference	High Price
Prices (2005 dollars per unit)										
Petroleum (dollars per barrel)										
Imported Low Sulfur Light Crude Oil Price[13]	56.76	49.21	57.47	69.21	34.10	52.04	89.12	35.68	59.12	100.14
Imported Crude Oil Price[13]	49.19	44.06	51.20	62.53	28.91	46.47	82.60	28.91	51.63	92.93
Natural Gas (dollars per million Btu)										
Price at Henry Hub	8.60	5.62	6.28	6.91	4.66	5.71	6.46	5.53	6.52	8.27
Wellhead Price[14]	7.29	4.99	5.59	6.16	4.12	5.07	5.75	4.91	5.80	7.41
Natural Gas (dollars per thousand cubic feet)										
Wellhead Price[14]	7.51	5.14	5.76	6.35	4.25	5.22	5.92	5.06	5.98	7.63
Coal (dollars per ton)										
Minemouth Price[15]	23.34	23.64	24.20	24.54	20.31	21.58	23.46	20.23	22.60	24.45
Coal (dollars per million Btu)										
Minemouth Price[15]	1.15	1.16	1.18	1.20	1.01	1.08	1.17	1.02	1.15	1.25
Average Delivered Price[16]	1.61	1.74	1.77	1.80	1.54	1.62	1.73	1.56	1.71	1.83
Average Electricity Price (cents per kilowatthour)	8.1	7.8	8.1	8.3	7.5	7.9	8.1	7.8	8.1	8.3

[1]Includes waste coal.
[2]Includes grid-connected electricity from wood and waste; biomass, such as corn, used for liquid fuels production; and non-electric energy demand from wood. Refer to Table A17 for details.
[3]Includes grid-connected electricity from landfill gas; municipal solid waste; wind; photovoltaic and solar thermal sources; and non-electric energy from renewable sources, such as active and passive solar systems. Excludes electricity imports using renewable sources and nonmarketed renewable energy. See Table A17 for selected nonmarketed residential and commercial renewable energy.
[4]Includes liquid hydrogen, methanol, and some domestic inputs to refineries.
[5]Includes imports of crude oil for the Strategic Petroleum Reserve.
[6]Includes imports of finished petroleum products, unfinished oils, alcohols, ethers, blending components, and renewable fuels such as ethanol.
[7]Includes coal, coal coke (net), and electricity (net).
[8]Includes crude oil and petroleum products.
[9]Balancing item. Includes unaccounted for supply, losses, gains, and net storage withdrawals.
[10]Includes petroleum-derived fuels and non-petroleum derived fuels, such as ethanol and biodiesel. Petroleum coke, which is a solid, is included. Also included are natural gas plant liquids, crude oil consumed as a fuel, and liquid hydrogen. Refer to Table A17 for detailed renewable liquid fuels consumption.
[11]Includes grid-connected electricity from wood and wood waste, non-electric energy from wood, and biofuels heat and coproducts used in the production of liquid fuels, but excludes the energy content of the liquid fuels.
[12]Includes net electricity imports.
[13]Weighted average price delivered to U.S. refiners.
[14]Represents lower 48 onshore and offshore supplies.
[15]Includes reported prices for both open market and captive mines.
[16]Prices weighted by consumption; weighted average excludes residential and commercial prices, and export free-alongside-ship (f.a.s.) prices.
Btu = British thermal unit.
N/A = Not applicable.
Note: Totals may not equal sum of components due to independent rounding. Data for 2005 are model results and may differ slightly from official EIA data reports.
Sources: 2005 natural gas supply values and natural gas wellhead price: EIA, *Natural Gas Monthly*, DOE/EIA-0130(2006/04) (Washington, DC, April 2006). 2005 coal minemouth and delivered coal prices: EIA, *Annual Coal Report 2005*, DOE/EIA-0584(2005) (Washington, DC, October 2006). 2005 petroleum supply values: EIA, *Petroleum Supply Annual 2005*, DOE/EIA-0340(2005)/1 (Washington, DC, October 2006). 2005 low sulfur light crude oil price: EIA, Form EIA-856, "Monthly Foreign Crude Oil Acquisition Report." Other 2005 coal values: *Quarterly Coal Report, October-December 2005*, DOE/EIA-0121(2005/4Q) (Washington, DC, March 2006). Other 2005 values: EIA, *Annual Energy Review 2005*, DOE/EIA-0384(2005) (Washington, DC, July 2006). **Projections:** EIA, AEO2007 National Energy Modeling System runs LP2007.D112106A, AEO2007.D112106A, and HP2007.D112106A.

Table C2. Energy Consumption by Sector and Source
(Quadrillion Btu per Year, Unless Otherwise Noted)

Sector and Source	2005	Projections								
		2010			2020			2030		
		Low Price	Reference	High Price	Low Price	Reference	High Price	Low Price	Reference	High Price
Energy Consumption										
Residential										
Liquefied Petroleum Gases	0.51	0.54	0.53	0.53	0.58	0.58	0.57	0.62	0.62	0.61
Kerosene	0.10	0.10	0.10	0.09	0.10	0.10	0.09	0.10	0.09	0.08
Distillate Fuel Oil	0.93	0.92	0.90	0.88	0.94	0.85	0.76	0.86	0.76	0.65
Liquid Fuels and Other Petroleum Subtotal	1.54	1.55	1.53	1.51	1.63	1.53	1.42	1.58	1.46	1.34
Natural Gas	4.98	5.23	5.18	5.14	5.53	5.43	5.37	5.56	5.47	5.35
Coal	0.01	0.01	0.01	0.01	0.01	0.01	0.01	0.01	0.01	0.01
Renewable Energy[1]	0.41	0.42	0.43	0.43	0.38	0.40	0.43	0.37	0.39	0.41
Electricity	4.66	5.08	5.06	5.04	5.85	5.80	5.77	6.50	6.47	6.46
Delivered Energy	**11.60**	**12.29**	**12.21**	**12.13**	**13.40**	**13.17**	**13.00**	**14.02**	**13.80**	**13.57**
Electricity Related Losses	10.15	10.86	10.90	10.91	11.88	12.08	12.05	12.67	12.89	12.54
Total	**21.75**	**23.15**	**23.11**	**23.04**	**25.28**	**25.26**	**25.05**	**26.69**	**26.70**	**26.11**
Commercial										
Liquefied Petroleum Gases	0.09	0.09	0.09	0.09	0.10	0.10	0.09	0.10	0.10	0.10
Motor Gasoline[2]	0.05	0.05	0.05	0.05	0.05	0.05	0.05	0.05	0.05	0.05
Kerosene	0.02	0.03	0.03	0.03	0.03	0.03	0.03	0.03	0.03	0.03
Distillate Fuel Oil	0.48	0.46	0.45	0.44	0.57	0.48	0.44	0.63	0.49	0.45
Residual Fuel Oil	0.14	0.14	0.14	0.14	0.16	0.14	0.13	0.16	0.14	0.13
Liquid Fuels and Other Petroleum Subtotal	0.77	0.78	0.75	0.74	0.91	0.80	0.74	0.98	0.81	0.76
Natural Gas	3.15	3.37	3.31	3.26	3.97	3.86	3.75	4.45	4.36	4.13
Coal	0.10	0.10	0.10	0.10	0.10	0.10	0.10	0.10	0.10	0.10
Renewable Energy[3]	0.12	0.12	0.12	0.12	0.12	0.12	0.12	0.12	0.12	0.12
Electricity	4.32	4.79	4.77	4.75	5.87	5.78	5.71	7.13	7.03	6.94
Delivered Energy	**8.46**	**9.15**	**9.05**	**8.97**	**10.97**	**10.66**	**10.41**	**12.77**	**12.43**	**12.05**
Electricity Related Losses	9.42	10.25	10.27	10.28	11.92	12.03	11.90	13.89	14.01	13.48
Total	**17.88**	**19.40**	**19.33**	**19.25**	**22.88**	**22.69**	**22.31**	**26.66**	**26.44**	**25.53**
Industrial[4]										
Liquefied Petroleum Gases	2.13	2.26	2.26	2.25	2.28	2.26	2.25	2.46	2.40	2.38
Motor Gasoline[2]	0.32	0.32	0.32	0.31	0.33	0.33	0.33	0.35	0.36	0.35
Distillate Fuel Oil	1.23	1.19	1.18	1.16	1.25	1.22	1.19	1.33	1.26	1.25
Residual Fuel Oil	0.23	0.18	0.18	0.17	0.20	0.17	0.15	0.22	0.18	0.15
Petrochemical Feedstocks	1.38	1.50	1.48	1.47	1.53	1.50	1.48	1.57	1.57	1.53
Other Petroleum[5]	4.45	4.17	4.05	4.10	4.51	4.34	4.16	5.06	4.78	4.53
Liquid Fuels and Other Petroleum Subtotal	9.73	9.62	9.47	9.47	10.10	9.82	9.56	10.99	10.55	10.19
Natural Gas	6.84	7.76	7.86	7.52	8.10	8.26	7.93	8.59	8.90	8.42
Natural-Gas-to-Liquids Heat and Power	0.00	0.00	0.00	0.00	0.00	0.00	0.15	0.00	0.00	0.15
Lease and Plant Fuel[6]	1.10	1.11	1.10	1.07	1.20	1.21	1.19	1.16	1.15	1.18
Natural Gas Subtotal	7.94	8.87	8.95	8.60	9.29	9.46	9.27	9.75	10.05	9.75
Metallurgical Coal	0.62	0.62	0.60	0.59	0.59	0.57	0.55	0.59	0.57	0.55
Other Industrial Coal	1.35	1.37	1.37	1.37	1.33	1.34	1.34	1.34	1.36	1.37
Coal-to-Liquids Heat and Power	0.00	0.00	0.00	0.00	0.00	0.21	1.26	0.00	0.93	3.49
Net Coal Coke Imports	0.04	0.02	0.02	0.02	0.02	0.02	0.02	0.02	0.02	0.02
Coal Subtotal	2.01	2.01	2.00	1.98	1.94	2.14	3.16	1.95	2.89	5.43
Biofuels Heat and Coproducts	0.24	0.59	0.69	0.71	0.60	0.78	0.78	0.62	0.88	0.96
Renewable Energy[7]	1.44	1.61	1.60	1.58	1.84	1.81	1.78	2.07	2.05	2.02
Electricity	3.48	3.66	3.63	3.58	3.87	3.83	3.81	4.00	4.09	4.22
Delivered Energy	**24.85**	**26.36**	**26.33**	**25.92**	**27.65**	**27.84**	**28.36**	**29.37**	**30.51**	**32.57**
Electricity Related Losses	7.60	7.82	7.81	7.75	7.85	7.98	7.95	7.80	8.15	8.20
Total	**32.45**	**34.18**	**34.14**	**33.67**	**35.50**	**35.82**	**36.32**	**37.17**	**38.66**	**40.77**

Table C2. Energy Consumption by Sector and Source (Continued)
(Quadrillion Btu per Year, Unless Otherwise Noted)

Sector and Source	2005	Projections								
		2010			2020			2030		
		Low Price	Reference	High Price	Low Price	Reference	High Price	Low Price	Reference	High Price
Transportation										
Liquefied Petroleum Gases	0.04	0.05	0.05	0.05	0.06	0.06	0.10	0.07	0.08	0.14
E85[8]	0.00	0.00	0.00	0.00	0.01	0.01	0.04	0.02	0.02	0.30
Motor Gasoline[2]	17.00	17.53	17.37	17.06	20.91	19.95	17.80	24.37	22.89	19.04
Jet Fuel[9]	3.37	4.07	4.04	4.00	4.59	4.54	4.48	4.73	4.70	4.26
Distillate Fuel Oil[10]	6.02	6.70	6.64	6.58	8.02	7.81	7.69	9.90	9.58	9.54
Residual Fuel Oil	0.81	0.82	0.82	0.82	0.85	0.85	0.84	0.87	0.87	0.87
Liquid Hydrogen	0.00	0.00	0.00	0.00	0.00	0.00	0.00	0.00	0.00	0.00
Other Petroleum[11]	0.18	0.18	0.18	0.18	0.19	0.19	0.19	0.19	0.19	0.20
Liquid Fuels and Other Petroleum Subtotal	27.42	29.36	29.11	28.71	34.62	33.41	31.14	40.16	38.34	34.35
Pipeline Fuel Natural Gas	0.58	0.67	0.66	0.64	0.81	0.79	0.74	0.83	0.79	0.75
Compressed Natural Gas	0.03	0.06	0.06	0.06	0.09	0.09	0.10	0.11	0.12	0.12
Electricity	0.02	0.03	0.03	0.03	0.03	0.03	0.03	0.04	0.04	0.04
Delivered Energy	**28.05**	**30.11**	**29.86**	**29.44**	**35.56**	**34.33**	**32.01**	**41.15**	**39.29**	**35.26**
Electricity Related Losses	0.05	0.06	0.06	0.06	0.07	0.07	0.07	0.08	0.08	0.07
Total	**28.11**	**30.17**	**29.92**	**29.50**	**35.63**	**34.40**	**32.08**	**41.23**	**39.37**	**35.34**
Delivered Energy Consumption for All Sectors										
Liquefied Petroleum Gases	2.77	2.94	2.93	2.93	3.02	2.99	3.02	3.25	3.19	3.23
E85[8]	0.00	0.00	0.00	0.00	0.01	0.01	0.04	0.02	0.02	0.30
Motor Gasoline[2]	17.37	17.90	17.74	17.43	21.30	20.34	18.18	24.78	23.30	19.44
Jet Fuel[9]	3.37	4.07	4.04	4.00	4.59	4.54	4.48	4.73	4.70	4.26
Kerosene	0.14	0.15	0.14	0.14	0.15	0.14	0.13	0.15	0.14	0.13
Distillate Fuel Oil	8.65	9.27	9.17	9.06	10.78	10.36	10.08	12.72	12.09	11.90
Residual Fuel Oil	1.17	1.15	1.13	1.13	1.20	1.16	1.12	1.24	1.19	1.15
Petrochemical Feedstocks	1.38	1.50	1.48	1.47	1.53	1.50	1.48	1.57	1.57	1.53
Liquid Hydrogen	0.00	0.00	0.00	0.00	0.00	0.00	0.00	0.00	0.00	0.00
Other Petroleum[12]	4.61	4.33	4.22	4.26	4.67	4.51	4.33	5.24	4.96	4.70
Liquid Fuels and Other Petroleum Subtotal	39.46	41.31	40.86	40.43	47.26	45.55	42.86	53.71	51.17	46.65
Natural Gas	15.01	16.41	16.41	15.98	17.69	17.65	17.14	18.71	18.86	18.02
Natural-Gas-to-Liquids Heat and Power	0.00	0.00	0.00	0.00	0.00	0.00	0.15	0.00	0.00	0.15
Lease and Plant Fuel[6]	1.10	1.11	1.10	1.07	1.20	1.21	1.19	1.16	1.15	1.18
Pipeline Natural Gas	0.58	0.67	0.66	0.64	0.81	0.79	0.74	0.83	0.79	0.75
Natural Gas Subtotal	16.68	18.19	18.17	17.70	19.70	19.64	19.22	20.70	20.80	20.10
Metallurgical Coal	0.62	0.62	0.60	0.59	0.59	0.57	0.55	0.59	0.57	0.55
Other Coal	1.46	1.48	1.48	1.48	1.44	1.45	1.45	1.45	1.47	1.47
Coal-to-Liquids Heat and Power	0.00	0.00	0.00	0.00	0.00	0.21	1.26	0.00	0.93	3.49
Net Coal Coke Imports	0.04	0.02	0.02	0.02	0.02	0.02	0.02	0.02	0.02	0.02
Coal Subtotal	2.12	2.12	2.11	2.08	2.05	2.24	3.27	2.06	2.99	5.54
Biofuels Heat and Coproducts	0.24	0.59	0.69	0.71	0.60	0.78	0.78	0.62	0.88	0.96
Renewable Energy[13]	1.97	2.15	2.14	2.13	2.35	2.34	2.33	2.56	2.56	2.55
Electricity	12.49	13.56	13.49	13.40	15.62	15.45	15.32	17.66	17.63	17.66
Delivered Energy	**72.97**	**77.91**	**77.46**	**76.46**	**87.58**	**86.00**	**83.78**	**97.31**	**96.03**	**93.45**
Electricity Related Losses	27.23	28.99	29.04	29.00	31.72	32.17	31.97	34.44	35.13	34.29
Total	**100.19**	**106.90**	**106.50**	**105.46**	**119.29**	**118.16**	**115.75**	**131.75**	**131.16**	**127.74**
Electric Power[14]										
Distillate Fuel Oil	0.19	0.23	0.24	0.23	0.57	0.25	0.26	1.10	0.28	0.28
Residual Fuel Oil	0.96	0.71	0.67	0.66	1.24	0.72	0.56	1.41	0.72	0.59
Liquid Fuels and Other Petroleum Subtotal	1.16	0.95	0.90	0.89	1.82	0.97	0.82	2.51	1.01	0.87
Natural Gas	5.95	6.92	6.56	6.13	9.08	7.40	5.48	9.92	6.09	4.50
Steam Coal	20.75	21.74	22.13	22.36	23.15	25.05	26.33	26.37	31.14	30.85
Nuclear Power	8.13	8.23	8.23	8.23	8.49	9.23	9.39	8.32	9.33	10.31
Renewable Energy[15]	3.64	4.68	4.67	4.74	4.76	4.93	5.24	4.94	5.15	5.33
Electricity Imports	0.08	0.04	0.04	0.05	0.04	0.04	0.04	0.04	0.04	0.08
Total	**39.71**	**42.54**	**42.53**	**42.40**	**47.34**	**47.62**	**47.30**	**52.10**	**52.77**	**51.95**

Table C2. Energy Consumption by Sector and Source (Continued)
(Quadrillion Btu per Year, Unless Otherwise Noted)

Sector and Source	2005	Projections								
		2010			2020			2030		
		Low Price	Reference	High Price	Low Price	Reference	High Price	Low Price	Reference	High Price
Total Energy Consumption										
Liquefied Petroleum Gases	2.77	2.94	2.93	2.93	3.02	2.99	3.02	3.25	3.19	3.23
E85[8]	0.00	0.00	0.00	0.00	0.01	0.01	0.04	0.02	0.02	0.30
Motor Gasoline[2]	17.37	17.90	17.74	17.43	21.30	20.34	18.18	24.78	23.30	19.44
Jet Fuel[9]	3.37	4.07	4.04	4.00	4.59	4.54	4.48	4.73	4.70	4.26
Kerosene	0.14	0.15	0.14	0.14	0.15	0.14	0.13	0.15	0.14	0.13
Distillate Fuel Oil	8.84	9.51	9.40	9.30	11.36	10.61	10.34	13.82	12.37	12.17
Residual Fuel Oil	2.14	1.86	1.80	1.79	2.44	1.88	1.68	2.65	1.91	1.74
Petrochemical Feedstocks	1.38	1.50	1.48	1.47	1.53	1.50	1.48	1.57	1.57	1.53
Liquid Hydrogen	0.00	0.00	0.00	0.00	0.00	0.00	0.00	0.00	0.00	0.00
Other Petroleum[12]	4.61	4.33	4.22	4.26	4.67	4.51	4.33	5.24	4.96	4.70
Liquid Fuels and Other Petroleum Subtotal	40.61	42.25	41.76	41.32	49.08	46.52	43.68	56.22	52.17	47.52
Natural Gas	20.96	23.33	22.97	22.12	26.77	25.05	22.62	28.63	24.95	22.52
Natural-Gas-to-Liquids Heat and Power	0.00	0.00	0.00	0.00	0.00	0.00	0.15	0.00	0.00	0.15
Lease and Plant Fuel[6]	1.10	1.11	1.10	1.07	1.20	1.21	1.19	1.16	1.15	1.18
Pipeline Natural Gas	0.58	0.67	0.66	0.64	0.81	0.79	0.74	0.83	0.79	0.75
Natural Gas Subtotal	22.63	25.11	24.73	23.83	28.78	27.04	24.70	30.62	26.89	24.60
Metallurgical Coal	0.62	0.62	0.60	0.59	0.59	0.57	0.55	0.59	0.57	0.55
Other Coal	22.21	23.22	23.61	23.84	24.59	26.50	27.78	27.82	32.61	32.32
Coal-to-Liquids Heat and Power	0.00	0.00	0.00	0.00	0.00	0.21	1.26	0.00	0.93	3.49
Net Coal Coke Imports	0.04	0.02	0.02	0.02	0.02	0.02	0.02	0.02	0.02	0.02
Coal Subtotal	22.87	23.86	24.24	24.45	25.20	27.29	29.60	28.43	34.14	36.39
Nuclear Power	8.13	8.23	8.23	8.23	8.49	9.23	9.39	8.32	9.33	10.31
Biofuels Heat and Coproducts	0.24	0.59	0.69	0.71	0.60	0.78	0.78	0.62	0.88	0.96
Renewable Energy[16]	5.61	6.83	6.81	6.87	7.10	7.27	7.57	7.50	7.71	7.89
Electricity Imports	0.08	0.04	0.04	0.05	0.04	0.04	0.04	0.04	0.04	0.08
Total	**100.19**	**106.90**	**106.50**	**105.46**	**119.29**	**118.16**	**115.75**	**131.75**	**131.16**	**127.74**
Energy Use and Related Statistics										
Delivered Energy Use	72.97	77.91	77.46	76.46	87.58	86.00	83.78	97.31	96.03	93.45
Total Energy Use	100.19	106.90	106.50	105.46	119.29	118.16	115.75	131.75	131.16	127.74
Ethanol Consumed in Motor Gasoline and E85	0.33	0.77	0.91	0.95	0.79	1.06	1.05	0.87	1.22	1.30
Population (millions)	296.94	310.26	310.26	310.26	337.13	337.13	337.13	364.94	364.94	364.94
Gross Domestic Product (billion 2000 dollars)	11049	12850	12790	12708	17129	17077	17027	22532	22494	22472
Carbon Dioxide Emissions (million metric tons)	5945.3	6241.2	6214.0	6155.6	7040.2	6944.5	6830.5	7928.6	7950.2	7701.0

[1]Includes wood used for residential heating. See Table A4 and/or Table A17 for estimates of nonmarketed renewable energy consumption for geothermal heat pumps, solar thermal hot water heating, and solar photovoltaic electricity generation.
[2]Includes ethanol (blends of 10 percent or less) and ethers blended into gasoline.
[3]Includes commercial sector consumption of wood and wood waste, landfill gas, municipal solid waste, and other biomass for combined heat and power. See Table A5 and/or Table A17 for estimates of nonmarketed renewable energy consumption for solar thermal hot water heating and solar photovoltaic electricity generation.
[4]Includes energy for combined heat and power plants, except those whose primary business is to sell electricity, or electricity and heat, to the public.
[5]Includes petroleum coke, asphalt, road oil, lubricants, still gas, tire-derived fuel, and miscellaneous petroleum products.
[6]Represents natural gas used in well, field, and lease operations, and in natural gas processing plant machinery.
[7]Includes consumption of energy produced from hydroelectric, wood and wood waste, municipal solid waste, and other biomass sources.
[8]E85 refers to a blend of 85 percent ethanol (renewable) and 15 percent motor gasoline (nonrenewable). To address cold starting issues, the percentage of ethanol actually varies seasonally. The annual average ethanol content of 74 percent is used for this forecast.
[9]Includes only kerosene type.
[10]Diesel fuel for on- and off- road use.
[11]Includes aviation gasoline and lubricants.
[12]Includes unfinished oils, natural gasoline, motor gasoline blending components, aviation gasoline, lubricants, still gas, asphalt, road oil, petroleum coke, tire-derived fuel, and miscellaneous petroleum products.
[13]Includes electricity generated for sale to the grid and for own use from renewable sources, and non-electric energy from renewable sources. Excludes nonmarketed renewable energy consumption for geothermal heat pumps, buildings photovoltaic systems, and solar thermal hot water heaters.
[14]Includes consumption of energy by electricity-only and combined heat and power plants whose primary business is to sell electricity, or electricity and heat, to the public. Includes small power producers and exempt wholesale generators.
[15]Includes conventional hydroelectric, geothermal, wood and wood waste, municipal solid waste, other biomass, petroleum coke, wind, photovoltaic and solar thermal sources. Excludes net electricity imports.
[16]Includes hydroelectric, geothermal, wood and wood waste, municipal solid waste, other biomass, wind, photovoltaic and solar thermal sources. Includes ethanol components of E85; excludes ethanol blends (10 percent or less) in motor gasoline. Excludes net electricity imports and nonmarketed renewable energy consumption for geothermal heat pumps, buildings photovoltaic systems, and solar thermal hot water heaters.
Btu = British thermal unit.
N/A = Not applicable.
Note: Totals may not equal sum of components due to independent rounding. Data for 2005 are model results and may differ slightly from official EIA data reports. Consumption values of 0.00 are values that round to 0.00, because they are less than 0.005.
Sources: 2005 consumption based on: Energy Information Administration (EIA), *Annual Energy Review 2005*, DOE/EIA-0384(2005) (Washington, DC, July 2006). 2005 population and gross domestic product: Global Insight macroeconomic model CTL0806. 2005 carbon dioxide emissions: EIA, *Emissions of Greenhouse Gases in the United States 2005*, DOE/EIA-0573(2005) (Washington, DC, November 2006). **Projections:** EIA, AEO2007 National Energy Modeling System runs LP2007.D112106A, AEO2007.D112106A, and HP2007.D112106A.

Table C3. Energy Prices by Sector and Source
(2005 Dollars per Million Btu, Unless Otherwise Noted)

Sector and Source	2005	Projections								
		2010			2020			2030		
		Low Price	Reference	High Price	Low Price	Reference	High Price	Low Price	Reference	High Price
Residential										
Liquefied Petroleum Gases	19.29	23.03	23.67	24.29	22.18	23.18	23.92	22.98	23.91	25.65
Distillate Fuel Oil	14.73	12.63	14.87	16.58	9.09	13.15	18.95	9.66	14.13	19.84
Natural Gas	12.43	10.33	10.98	11.59	9.50	10.54	11.29	10.40	11.43	13.12
Electricity	27.59	26.24	26.91	27.58	25.24	26.37	26.96	26.08	26.76	27.24
Commercial										
Distillate Fuel Oil	12.68	10.58	12.72	14.58	7.15	11.35	16.85	7.88	12.45	18.06
Residual Fuel Oil	8.41	6.61	7.54	7.31	4.39	7.07	11.35	4.59	7.31	12.07
Natural Gas	11.20	8.70	9.34	9.94	7.65	8.67	9.41	8.33	9.30	10.99
Electricity	25.25	23.69	24.50	25.28	22.58	23.95	24.67	23.39	24.27	25.26
Industrial[1]										
Liquefied Petroleum Gases	16.96	15.79	16.42	17.09	14.89	15.91	16.66	15.75	16.55	18.25
Distillate Fuel Oil	13.08	10.93	12.95	14.94	7.90	12.04	17.30	8.84	13.25	18.79
Residual Fuel Oil	7.77	8.42	9.50	10.74	6.15	8.91	14.15	6.41	9.58	15.36
Natural Gas[2]	8.16	5.83	6.43	7.03	4.92	5.90	6.61	5.65	6.56	8.21
Metallurgical Coal	3.06	3.08	3.09	3.11	2.69	2.71	2.76	2.73	2.75	2.78
Other Industrial Coal	2.15	2.22	2.26	2.30	2.07	2.18	2.32	2.12	2.29	2.43
Coal to Liquids	0.00	0.00	0.00	0.00	0.00	0.97	1.42	0.00	1.33	1.78
Electricity	16.69	17.35	18.01	18.64	16.11	17.07	17.41	16.95	17.43	17.72
Transportation										
Liquefied Petroleum Gases[3]	23.92	23.72	24.34	24.88	22.68	23.66	24.20	23.35	24.29	25.85
E85[4]	23.10	22.00	21.29	22.41	18.58	20.61	23.33	19.38	21.50	25.54
Motor Gasoline[5]	18.64	16.55	17.90	20.64	13.97	16.63	23.62	14.39	17.76	26.42
Jet Fuel[6]	13.14	9.66	10.91	12.84	7.43	10.51	15.86	8.21	11.75	17.44
Distillate Fuel Oil[7]	17.52	14.79	16.81	18.76	11.24	15.42	20.72	12.02	16.47	22.20
Residual Fuel Oil	5.51	6.89	8.05	9.83	4.76	7.36	12.85	5.03	8.27	14.38
Natural Gas[8]	14.76	13.27	13.97	14.90	11.66	12.98	14.98	11.99	13.45	16.55
Electricity	25.22	23.98	24.86	25.66	22.90	24.22	24.77	23.72	24.46	25.19
Electric Power[9]										
Distillate Fuel Oil	11.38	9.44	11.71	13.59	5.07	9.84	15.34	5.43	10.79	15.94
Residual Fuel Oil	6.96	5.64	6.58	7.59	3.51	6.08	11.24	3.68	6.85	12.57
Natural Gas	8.18	5.64	6.22	6.75	4.93	5.76	6.32	5.71	6.33	7.79
Steam Coal	1.53	1.67	1.71	1.74	1.48	1.58	1.71	1.51	1.69	1.80
Average Price to All Users[10]										
Liquefied Petroleum Gases	17.48	17.39	18.02	18.68	16.60	17.62	18.44	17.44	18.30	20.13
E85[4]	23.10	22.00	21.29	22.41	18.58	20.61	23.33	19.38	21.50	25.54
Motor Gasoline[5]	18.60	16.55	17.90	20.64	13.97	16.63	23.61	14.39	17.75	26.42
Jet Fuel	13.14	9.66	10.91	12.84	7.43	10.51	15.86	8.21	11.75	17.44
Distillate Fuel Oil	16.22	13.66	15.70	17.64	10.18	14.53	19.89	10.85	15.70	21.45
Residual Fuel Oil	6.59	6.54	7.61	8.90	4.22	7.00	12.31	4.39	7.79	13.68
Natural Gas	9.65	7.22	7.83	8.46	6.30	7.32	8.15	7.04	8.09	9.85
Metallurgical Coal	3.06	3.08	3.09	3.11	2.69	2.71	2.76	2.73	2.75	2.78
Other Coal	1.57	1.70	1.74	1.77	1.52	1.61	1.74	1.54	1.72	1.83
Coal to Liquids	0.00	0.00	0.00	0.00	0.00	0.97	1.42	0.00	1.33	1.78
Electricity	23.73	22.94	23.66	24.37	21.98	23.15	23.73	22.92	23.60	24.18

Table C3. Energy Prices by Sector and Source (Continued)
(2005 Dollars per Million Btu, Unless Otherwise Noted)

Sector and Source	2005	Projections								
		2010			2020			2030		
		Low Price	Reference	High Price	Low Price	Reference	High Price	Low Price	Reference	High Price
Non-Renewable Energy Expenditures by Sector (billion 2005 dollars)										
Residential	215.13	212.43	220.44	227.60	222.50	236.03	246.06	250.70	262.21	276.15
Commercial	154.38	151.98	157.97	163.50	170.74	181.74	188.76	212.65	222.08	234.83
Industrial	196.07	187.02	200.48	213.85	169.84	194.88	224.73	193.83	222.08	269.71
Transportation	474.66	437.09	476.38	538.62	424.57	511.07	668.16	517.92	632.79	811.45
Total Non-Renewable Expenditures	1040.25	988.51	1055.27	1143.57	987.65	1123.73	1327.72	1175.11	1339.16	1592.14
Transportation Renewable Expenditures	0.03	0.05	0.06	0.09	0.13	0.15	0.89	0.40	0.51	7.62
Total Expenditures	**1040.29**	**988.56**	**1055.33**	**1143.66**	**987.78**	**1123.89**	**1328.60**	**1175.51**	**1339.68**	**1599.77**

[1]Includes energy for combined heat and power plants, except those whose primary business is to sell electricity, or electricity and heat, to the public.
[2]Excludes use for lease and plant fuel.
[3]Includes Federal and State taxes while excluding county and local taxes.
[4]E85 refers to a blend of 85 percent ethanol (renewable) and 15 percent motor gasoline (nonrenewable). To address cold starting issues, the percentage of ethanol actually varies seasonally. The annual average ethanol content of 74 percent is used for this forecast.
[5]Sales weighted-average price for all grades. Includes Federal, State and local taxes.
[6]Kerosene-type jet fuel. Includes Federal and State taxes while excluding county and local taxes.
[7]Diesel fuel for on-road use. Includes Federal and State taxes while excluding county and local taxes.
[8]Compressed natural gas used as a vehicle fuel. Includes estimated motor vehicle fuel taxes and estimated dispensing costs or charges.
[9]Includes electricity-only and combined heat and power plants whose primary business is to sell electricity, or electricity and heat, to the public.
[10]Weighted averages of end-use fuel prices are derived from the prices shown in each sector and the corresponding sectoral consumption.
Btu = British thermal unit.
N/A = Not applicable.
Note: Data for 2005 are model results and may differ slightly from official EIA data reports.
Sources: 2005 prices for motor gasoline, distillate fuel oil, and jet fuel are based on prices in the Energy Information Administration (EIA), *Petroleum Marketing Annual 2005*, DOE/EIA-0487(2005) (Washington, DC, August 2006). 2005 residential and commercial natural gas delivered prices: EIA, *Natural Gas Monthly*, DOE/EIA-0130(2006/04) (Washington, DC, April 2006). 2005 industrial natural gas delivered prices are estimated based on: EIA, *Manufacturing Energy Consumption Survey 1994* and industrial and wellhead prices from the *Natural Gas Annual 2004*, DOE/EIA-0131(2004) (Washington, DC, December 2005) and the *Natural Gas Monthly*, DOE/EIA-0130(2006/04) (Washington, DC, April 2006). 2005 transportation sector natural gas delivered prices are model results. 2005 electric power sector natural gas prices: EIA, *Electric Power Monthly*, DOE/EIA-0226, May 2003 through April 2004, Table 4.11.A. 2005 coal prices based on: EIA, *Quarterly Coal Report, October-December 2005*, DOE/EIA-0121(2005/4Q) (Washington, DC, March 2006) and EIA, AEO2007 National Energy Modeling System run AEO2007.D112106A. 2005 electricity prices: EIA, *Annual Energy Review 2005*, DOE/EIA-0384(2005) (Washington, DC, July 2006). 2005 ethanol prices derived from weekly spot prices in the Oxy Fuel News. **Projections:** EIA, AEO2007 National Energy Modeling System runs LP2007.D112106A, AEO2007.D112106A, and HP2007.D112106A.

Price Case Comparisons

Table C4. Liquid Fuels Supply and Disposition
(Million Barrels per Day, Unless Otherwise Noted)

Supply and Disposition	2005	Projections								
		2010			2020			2030		
		Low Price	Reference	High Price	Low Price	Reference	High Price	Low Price	Reference	High Price
Crude Oil										
Domestic Crude Production[1]	5.18	5.81	5.67	5.51	5.95	5.89	5.98	5.25	5.39	6.04
Alaska	0.86	0.69	0.69	0.69	0.69	0.74	0.71	0.25	0.27	0.42
Lower 48 States	4.31	5.12	4.98	4.82	5.25	5.15	5.28	4.99	5.12	5.62
Net Imports	10.09	9.94	9.99	10.01	11.69	11.29	10.02	14.35	13.09	10.59
Gross Imports	10.12	9.98	10.03	10.05	11.73	11.33	10.06	14.38	13.12	10.63
Exports	0.03	0.04	0.04	0.03	0.04	0.04	0.04	0.03	0.03	0.04
Other Crude Supply[2]	-0.06	0.00	0.00	0.00	0.00	0.00	0.00	0.00	0.00	0.00
Total Crude Supply	**15.22**	**15.75**	**15.66**	**15.52**	**17.64**	**17.19**	**16.00**	**19.60**	**18.47**	**16.63**
Other Supply										
Natural Gas Plant Liquids	1.72	1.82	1.80	1.76	1.77	1.76	1.69	1.73	1.72	1.71
Net Product Imports	2.48	1.99	1.80	1.72	3.46	2.27	1.53	4.95	3.28	1.45
Gross Refined Product Imports[3]	2.45	1.99	1.78	1.76	2.83	1.98	1.52	3.73	2.52	1.43
Unfinished Oil Imports	0.58	0.42	0.41	0.36	0.66	0.51	0.42	0.90	0.67	0.43
Ethanol Imports	0.01	0.01	0.02	0.02	0.01	0.04	0.04	0.05	0.05	0.05
Blending Component Imports	0.54	0.82	0.82	0.80	1.29	1.03	0.79	1.72	1.36	0.81
Exports	1.07	1.25	1.23	1.22	1.33	1.29	1.25	1.44	1.33	1.28
Refinery Processing Gain[4]	0.99	1.18	1.21	1.16	1.18	1.41	1.28	1.22	1.49	1.37
Other Inputs	0.39	0.96	1.02	1.13	1.07	1.31	1.98	1.19	1.88	3.30
Ethanol	0.26	0.59	0.69	0.72	0.60	0.79	0.78	0.63	0.90	0.96
Liquids from Gas	0.00	0.00	0.00	0.00	0.00	0.00	0.10	0.00	0.00	0.10
Liquids from Coal	0.00	0.00	0.00	0.00	0.00	0.10	0.60	0.00	0.44	1.65
Other[5]	0.13	0.36	0.33	0.41	0.47	0.43	0.50	0.57	0.53	0.59
Total Primary Supply[6]	**20.79**	**21.70**	**21.49**	**21.28**	**25.13**	**23.94**	**22.48**	**28.70**	**26.84**	**24.46**
Liquid Fuels Consumption										
by Fuel										
Liquefied Petroleum Gases	2.03	2.22	2.22	2.21	2.28	2.26	2.29	2.46	2.42	2.45
E85[7]	0.00	0.00	0.00	0.00	0.01	0.01	0.03	0.01	0.02	0.21
Motor Gasoline[8]	9.16	9.58	9.53	9.38	11.37	10.93	9.79	13.23	12.53	10.47
Jet Fuel[9]	1.68	1.97	1.95	1.94	2.22	2.19	2.16	2.29	2.27	2.06
Distillate Fuel Oil[10]	4.12	4.58	4.53	4.48	5.47	5.11	4.98	6.64	5.95	5.85
Residual Fuel Oil	0.92	0.81	0.79	0.78	1.06	0.82	0.73	1.16	0.83	0.76
Other[11]	2.84	2.63	2.57	2.58	2.79	2.70	2.61	3.06	2.93	2.80
by Sector										
Residential and Commercial	1.26	1.28	1.25	1.24	1.39	1.29	1.21	1.41	1.28	1.19
Industrial[12]	5.07	5.08	5.01	5.00	5.30	5.16	5.04	5.75	5.53	5.37
Transportation	13.87	15.01	14.93	14.74	17.71	17.15	15.97	20.55	19.69	17.64
Electric Power[13]	0.51	0.42	0.40	0.40	0.81	0.43	0.37	1.13	0.45	0.39
Total	**20.75**	**21.79**	**21.59**	**21.37**	**25.21**	**24.03**	**22.59**	**28.84**	**26.95**	**24.60**
Discrepancy[14]	**0.04**	**-0.09**	**-0.10**	**-0.09**	**-0.08**	**-0.09**	**-0.11**	**-0.14**	**-0.11**	**-0.13**

Table C4. Liquid Fuels Supply and Disposition (Continued)
(Million Barrels per Day, Unless Otherwise Noted)

Supply and Disposition	2005	Projections								
		2010			2020			2030		
		Low Price	Reference	High Price	Low Price	Reference	High Price	Low Price	Reference	High Price
Domestic Refinery Distillation Capacity[15]	17.1	17.9	17.8	18.0	19.0	18.7	18.1	21.0	20.0	18.5
Capacity Utilization Rate (percent)[16]	91.0	89.4	89.1	87.5	94.2	93.4	89.6	94.7	93.5	91.1
Net Import Share of Product Supplied (percent)	60.5	55.0	54.9	55.1	60.3	56.6	51.4	67.3	61.0	49.2
Net Expenditures for Imported Crude Oil and Petroleum Products (billion 2005 dollars)	236.65	194.59	222.76	268.39	161.96	229.80	344.40	211.83	300.51	404.77

[1]Includes lease condensate.
[2]Strategic petroleum reserve stock additions plus unaccounted for crude oil and crude stock withdrawals minus crude product supplied.
[3]Includes other hydrocarbons and alcohols.
[4]Represents volumetric gain in refinery distillation and cracking processes.
[5]Includes petroleum product stock withdrawals, domestic sources of blending components, other hydrocarbons, ethers, and renewable fuels such as biodiesel.
[6]Total crude supply plus natural gas plant liquids, other inputs, refinery processing gain, and net product imports.
[7]E85 refers to a blend of 85 percent ethanol (renewable) and 15 percent motor gasoline (nonrenewable). To address cold starting issues, the percentage of ethanol actually varies seasonally. The annual average ethanol content of 74 percent is used for this forecast.
[8]Includes ethanol and ethers blended into gasoline.
[9]Includes only kerosene type.
[10]Includes distillate fuel oil and kerosene from petroleum and biomass feedstocks.
[11]Includes aviation gasoline, petrochemical feedstocks, lubricants, waxes, asphalt, road oil, still gas, special naphthas, petroleum coke, crude oil product supplied, tire-derived fuel, methanol, liquid hydrogen,and miscellaneous petroleum products.
[12]Includes consumption for combined heat and power, which produces electricity and other useful thermal energy.
[13]Includes consumption of energy by electricity-only and combined heat and power plants whose primary business is to sell electricity, or electricity and heat, to the public. Includes small power producers and exempt wholesale generators.
[14]Balancing item. Includes unaccounted for supply, losses, and gains.
[15]End-of-year operable capacity.
[16]Rate is calculated by dividing the gross annual input to atmospheric crude oil distillation units by their operable refining capacity in barrels per calendar day.
N/A = Not applicable.
Note: Totals may not equal sum of components due to independent rounding. Data for 2005 are model results and may differ slightly from official EIA data reports.
Sources: 2005 imported crude oil price and petroleum product supplied based on: Energy Information Administration (EIA), *Annual Energy Review 2005*, DOE/EIA-0384(2005) (Washington, DC, July 2006). 2005 imported low sulfur light crude oil price: EIA, Form EIA-856, "Monthly Foreign Crude Oil Acquisition Report." Other 2005 data: EIA, *Petroleum Supply Annual 2005,* DOE/EIA-0340(2005)/1 (Washington, DC, October 2006). **Projections:** EIA, AEO2007 National Energy Modeling System runs LP2007.D112106A, AEO2007.D112106A, and HP2007.D112106A.

Table C5. Petroleum Product Prices
(2005 Cents per Gallon, Unless Otherwise Noted)

Sector and Fuel	2005	Projections								
		2010			2020			2030		
		Low Price	Reference	High Price	Low Price	Reference	High Price	Low Price	Reference	High Price
Crude Oil Prices (2005 dollars per barrel)										
Imported Low Sulfur Light Crude Oil[1]	56.76	49.21	57.47	69.21	34.10	52.04	89.12	35.68	59.12	100.14
Imported Crude Oil[1]	49.19	44.06	51.20	62.53	28.91	46.47	82.60	28.91	51.63	92.93
Delivered Sector Product Prices										
Residential										
Liquefied Petroleum Gases	166.3	198.5	204.0	209.4	191.1	199.8	206.2	198.0	206.1	221.0
Distillate Fuel Oil	204.3	175.2	206.3	229.9	126.1	182.3	262.8	134.0	195.9	275.1
Commercial										
Distillate Fuel Oil	175.4	145.9	175.5	201.2	98.6	156.5	232.3	108.6	171.7	248.9
Residual Fuel Oil	126.0	99.0	112.8	109.5	65.8	105.8	169.8	68.7	109.4	180.7
Residual Fuel Oil (2005 dollars per barrel)	52.90	41.56	47.39	45.99	27.62	44.44	71.33	28.85	45.97	75.89
Industrial[2]										
Liquefied Petroleum Gases	146.2	136.1	141.5	147.3	128.3	137.1	143.6	135.7	142.6	157.3
Distillate Fuel Oil	181.1	150.3	178.1	205.4	108.5	165.3	237.5	121.4	181.8	257.9
Residual Fuel Oil	116.3	126.0	142.2	160.8	92.1	133.4	211.9	95.9	143.5	230.0
Residual Fuel Oil (2005 dollars per barrel)	48.86	52.94	59.74	67.52	38.68	56.04	88.98	40.29	60.25	96.58
Transportation										
Liquefied Petroleum Gases	206.1	204.4	209.8	214.4	195.5	203.9	208.6	201.2	209.3	222.8
Ethanol (E85)[3]	217.1	205.4	198.5	208.8	173.6	192.1	217.3	181.0	200.4	237.9
Ethanol Wholesale Price	180.4	174.8	181.4	184.0	152.3	168.2	174.0	155.3	170.2	183.5
Motor Gasoline[4]	231.6	201.8	217.3	250.1	170.7	201.9	286.1	175.8	215.4	320.1
Jet Fuel[5]	177.4	130.5	147.2	173.3	100.3	141.8	214.1	110.8	158.6	235.5
Diesel Fuel (distillate fuel oil)[6]	241.3	202.7	230.4	257.2	154.1	211.2	283.9	164.7	225.7	304.2
Residual Fuel Oil	82.4	103.2	120.5	147.2	71.2	110.2	192.4	75.2	123.8	215.2
Residual Fuel Oil (2005 dollars per barrel)	34.62	43.33	50.60	61.81	29.91	46.27	80.79	31.60	52.02	90.40
Electric Power[7]										
Distillate Fuel Oil	157.9	130.9	162.3	188.5	70.3	136.5	212.7	75.4	149.6	221.1
Residual Fuel Oil	104.2	84.4	98.5	113.7	52.6	91.0	168.3	55.0	102.5	188.1
Residual Fuel Oil (2005 dollars per barrel)	43.76	35.44	41.37	47.75	22.10	38.24	70.69	23.11	43.05	79.02
Refined Petroleum Product Prices[8]										
Liquefied Petroleum Gases	150.7	149.9	155.3	161.0	143.1	151.9	158.9	150.3	157.7	173.5
Motor Gasoline[4]	231.1	201.8	217.3	250.1	170.6	201.8	286.0	175.8	215.4	320.1
Jet Fuel[5]	177.4	130.5	147.2	173.3	100.3	141.8	214.1	110.8	158.6	235.5
Distillate Fuel Oil	223.9	187.8	215.9	242.5	139.8	199.4	273.0	149.0	215.5	294.4
Residual Fuel Oil	98.6	97.9	113.9	133.3	63.1	104.7	184.3	65.8	116.6	204.8
Residual Fuel Oil (2005 dollars per barrel)	41.42	41.10	47.84	55.97	26.50	43.98	77.40	27.63	48.96	86.01
Average	**204.5**	**177.5**	**195.0**	**221.3**	**145.8**	**183.4**	**253.3**	**153.5**	**198.1**	**281.4**

[1]Weighted average price delivered to U.S. refiners.
[2]Includes energy for combined heat and power plants, except those whose primary business is to sell electricity, or electricity and heat, to the public.
[3]E85 refers to a blend of 85 percent ethanol (renewable) and 15 percent motor gasoline (nonrenewable). To address cold starting issues, the percentage of ethanol actually varies seasonally. The annual average ethanol content of 74 percent is used for this forecast.
[4]Sales weighted-average price for all grades. Includes Federal, State and local taxes.
[5]Includes only kerosene type.
[6]Diesel fuel for on-road use. Includes Federal and State taxes while excluding county and local taxes.
[7]Includes electricity-only and combined heat and power plants whose primary business is to sell electricity, or electricity and heat, to the public. Includes small power producers and exempt wholesale generators.
[8]Weighted averages of end-use fuel prices are derived from the prices in each sector and the corresponding sectoral consumption.
Note: Data for 2005 are model results and may differ slightly from official EIA data reports.
Sources: 2005 imported low sulfur light crude oil price: Energy Information Administration (EIA), Form EIA-856, "Monthly Foreign Crude Oil Acquisition Report." 2005 imported crude oil price: EIA, *Annual Energy Review 2005*, DOE/EIA-0384(2005) (Washington, DC, July 2006). 2005 prices for motor gasoline, distillate fuel oil, and jet fuel are based on: EIA, *Petroleum Marketing Annual 2005*, DOE/EIA-0487(2005) (Washington, DC, August 2006). 2005 residential, commercial, industrial, and transportation sector petroleum product prices are derived from: EIA, Form EIA-782A, "Refiners'/Gas Plant Operators' Monthly Petroleum Product Sales Report." 2005 electric power prices based on: Federal Energy Regulatory Commission, FERC Form 423, "Monthly Report of Cost and Quality of Fuels for Electric Plants." 2005 ethanol prices derived from weekly spot prices in the Oxy Fuel News. 2005 wholesale ethanol prices derived from Bloomburg U.S. average rack price. **Projections:** EIA, AEO2007 National Energy Modeling System runs LP2007.D112106A, AEO2007.D112106A, and HP2007.D112106A.

Table C6. International Petroleum Supply and Disposition Summary
(Million Barrels per Day, Unless Otherwise Noted)

Supply and Disposition	2005	Projections								
		2010			2020			2030		
		Low Price	Reference	High Price	Low Price	Reference	High Price	Low Price	Reference	High Price
Crude Oil Prices (2005 dollars per barrel)										
Imported Low Sulfur Light Crude Oil Price[1]	56.76	49.21	57.47	69.21	34.10	52.04	89.12	35.68	59.12	100.14
Imported Crude Oil Price[1]	49.19	44.06	51.20	62.53	28.91	46.47	82.60	28.91	51.63	92.93
Conventional Production (Conventional)[2]										
OPEC[3]										
Asia	1.17	1.11	1.11	0.96	1.17	1.09	0.76	1.30	1.10	0.70
Middle East	22.96	21.81	22.23	20.37	29.34	26.60	17.99	39.85	33.20	21.34
North Africa	3.78	4.33	4.29	3.55	4.57	4.24	2.87	4.48	3.93	2.48
West Africa	2.78	3.10	3.07	2.38	4.22	4.10	2.40	4.61	4.48	2.21
South America	2.71	2.95	2.59	2.57	3.10	2.30	2.23	3.02	2.24	2.15
Total OPEC	**33.41**	**33.30**	**33.30**	**29.83**	**42.40**	**38.33**	**26.25**	**53.26**	**44.95**	**28.88**
Non-OPEC										
OECD										
United States (50 states)	8.03	9.15	8.98	8.81	9.37	9.48	9.29	8.80	9.12	9.19
Canada	2.12	1.95	1.93	1.80	2.18	1.89	1.61	2.11	1.62	1.25
Mexico	3.78	3.23	3.15	3.08	3.51	3.18	3.04	4.05	3.52	2.99
OECD Europe[4]	5.96	5.92	5.73	5.64	4.74	4.22	4.01	3.65	3.16	2.62
Japan	0.10	0.10	0.10	0.10	0.10	0.10	0.10	0.10	0.10	0.10
Australia and New Zealand	0.60	0.56	0.56	0.56	0.56	0.51	0.51	0.64	0.60	0.56
Total OECD	**20.59**	**20.90**	**20.45**	**19.99**	**20.46**	**19.39**	**18.56**	**19.34**	**18.12**	**16.71**
Non-OECD										
Russia	9.51	10.30	9.98	9.88	11.80	10.79	10.39	13.18	11.54	9.60
Other Eurasia[5]	2.48	4.10	3.98	3.88	6.01	5.41	5.22	7.50	6.55	5.53
China	3.74	3.63	3.53	3.42	3.70	3.30	3.20	3.80	3.20	2.60
Other Asia[6]	2.53	2.39	2.29	2.19	2.90	2.60	2.50	2.90	2.50	2.10
Middle East[7]	1.67	2.10	2.00	2.00	2.70	2.40	2.30	3.30	2.90	2.30
Africa	3.59	5.33	5.19	5.19	7.73	7.38	6.95	10.72	9.83	8.10
Brazil	1.76	2.49	2.39	2.39	3.50	3.20	3.10	4.40	3.90	3.20
Other Central and South America	2.31	2.34	2.32	2.20	2.96	2.66	2.56	3.26	2.90	2.46
Total Non-OECD	**27.59**	**32.68**	**31.67**	**31.14**	**41.31**	**37.75**	**36.22**	**49.05**	**43.32**	**35.89**
Total Conventional Production	**81.59**	**86.88**	**85.42**	**80.97**	**104.16**	**95.47**	**81.03**	**121.66**	**106.40**	**81.48**
Unconventional Production[8]										
United States (50 states)	0.25	0.61	0.71	0.74	0.61	0.91	1.62	0.64	1.37	3.20
Other North America	1.09	1.90	1.91	1.87	2.10	2.74	3.57	2.32	3.66	4.91
OECD Europe[3]	0.08	0.14	0.15	0.17	0.18	0.19	0.36	0.24	0.27	0.53
Middle East[7]	0.02	0.57	0.57	0.57	0.47	0.75	0.97	0.28	1.11	1.28
Africa.	0.16	0.30	0.32	0.42	0.39	0.52	1.26	0.52	0.73	2.36
Central and South America	0.93	1.35	1.35	1.54	1.64	1.81	2.99	1.82	2.40	4.15
Other	0.28	0.50	0.62	0.60	0.57	0.90	1.55	0.60	1.41	3.69
Total Unconventional Production	**2.80**	**5.36**	**5.63**	**5.91**	**5.95**	**7.83**	**12.33**	**6.41**	**10.93**	**20.12**
Total Production	**84.39**	**92.24**	**91.05**	**86.88**	**110.11**	**103.29**	**93.36**	**128.07**	**117.33**	**101.60**

Table C6. International Petroleum Supply and Disposition Summary (Continued)
(Million Barrels per Day, Unless Otherwise Noted)

Supply and Disposition	2005	2010 Low Price	2010 Reference	2010 High Price	2020 Low Price	2020 Reference	2020 High Price	2030 Low Price	2030 Reference	2030 High Price
Consumption[8]										
OECD										
United States (50 states)	20.75	21.78	21.59	21.37	25.20	24.02	22.57	28.83	26.93	24.44
United States Territories	0.38	0.45	0.43	0.39	0.58	0.51	0.48	0.68	0.59	0.56
Canada	2.28	2.44	2.42	2.35	2.55	2.49	2.35	2.68	2.59	2.34
Mexico	2.09	2.26	2.22	2.06	2.92	2.68	2.34	3.57	3.19	2.67
OECD Europe[3]	15.73	15.81	15.82	15.42	16.03	15.76	15.02	16.61	16.26	14.84
Japan	5.58	5.43	5.42	5.28	5.52	5.43	5.17	5.57	5.45	4.97
South Korea	2.30	2.58	2.58	2.50	3.12	3.04	2.87	3.57	3.45	3.12
Australia and New Zealand	1.05	1.09	1.08	1.04	1.16	1.13	1.07	1.26	1.22	1.10
Total OECD	**50.16**	**51.84**	**51.54**	**50.41**	**57.08**	**55.05**	**51.86**	**62.74**	**59.69**	**54.04**
Non-OECD										
Russia	2.75	2.88	2.85	2.68	3.32	3.11	2.79	3.69	3.39	2.91
Other Non-OECD Eurasia[5]	2.33	2.66	2.63	2.48	3.39	3.18	2.85	4.08	3.75	3.22
China	6.86	8.93	8.70	7.97	12.95	11.66	9.92	17.25	15.05	12.25
India	2.52	3.02	2.94	2.69	4.06	3.66	3.11	5.10	4.45	3.63
Other Non-OECD Asia	6.02	7.07	6.89	6.31	9.45	8.51	7.24	11.79	10.29	8.37
Middle East[7]	5.56	6.17	6.06	5.63	7.62	7.00	6.11	8.72	7.81	6.52
Africa	3.01	3.80	3.70	3.39	4.77	4.30	3.66	5.65	4.93	4.02
Brazil	2.20	2.44	2.39	2.23	3.07	2.82	2.46	3.68	3.29	2.75
Other Central and South America	2.99	3.44	3.36	3.08	4.43	4.00	3.36	5.36	4.68	3.73
Total Non-OECD	**34.23**	**40.40**	**39.52**	**36.46**	**53.08**	**48.23**	**41.49**	**65.33**	**57.64**	**47.40**
Total Consumption	**84.39**	**92.24**	**91.05**	**86.88**	**110.16**	**103.29**	**93.36**	**128.07**	**117.33**	**101.44**
OPEC Production[10]	34.04	34.72	34.72	31.25	43.89	40.19	29.26	54.68	47.65	33.29
Non-OPEC Production[10]	50.35	57.52	56.34	55.63	66.22	63.10	64.10	73.39	69.68	68.30
Net Eurasia Exports	8.67	11.35	10.87	10.98	14.60	13.12	13.07	17.31	14.85	12.20
OPEC Market Share	40.3	37.6	38.1	36.0	39.9	38.9	31.3	42.7	40.6	32.8

[1]Weighted average price delivered to U.S. refiners.

[2]Includes production of crude oil (including lease condensates), natural gas plant liquids, other hydrogen and hydrocarbons for refinery feedstocks, alcohol and other sources, and refinery gains.

[3]OPEC = Organization of Petroleum Exporting Countries - Algeria, Indonesia, Iran, Iraq, Kuwait, Libya, Nigeria, Qatar, Saudi Arabia, the United Arab Emirates, and Venezuela. Does not include Angola, which was admitted as a full member to OPEC on December 14, 2006.

[4]OECD Europe = Organization for Economic Cooperation and Development - Austria, Belgium, Czech Republic, Denmark, Finland, France, Germany, Greece, Hungary, Iceland, Ireland, Italy, Luxembourg, the Netherlands, Norway, Poland, Portugal, Slovakia, Spain, Sweden, Switzerland, Turkey, and the United Kingdom.

[5]Eurasia consists of Armenia, Azerbaijan, Belarus, Estonia, Georgia, Kazakhstan, Kyrgyzstan, Latvia, Lithuania, Moldova, Russia, Tajikistan, Turkmenistan, Ukraine, and Uzbekistan.

[6]Other Asia = Afghanistan, Bangladesh, Bhutan, Brunei, Cambodia (Kampuchea), Fiji, French Polynesia, Guam, Hong Kong, Indonesia, Kiribati, Laos, Malaysia, Macau, Maldives, Mongolia, Myanmar (Burma), Nauru, Nepal, New Caledonia, Niue, North Korea, Pakistan, Papua New Guinea, Philippines, Samoa, Singapore, Solomon Islands, Sri Lanka, Taiwan, Thailand, Tonga, Vanuatu, and Vietnam.

[7]Non-OPEC Middle East includes Turkey.

[8]Includes liquids produced from energy crops, natural gas, coal, oil sands, and shale. Includes both OPEC and non-OPEC producers in the regional breakdown.

[9]Includes both OPEC and non-OPEC consumers in the regional breakdown.

[10]Includes both conventional and nonconventional liquids production.

Note: Totals may not equal sum of components due to independent rounding. Data for 2005 are model results and may differ slightly from official EIA data reports.

Sources: 2005 low sulfur light crude oil price: Energy Information Administration (EIA), Form EIA-856, "Monthly Foreign Crude Oil Acquisition Report." 2005 imported crude oil price: EIA, *Annual Energy Review 2005*, DOE/EIA-0384(2005) (Washington, DC, July 2006). **2005 quantities and projections:** Energy Information Administration, AEO2007 National Energy Modeling System runs LP2007.D112106A, AEO2007.D112106A, and HP2007.D112106A.

Results from Side Cases

Table D1. Key Results for Residential and Commercial Sector Technology Cases

Energy Consumption	2005	2010				2020			
		2006 Technology	Reference	High Technology	Best Available Technology	2006 Technology	Reference	High Technology	Best Available Technology
Residential									
Energy Consumption (quadrillion Btu)									
Liquefied Petroleum Gases	0.51	0.54	0.53	0.53	0.52	0.58	0.58	0.56	0.54
Kerosene	0.10	0.10	0.10	0.10	0.09	0.10	0.10	0.09	0.09
Distillate Fuel Oil	0.93	0.91	0.90	0.89	0.87	0.88	0.85	0.83	0.78
Liquid Fuels and Other Petroleum	1.54	1.54	1.53	1.52	1.49	1.56	1.53	1.48	1.41
Natural Gas	4.98	5.21	5.18	5.15	4.88	5.55	5.43	5.23	4.38
Coal	0.01	0.01	0.01	0.01	0.01	0.01	0.01	0.01	0.01
Renewable Energy[1]	0.41	0.43	0.43	0.42	0.41	0.42	0.40	0.39	0.38
Electricity	4.66	5.08	5.06	5.02	4.26	5.95	5.80	5.54	4.27
Delivered Energy	**11.60**	**12.27**	**12.21**	**12.11**	**11.05**	**13.48**	**13.17**	**12.66**	**10.45**
Electricity Related Losses	10.15	10.95	10.90	10.80	9.18	12.39	12.08	11.53	8.88
Total	**21.75**	**23.22**	**23.11**	**22.91**	**20.23**	**25.87**	**25.26**	**24.18**	**19.33**
Delivered Energy Intensity (million Btu per household)	**102.3**	**101.7**	**101.1**	**100.3**	**91.6**	**100.1**	**97.8**	**94.0**	**77.5**
Nonmarketed Renewables Consumption (quadrillion Btu)	**0.03**	**0.04**	**0.04**	**0.04**	**0.04**	**0.06**	**0.06**	**0.09**	**0.06**
Commercial									
Energy Consumption (quadrillion Btu)									
Liquefied Petroleum Gases	0.09	0.09	0.09	0.09	0.09	0.10	0.10	0.10	0.10
Motor Gasoline[2]	0.05	0.05	0.05	0.05	0.05	0.05	0.05	0.05	0.05
Kerosene	0.02	0.03	0.03	0.03	0.03	0.03	0.03	0.03	0.03
Distillate Fuel Oil	0.48	0.45	0.45	0.45	0.46	0.52	0.48	0.48	0.55
Residual Fuel Oil	0.14	0.14	0.14	0.14	0.14	0.14	0.14	0.14	0.14
Liquid Fuels and Other Petroleum	0.77	0.75	0.75	0.75	0.76	0.84	0.80	0.79	0.86
Natural Gas	3.15	3.32	3.31	3.31	3.26	3.86	3.86	3.86	3.67
Coal	0.10	0.10	0.10	0.10	0.10	0.10	0.10	0.10	0.10
Renewable Energy[3]	0.12	0.12	0.12	0.12	0.12	0.12	0.12	0.12	0.12
Electricity	4.32	4.83	4.77	4.64	4.37	6.06	5.78	5.44	4.78
Delivered Energy	**8.46**	**9.12**	**9.05**	**8.92**	**8.61**	**10.98**	**10.66**	**10.31**	**9.53**
Electricity Related Losses	9.42	10.40	10.27	9.98	9.42	12.63	12.03	11.32	9.96
Total	**17.88**	**19.52**	**19.33**	**18.90**	**18.03**	**23.60**	**22.69**	**21.63**	**19.48**
Delivered Energy Intensity (thousand Btu per square foot)	**113.9**	**113.5**	**112.6**	**110.9**	**107.1**	**118.2**	**114.7**	**111.0**	**102.6**
Commercial Sector Generation									
Net Summer Generation Capacity (megawatts)									
Natural Gas	588	591	593	591	591	607	622	658	666
Solar Photovoltaic	159	209	487	487	487	210	617	622	817
Electricity Generation (billion kilowatthours)									
Natural Gas	4.23	4.26	4.27	4.26	4.26	4.37	4.48	4.74	4.80
Solar Photovoltaic	0.30	0.40	0.93	0.93	0.93	0.40	1.18	1.18	1.55
Nonmarketed Renewables Consumption (quadrillion Btu)	**0.03**	**0.03**	**0.03**	**0.03**	**0.03**	**0.03**	**0.03**	**0.03**	**0.03**

[1]Includes wood used for residential heating. See Table A4 and/or Table A17 for estimates of nonmarketed renewable energy consumption for geothermal heat pumps, solar thermal hot water heating, and solar photovoltaic electricity generation.
[2]Includes ethanol (blends of 10 percent or less) and ethers blended into gasoline.
[3]Includes commercial sector consumption of wood and wood waste, landfill gas, municipal solid waste, and other biomass for combined heat and power.
Btu = British thermal unit.
Note: Totals may not equal sum of components due to independent rounding. Data for 2005 are model results and may differ slightly from official EIA data reports. Side cases were run without the fully integrated modeling system, so not all feedbacks are captured. The reference case ratio of electricity losses to electricity use was used to compute electricity losses for the technology cases.
Source: Energy Information Administration, AEO2007 National Energy Modeling System, runs BLDFRZN.D112206A, AEO2007.D112106A, BLDHIGH.D112206A, and BLDBEST.D112206A.

2030				Annual Growth 2005-2030 (percent)			
2006 Technology	Reference	High Technology	Best Available Technology	2006 Technology	Reference	High Technology	Best Available Technology
0.62	0.62	0.59	0.57	0.8%	0.8%	0.6%	0.5%
0.09	0.09	0.08	0.08	-0.1%	-0.3%	-0.5%	-0.7%
0.81	0.76	0.72	0.66	-0.6%	-0.8%	-1.0%	-1.4%
1.52	1.46	1.40	1.32	-0.0%	-0.2%	-0.4%	-0.6%
5.70	5.47	5.15	4.22	0.5%	0.4%	0.1%	-0.7%
0.01	0.01	0.01	0.01	-0.9%	-1.2%	-1.3%	-1.4%
0.42	0.39	0.38	0.36	0.1%	-0.2%	-0.3%	-0.5%
6.75	6.47	5.91	4.54	1.5%	1.3%	1.0%	-0.1%
14.39	**13.80**	**12.85**	**10.44**	**0.9%**	**0.7%**	**0.4%**	**-0.4%**
13.44	12.89	11.77	9.04	1.1%	1.0%	0.6%	-0.5%
27.84	**26.70**	**24.62**	**19.49**	**1.0%**	**0.8%**	**0.5%**	**-0.4%**
97.6	**93.6**	**87.1**	**70.8**	**-0.2%**	**-0.4%**	**-0.6%**	**-1.5%**
0.08	**0.08**	**0.14**	**0.10**	**4.0%**	**4.0%**	**6.4%**	**5.1%**
0.10	0.10	0.10	0.10	0.4%	0.4%	0.4%	0.4%
0.05	0.05	0.05	0.05	0.6%	0.6%	0.6%	0.6%
0.03	0.03	0.03	0.03	0.4%	0.4%	0.4%	0.4%
0.56	0.49	0.49	0.61	0.7%	0.1%	0.1%	1.0%
0.14	0.14	0.14	0.14	0.2%	0.2%	0.2%	0.2%
0.89	0.81	0.81	0.94	0.6%	0.2%	0.2%	0.8%
4.35	4.36	4.38	4.10	1.3%	1.3%	1.3%	1.1%
0.10	0.10	0.10	0.10	-0.1%	-0.1%	-0.1%	-0.1%
0.12	0.12	0.12	0.12	0.0%	0.0%	0.0%	0.0%
7.56	7.03	6.54	5.53	2.3%	2.0%	1.7%	1.0%
13.02	**12.43**	**11.95**	**10.79**	**1.7%**	**1.6%**	**1.4%**	**1.0%**
15.06	14.01	13.03	11.03	1.9%	1.6%	1.3%	0.6%
28.08	**26.44**	**24.98**	**21.81**	**1.8%**	**1.6%**	**1.3%**	**0.8%**
120.5	**115.1**	**110.6**	**99.8**	**0.2%**	**0.0%**	**-0.1%**	**-0.5%**
734	979	1571	1808	0.9%	2.1%	4.0%	4.6%
212	2292	3204	10850	1.1%	11.3%	12.8%	18.4%
5.29	7.08	11.38	13.11	0.9%	2.1%	4.0%	4.6%
0.40	4.34	6.04	19.82	1.2%	11.3%	12.7%	18.2%
0.03	**0.04**	**0.05**	**0.10**	**0.6%**	**2.1%**	**2.6%**	**5.4%**

Table D2. Key Results for Industrial Sector Technology Cases, Excluding Refining

Consumption	2005	2010			2020			2030		
		2006 Technology	Reference	High Technology	2006 Technology	Reference	High Technology	2006 Technology	Reference	High Technology
Value of Shipments (billion 2000 dollars)										
Manufacturing	4002	4462	4462	4462	5666	5666	5666	7183	7183	7183
Nonmanufacturing	1538	1596	1596	1596	1846	1846	1846	2023	2023	2023
Total	**5540**	**6059**	**6059**	**6059**	**7513**	**7513**	**7513**	**9207**	**9207**	**9207**
Energy Consumption excluding Refining[1] (quadrillion Btu)										
Liquefied Petroleum Gases	2.12	2.31	2.26	2.21	2.43	2.26	2.13	2.61	2.37	2.22
Heat and Power	0.13	0.09	0.08	0.08	0.09	0.08	0.08	0.10	0.08	0.08
Feedstocks	1.98	2.22	2.17	2.12	2.33	2.18	2.05	2.51	2.29	2.15
Motor Gasoline	0.32	0.33	0.32	0.31	0.37	0.33	0.31	0.40	0.36	0.32
Distillate Fuel Oil	1.22	1.21	1.18	1.14	1.35	1.22	1.11	1.45	1.26	1.13
Residual Fuel Oil	0.22	0.15	0.14	0.14	0.16	0.14	0.13	0.17	0.14	0.14
Petrochemical Feedstocks	1.38	1.52	1.48	1.46	1.59	1.50	1.43	1.69	1.57	1.48
Petroleum Coke	0.33	0.33	0.31	0.30	0.36	0.31	0.29	0.41	0.34	0.30
Asphalt and Road Oil	1.31	1.32	1.24	1.17	1.56	1.29	1.08	1.71	1.37	1.12
Miscellaneous Petroleum[2]	0.59	0.49	0.45	0.44	0.51	0.38	0.37	0.55	0.38	0.35
Petroleum Subtotal	7.48	7.66	7.38	7.17	8.32	7.43	6.84	8.98	7.79	7.07
Natural Gas Heat and Power	5.30	6.12	5.83	5.75	7.06	6.22	5.99	7.96	6.97	6.61
Natural Gas Feedstocks	0.57	0.59	0.58	0.57	0.60	0.57	0.54	0.62	0.58	0.54
Lease and Plant Fuel[3]	1.10	1.10	1.10	1.10	1.21	1.21	1.21	1.15	1.15	1.15
Natural Gas Subtotal	6.97	7.81	7.51	7.41	8.87	7.99	7.73	9.74	8.70	8.30
Metallurgical Coal and Coke[4]	0.66	0.65	0.63	0.59	0.65	0.59	0.50	0.67	0.59	0.46
Other Industrial Coal	1.23	1.29	1.26	1.25	1.32	1.23	1.19	1.36	1.25	1.18
Coal Subtotal	1.89	1.94	1.88	1.84	1.97	1.81	1.68	2.03	1.84	1.64
Renewables[5]	1.44	1.60	1.60	1.62	1.81	1.81	1.94	2.03	2.05	2.32
Purchased Electricity	3.35	3.53	3.44	3.36	3.91	3.63	3.40	4.35	3.87	3.49
Delivered Energy	**21.14**	**22.54**	**21.81**	**21.41**	**24.88**	**22.67**	**21.59**	**27.12**	**24.24**	**22.82**
Electricity Related Losses	7.31	7.59	7.41	7.24	8.14	7.56	7.08	8.67	7.70	6.96
Total	**28.45**	**30.13**	**29.22**	**28.65**	**33.01**	**30.23**	**28.67**	**35.79**	**31.94**	**29.78**
Delivered Energy Use per Dollar of Shipments (thousand Btu per 2000 dollar)	**3.81**	**3.72**	**3.60**	**3.53**	**3.31**	**3.02**	**2.87**	**2.95**	**2.63**	**2.48**
Onsite Industrial Combined Heat and Power										
Capacity (gigawatts)	21.43	23.48	23.46	23.77	29.71	29.87	31.72	36.15	37.95	42.17
Generation (billion kilowatthours)	114.89	129.87	129.75	131.74	175.24	176.16	188.16	222.07	234.87	261.67

[1]Fuel consumption includes energy for combined heat and power plants, except those whose primary business is to sell electricity, or electricity and heat, to the public.
[2]Includes lubricants and miscellaneous petroleum products.
[3]Represents natural gas used in the field gathering and processing plant machinery.
[4]Includes net coal coke imports.
[5]Includes consumption of energy from hydroelectric, wood and wood waste, municipal solid waste, and other biomass.
Btu = British thermal unit.
Note: Totals may not equal sum of components due to independent rounding. Data for 2005 are model results and may differ slightly from official EIA data reports. Side cases were run without the fully integrated modeling system, so not all potential feedbacks were captured. The reference case ratio of electricity losses to electricity use was used to compute electricity losses for the technology cases.
Source: Energy Information Administration, AEO2007 National Energy Modeling System runs INDFRZN.D112406A, AEO2007.D112106A, and INDHIGH.D112406A.

Table D3. Key Results for Transportation Sector Technology Cases

Consumption and Indicators	2005	2010			2020			2030		
		2006 Technology	Reference	High Technology	2006 Technology	Reference	High Technology	2006 Technology	Reference	High Technology
Level of Travel										
(billion vehicle miles traveled)										
Light-Duty Vehicles less than 8,500	2655	2796	2799	2799	3461	3474	3501	4186	4226	4290
Commercial Light Trucks[1]	67	72	72	72	89	89	89	109	110	110
Freight Trucks greater than 10,000	230	255	255	255	318	318	319	398	397	398
(billion seat miles available)										
Air	1027	1172	1172	1172	1421	1410	1421	1555	1544	1555
(billion ton miles traveled)										
Rail	1590	1715	1714	1717	2002	2000	2004	2446	2445	2449
Domestic Shipping	613	662	661	662	731	730	731	775	775	776
Energy Efficiency Indicators										
(miles per gallon)										
New Light-Duty Vehicle[2]	25.2	26.8	27.3	28.6	26.9	28.2	30.7	27.1	29.2	32.1
New Car[2]	30.0	30.4	31.7	33.9	30.2	32.8	35.7	30.3	33.7	36.9
New Light Truck[2]	21.8	23.7	23.7	24.5	24.5	25.3	27.5	25.0	26.5	29.2
Light-Duty Stock[3]	19.6	19.6	19.8	19.9	20.6	21.2	22.4	20.9	22.2	24.1
New Commercial Light Truck[1]	14.6	15.8	15.8	16.4	16.1	16.7	18.3	16.1	17.4	19.4
Stock Commercial Light Truck[1]	14.1	14.7	14.7	14.8	16.0	16.2	17.2	16.1	17.0	18.8
Freight Truck	6.0	6.0	6.0	6.1	6.1	6.4	6.5	6.1	6.7	6.8
(seat miles per gallon)										
Aircraft	55.7	57.7	58.2	62.5	60.5	66.4	80.0	61.5	75.6	99.2
(ton miles per thousand Btu)										
Rail	2.9	2.9	2.9	3.0	2.9	3.0	3.3	2.9	3.0	3.5
Domestic Shipping	2.4	2.4	2.4	2.5	2.4	2.4	2.6	2.4	2.5	2.7
Energy Use (quadrillion Btu)										
by Mode										
Light-Duty Vehicles	16.36	16.79	16.76	16.58	19.83	19.44	18.45	23.72	22.66	21.06
Commercial Light Trucks[1]	0.59	0.61	0.61	0.61	0.70	0.69	0.65	0.85	0.81	0.74
Bus Transportation	0.26	0.27	0.27	0.27	0.30	0.28	0.28	0.33	0.30	0.30
Freight Trucks	4.77	5.29	5.29	5.26	6.55	6.18	6.11	8.15	7.40	7.32
Rail, Passenger	0.04	0.05	0.05	0.05	0.05	0.05	0.05	0.06	0.06	0.06
Rail, Freight	0.55	0.59	0.59	0.57	0.69	0.68	0.61	0.84	0.82	0.69
Shipping, Domestic	0.26	0.28	0.27	0.27	0.30	0.30	0.28	0.32	0.32	0.29
Shipping, International	0.76	0.77	0.77	0.77	0.79	0.79	0.78	0.81	0.80	0.80
Recreational Boats	0.18	0.20	0.20	0.20	0.24	0.24	0.24	0.27	0.27	0.27
Air	2.84	3.53	3.50	3.26	4.38	3.97	3.31	5.07	4.11	3.14
Military Use	0.71	0.73	0.73	0.73	0.77	0.77	0.77	0.80	0.80	0.80
Lubricants	0.15	0.15	0.15	0.15	0.16	0.16	0.16	0.16	0.16	0.16
Pipeline Fuel	0.58	0.66	0.66	0.66	0.79	0.79	0.79	0.79	0.79	0.79
Total	**28.05**	**29.92**	**29.86**	**29.37**	**35.55**	**34.33**	**32.50**	**42.17**	**39.29**	**36.41**
by Fuel										
Liquefied Petroleum Gases	0.04	0.05	0.05	0.05	0.06	0.06	0.06	0.08	0.08	0.07
E85[4]	0.00	0.00	0.00	0.00	0.01	0.01	0.01	0.02	0.02	0.02
Motor Gasoline[5]	17.00	17.39	17.37	17.18	20.37	19.95	18.97	24.06	22.89	21.33
Jet Fuel[6]	3.37	4.07	4.04	3.80	4.96	4.54	3.89	5.67	4.70	3.74
Distillate Fuel Oil[7]	6.02	6.65	6.64	6.59	8.20	7.81	7.63	10.32	9.58	9.25
Residual Fuel Oil	0.81	0.82	0.82	0.82	0.85	0.85	0.84	0.88	0.87	0.85
Liquid Hydrogen	0.00	0.00	0.00	0.00	0.00	0.00	0.00	0.00	0.00	0.00
Other Petroleum[8]	0.18	0.18	0.18	0.18	0.19	0.19	0.19	0.19	0.19	0.19
Liquid Fuels and Other Petroleum	27.42	29.17	29.11	28.63	34.64	33.41	31.58	41.22	38.34	35.47
Pipeline Fuel Natural Gas	0.58	0.66	0.66	0.66	0.79	0.79	0.79	0.79	0.79	0.79
Compressed Natural Gas	0.03	0.06	0.06	0.06	0.09	0.09	0.09	0.12	0.12	0.12
Electricity	0.02	0.03	0.03	0.03	0.03	0.03	0.03	0.04	0.04	0.04
Delivered Energy	**28.05**	**29.92**	**29.86**	**29.37**	**35.55**	**34.33**	**32.50**	**42.17**	**39.29**	**36.41**
Electricity Related Losses	0.05	0.06	0.06	0.06	0.07	0.07	0.07	0.08	0.08	0.08
Total	**28.11**	**29.98**	**29.92**	**29.43**	**35.63**	**34.40**	**32.57**	**42.25**	**39.37**	**36.49**

[1]Commercial trucks 8,500 to 10,000 pounds.
[2]Environmental Protection Agency rated miles per gallon.
[3]Combined car and light truck "on-the-road" estimate.
[4]E85 refers to a blend of 85 percent ethanol (renewable) and 15 percent motor gasoline (nonrenewable). To address cold starting issues, the percentage of ethanol actually varies seasonally. The annual average ethanol content of 74 percent is used for this forecast.
[5]Includes ethanol (blends of 10 percent or less) and ethers blended into gasoline.
[6]Includes only kerosene type.
[7]Diesel fuel for on- and off- road use.
[8]Includes aviation gasoline and lubricants.
Btu = British thermal unit.
Note: Totals may not equal sum of components due to independent rounding. Data for 2005 are model results and may differ slightly from official EIA data reports. Side cases were run without the fully integrated modeling system, so not all potential feedbacks were captured. The reference case ratio of electricity losses to electricity use was used to compute electricity losses for the technology cases.
Source: Energy Information Administration, AEO2007 National Energy Modeling System runs TRNFRZN.D120806A, AEO2007.D112106A, and TRNHIGH.D120806A.

Table D4. Key Results for Integrated Technology Cases

Consumption and Emissions	2005	2010			2020			2030		
		2006 Technology	Reference	High Technology	2006 Technology	Reference	High Technology	2006 Technology	Reference	High Technology
Energy Consumption by Sector (quadrillion Btu)										
Residential	11.60	12.25	12.21	12.14	13.44	13.17	12.69	14.30	13.80	12.87
Commercial	8.46	9.09	9.05	8.95	10.89	10.66	10.36	12.84	12.43	12.00
Industrial [1]	24.85	27.03	26.33	25.95	30.09	27.84	26.71	33.54	30.51	28.90
Transportation	28.05	29.93	29.86	29.41	35.51	34.33	32.60	42.01	39.29	36.64
Electric Power [2]	39.71	42.89	42.53	41.97	49.24	47.62	45.28	55.63	52.77	48.34
Total	**100.19**	**107.57**	**106.50**	**105.14**	**123.13**	**118.16**	**112.93**	**139.72**	**131.16**	**122.41**
Energy Consumption by Fuel (quadrillion Btu)										
Liquid Fuels and Other Petroleum [3]	40.61	42.14	41.76	41.08	48.62	46.52	44.08	56.34	52.17	48.70
Natural Gas	22.63	25.24	24.73	24.29	28.29	27.04	26.33	29.24	26.89	26.33
Coal	22.87	24.38	24.24	23.96	28.98	27.29	25.47	36.09	34.14	29.81
Nuclear Power	8.13	8.23	8.23	8.23	8.99	9.23	8.93	9.06	9.33	8.81
Renewable Energy [4]	5.86	7.53	7.50	7.55	8.22	8.05	8.09	8.89	8.59	8.72
Electricity Imports	0.08	0.05	0.04	0.04	0.04	0.04	0.04	0.08	0.04	0.04
Total	**100.19**	**107.57**	**106.50**	**105.14**	**123.13**	**118.16**	**112.93**	**139.72**	**131.16**	**122.41**
Energy Intensity (thousand Btu per 2000 dollar of GDP)	**9.07**	**8.42**	**8.33**	**8.21**	**7.22**	**6.92**	**6.60**	**6.23**	**5.83**	**5.43**
Carbon Dioxide Emissions by Sector (million metric tons)										
Residential	368	381	380	378	400	393	380	404	390	369
Commercial	230	238	238	239	271	270	271	300	298	300
Industrial [1]	1020	1090	1058	1043	1218	1115	1065	1393	1250	1171
Transportation	1953	2037	2032	2001	2416	2335	2217	2861	2674	2493
Electric Power [5]	2375	2527	2505	2464	2981	2832	2658	3555	3338	2970
Total	**5945**	**6273**	**6214**	**6126**	**7286**	**6944**	**6591**	**8512**	**7950**	**7303**
Carbon Dioxide Emissions by Fuel (million metric tons)										
Petroleum	2614	2647	2629	2590	3066	2947	2802	3573	3318	3105
Natural Gas	1178	1325	1298	1275	1486	1420	1383	1536	1412	1383
Coal	2142	2288	2275	2249	2721	2563	2392	3389	3206	2801
Other [6]	12	13	12	12	14	14	13	14	14	14
Total	**5945**	**6273**	**6214**	**6126**	**7286**	**6944**	**6591**	**8512**	**7950**	**7303**
Carbon Dioxide Emissions (tons per person)	**20.0**	**20.2**	**20.0**	**19.7**	**21.6**	**20.6**	**19.5**	**23.3**	**21.8**	**20.0**

[1]Includes energy for combined heat and power plants, except those whose primary business is to sell electricity, or electricity and heat, to the public.
[2]Includes electricity-only and combined heat and power plants whose primary business is to sell electricity, or electricity and heat, to the public.
[3]Includes petroleum-derived fuels and non-petroleum derived fuels, such as ethanol and biodiesel. Petroleum coke, which is a solid, is included. Also included are natural gas plant liquids, crude oil consumed as a fuel, and liquid hydrogen.
[4]Includes grid-connected electricity from conventional hydroelectric; wood and wood waste; landfill gas; municipal solid waste; other biomass wind; photovoltaic and solar thermal sources; and non-electric energy from renewable sources, such as active and passive solar systems, and wood; and both the ethanol and gasoline components of E85, but not the ethanol component of blends less than 85 percent. Excludes electricity imports using renewable sources and nonmarketed renewable energy.
[5]Includes electricity-only and combined heat and power plants whose primary business is to sell electricity, or electricity and heat, to the public. Does not include emissions from the nonbiogenic component of municipal solid waste because under international guidelines these are accounted for as waste, not energy.
[6]Includes emissions from geothermal power and nonbiogenic emissions from municipal solid waste.
Btu = British thermal unit.
GDP = Gross domestic product.
Note: Includes end-use, fossil electricity, and renewable technology assumptions. Totals may not equal sum of components due to independent rounding. Data for 2005 are model results and may differ slightly from official EIA data reports.
Source: Energy Information Administration, AEO2007 National Energy Modeling System runs LTRKITEN.D121106A, AEO2007.D112106A, and HTRKITEN.D121106A.

Table D5. Key Results for Advanced Nuclear Cost Cases
(Gigawatts, Unless Otherwise Noted)

Net Summer Capacity, Generation, Emissions, and Fuel Prices	2005	2010			2020			2030		
		High Cost	Reference	Low Cost	High Cost	Reference	Low Cost	High Cost	Reference	Low Cost
Capacity										
Coal Steam	310.6	320.8	320.9	320.9	350.5	347.2	348.2	457.1	449.9	444.5
Other Fossil Steam	121.3	119.5	119.5	119.5	89.8	89.3	90.0	86.9	87.5	85.6
Combined Cycle	176.6	193.3	193.3	193.3	202.9	203.9	203.3	213.0	211.6	209.0
Combustion Turbine/Diesel	133.2	137.0	137.0	137.0	127.7	127.2	127.6	152.9	155.1	152.1
Nuclear Power	100.0	100.5	100.5	100.5	108.5	111.7	111.7	105.9	112.6	128.7
Pumped Storage	20.8	20.8	20.8	20.8	20.8	20.8	20.8	20.8	20.8	20.8
Fuel Cells	0.0	0.0	0.0	0.0	0.0	0.0	0.0	0.0	0.0	0.0
Renewable Sources	97.6	105.7	105.7	105.7	107.8	107.7	107.7	110.2	110.0	109.5
Distributed Generation (Natural Gas)	0.0	0.2	0.2	0.2	2.4	2.1	2.4	12.3	11.4	11.6
Combined Heat and Power[1]	27.5	31.7	31.7	31.7	41.7	41.7	41.5	61.3	61.3	60.3
Total	**987.6**	**1029.5**	**1029.5**	**1029.5**	**1052.0**	**1051.6**	**1053.0**	**1220.4**	**1220.2**	**1222.1**
Cumulative Additions										
Coal Steam	0.0	11.5	11.5	11.5	45.8	42.3	43.3	152.4	145.1	139.7
Other Fossil Steam	0.0	0.0	0.0	0.0	0.0	0.0	0.0	0.0	0.0	0.0
Combined Cycle	0.0	16.7	16.7	16.7	26.3	27.3	26.7	36.4	35.1	32.4
Combustion Turbine/Diesel	0.0	4.4	4.4	4.4	14.6	13.8	13.9	39.8	41.8	38.5
Nuclear Power	0.0	0.0	0.0	0.0	5.8	9.0	9.0	5.8	12.5	28.5
Pumped Storage	0.0	0.0	0.0	0.0	0.0	0.0	0.0	0.0	0.0	0.0
Fuel Cells	0.0	0.0	0.0	0.0	0.0	0.0	0.0	0.0	0.0	0.0
Renewable Sources	0.0	8.1	8.1	8.1	10.2	10.1	10.1	12.6	12.4	11.9
Distributed Generation	0.0	0.2	0.2	0.2	2.4	2.1	2.4	12.3	11.4	11.6
Combined Heat and Power[1]	0.0	4.2	4.2	4.2	14.2	14.2	14.0	33.8	33.8	32.8
Total	**0.0**	**45.1**	**45.1**	**45.1**	**119.2**	**118.9**	**119.4**	**293.1**	**291.9**	**295.4**
Cumulative Retirements	**0.0**	**3.7**	**3.7**	**3.7**	**57.6**	**57.6**	**56.7**	**63.1**	**62.0**	**63.7**
Generation by Fuel (billion kilowatthours)										
Coal	1993	2125	2121	2124	2473	2447	2450	3262	3220	3172
Petroleum	116	83	84	83	90	90	90	94	94	95
Natural Gas	675	795	795	794	912	918	916	733	732	672
Nuclear Power	780	789	789	789	861	885	886	844	896	1018
Pumped Storage	-7	-9	-9	-9	-9	-9	-9	-9	-9	-9
Renewable Sources	323	427	429	428	452	449	451	471	465	459
Distributed Generation	0	0	0	0	1	1	1	5	5	5
Combined Heat and Power[1]	155	183	183	183	256	256	255	396	395	391
Total	**4035**	**4393**	**4392**	**4393**	**5036**	**5037**	**5039**	**5797**	**5797**	**5805**
Carbon Dioxide Emissions by the Electric Power Sector (million metric tons)[2]										
Petroleum	100	69	69	69	74	74	74	77	77	78
Natural Gas	319	346	346	345	388	390	389	321	321	299
Coal	1944	2083	2078	2083	2374	2354	2354	2953	2927	2886
Other[3]	12	12	12	12	14	14	14	14	14	14
Total	**2375**	**2509**	**2505**	**2509**	**2850**	**2832**	**2831**	**3365**	**3338**	**3277**
Prices to the Electric Power Sector[2] (2005 dollars per million Btu)										
Petroleum	7.70	7.91	7.92	7.91	7.08	7.07	7.06	7.97	7.96	7.89
Natural Gas	8.18	6.21	6.22	6.22	5.75	5.76	5.75	6.31	6.33	6.15
Coal	1.53	1.71	1.71	1.71	1.59	1.58	1.58	1.70	1.69	1.67

[1]Includes combined heat and power plants and electricity-only plants in commercial and industrial sectors. Includes small on-site generating systems in the residential, commercial, and industrial sectors used primarily for own-use generation, but which may also sell some power to the grid. Excludes off-grid photovoltaics and other generators not connected to the distribution or transmission systems.

[2]Includes electricity-only and combined heat and power plants whose primary business to sell electricity, or electricity and heat, to the public.

[3]Includes emissions from geothermal power and nonbiogenic emissions from municipal solid waste.

Btu = British thermal unit.

Note: Totals may not equal sum of components due to independent rounding. Data for 2005 are model results and may differ slightly from official EIA data reports.

Source: Energy Information Administration, AEO2007 National Energy Modeling System runs LONUC07.D112706A, AEO2007.D112106A, and ADVNUC07.D112906A.

Table D6. Key Results for Electric Power Sector Fossil Technology Cases
(Gigawatts, Unless Otherwise Noted)

Net Summer Capacity, Generation Consumption, and Emissions	2005	2010			2020			2030		
		Low Fossil	Reference	High Fossil	Low Fossil	Reference	High Fossil	Low Fossil	Reference	High Fossil
Capacity										
Pulverized Coal	310.1	320.3	320.3	320.3	346.0	339.7	328.5	430.5	382.7	341.0
Coal Gasification Combined-Cycle	0.5	0.5	0.5	1.6	3.5	7.5	15.1	3.9	67.2	112.0
Conventional Natural Gas Combined-Cycle	176.6	193.3	193.3	193.3	194.2	194.2	194.2	194.4	194.2	194.2
Advanced Natural Gas Combined-Cycle	0.0	0.0	0.0	0.0	0.3	9.7	29.7	2.0	17.5	58.4
Conventional Combustion Turbine	133.2	136.5	136.5	136.4	118.2	117.6	117.6	122.8	118.2	117.6
Advanced Combustion Turbine	0.0	0.6	0.6	0.6	11.6	9.6	6.5	37.3	37.0	29.9
Fuel Cells	0.0	0.0	0.0	0.0	0.0	0.0	0.0	0.0	0.0	0.0
Nuclear	100.0	100.5	100.5	100.5	111.7	111.7	108.7	123.4	112.6	106.1
Oil and Gas Steam	121.3	119.5	119.5	119.5	89.8	89.3	85.4	88.0	87.5	77.8
Renewable Sources/Pumped Storage	118.3	126.4	126.4	126.4	129.9	128.5	128.1	133.3	130.8	128.8
Distributed Generation	0.0	0.2	0.2	0.2	2.8	2.1	1.1	14.6	11.4	7.3
Combined Heat and Power[1]	27.5	31.7	31.7	31.7	41.7	41.7	41.2	63.3	61.3	58.6
Total	**987.6**	**1029.5**	**1029.5**	**1030.6**	**1049.7**	**1051.6**	**1056.2**	**1213.6**	**1220.2**	**1231.7**
Cumulative Additions										
Pulverized Coal	0.0	11.5	11.5	11.5	42.0	35.4	24.2	126.5	78.4	36.9
Coal Gasification Combined-Cycle	0.0	0.0	0.0	1.0	3.0	6.9	14.6	3.4	66.7	111.5
Conventional Natural Gas Combined-Cycle	0.0	16.7	16.7	16.7	17.6	17.6	17.6	17.8	17.6	17.6
Advanced Natural Gas Combined-Cycle	0.0	0.0	0.0	0.0	0.3	9.7	29.7	2.0	17.5	58.4
Conventional Combustion Turbine	0.0	3.9	3.9	3.8	4.5	4.3	3.8	9.2	4.8	4.1
Advanced Combustion Turbine	0.0	0.6	0.6	0.6	11.6	9.6	6.5	37.3	37.0	29.9
Fuel Cells	0.0	0.0	0.0	0.0	0.0	0.0	0.0	0.0	0.0	0.0
Nuclear	0.0	0.0	0.0	0.0	9.0	9.0	6.0	23.2	12.5	6.0
Oil and Gas Steam	0.0	0.0	0.0	0.0	0.0	0.0	0.0	0.0	0.0	0.0
Renewable Sources	0.0	8.1	8.1	8.1	11.5	10.1	9.8	15.0	12.4	10.5
Distributed Generation	0.0	0.2	0.2	0.2	2.8	2.1	1.1	14.6	11.4	7.3
Combined Heat and Power[1]	0.0	4.2	4.2	4.2	14.2	14.2	13.7	35.8	33.8	31.1
Total	**0.0**	**45.1**	**45.1**	**46.1**	**116.7**	**118.9**	**127.1**	**284.8**	**291.9**	**313.2**
Cumulative Retirements	**0.0**	**3.7**	**3.7**	**3.7**	**57.3**	**57.6**	**61.2**	**61.6**	**62.0**	**71.9**
Generation by Fuel (billion kilowatthours)										
Coal	1992.5	2120.4	2120.7	2128.3	2463.5	2447.3	2420.4	3092.9	3219.8	3235.3
Petroleum	116.0	82.9	83.8	83.2	92.4	89.8	89.6	100.7	93.6	91.6
Natural Gas	675.1	795.8	795.1	789.2	886.1	917.7	977.6	720.4	731.9	818.0
Nuclear Power	780.5	789.3	789.3	789.3	884.9	885.2	862.4	978.0	895.7	845.9
Renewable Sources/Pumped Storage	316.1	421.3	420.6	419.9	449.0	440.0	441.7	474.0	455.9	441.2
Distributed Generation	0.0	0.1	0.1	0.1	1.2	0.9	0.5	6.4	5.0	3.2
Combined Heat and Power[1]	155.0	182.7	182.7	182.7	256.8	256.3	253.0	409.2	395.0	377.8
Total	**4035.1**	**4392.6**	**4392.2**	**4392.7**	**5033.9**	**5037.2**	**5045.1**	**5781.6**	**5796.9**	**5812.9**
Fuel Consumption by the Electric Power Sector (quadrillion Btu)[2]										
Coal	20.75	22.13	22.13	22.20	25.25	25.05	24.68	30.73	31.14	30.18
Petroleum	1.16	0.90	0.90	0.90	0.99	0.97	0.97	1.07	1.01	0.99
Natural Gas	5.95	6.57	6.56	6.52	7.24	7.40	7.55	6.10	6.09	6.29
Nuclear Power	8.13	8.23	8.23	8.23	9.22	9.23	8.99	10.19	9.33	8.81
Renewable Sources	3.64	4.68	4.67	4.67	5.01	4.93	4.94	5.33	5.15	4.97
Total	**39.63**	**42.50**	**42.49**	**42.51**	**47.71**	**47.58**	**47.13**	**53.42**	**52.72**	**51.25**
Carbon Dioxide Emissions by the Electric Power Sector (million metric tons)[2]										
Coal	1944	2078	2078	2085	2372	2354	2319	2889	2927	2837
Petroleum	100	69	69	69	76	74	74	82	77	76
Natural Gas	319	346	346	343	381	390	398	321	321	331
Other[1]	12	12	12	12	14	14	14	14	14	14
Total	**2375**	**2505**	**2505**	**2509**	**2843**	**2832**	**2805**	**3305**	**3338**	**3258**

[1]Includes combined heat and power plants and electricity-only plants in the commercial and industrial sectors. Includes small on-site generating systems in the residential, commercial, and industrial sectors used primarily for own-use generation, but which may also sell some power to the grid. Excludes off-grid photovoltaics and other generators not connected to the distribution or transmission systems.

[2]Includes electricity-only and combined heat and power plants whose primary business to sell electricity, or electricity and heat, to the public.

[3]Includes emissions from geothermal power and nonbiogenic emissions from municipal solid waste.

Note: Totals may not equal sum of components due to independent rounding. Data for 2005 are model results and may differ slightly from official EIA data reports.

Source: Energy Information Administration, AEO2007 National Energy Modeling System runs LFOSS07.D112706A, AEO2007.D112106A, and HFOSS07.D112706A.

Table D7. Key Results for Renewable Technology Cases

Capacity, Generation, and Emissions	2005	2010 Low Technology	2010 Reference	2010 High Technology	2020 Low Technology	2020 Reference	2020 High Technology	2030 Low Technology	2030 Reference	2030 High Technology
Net Summer Capacity (gigawatts)										
Electric Power Sector[1]										
Conventional Hydropower	79.97	79.99	79.99	79.99	80.12	80.12	80.19	80.12	80.18	80.41
Geothermal[2]	2.28	2.46	2.46	2.46	2.67	2.79	2.84	2.84	3.15	3.20
Municipal Waste[3]	3.23	3.43	3.43	3.63	3.79	3.80	3.85	3.81	3.87	3.87
Wood and Other Biomass[4]	2.06	2.22	2.22	2.22	2.27	2.37	3.30	2.67	3.80	9.64
Solar Thermal	0.40	0.54	0.54	0.54	0.58	0.58	0.58	0.63	0.63	0.63
Solar Photovoltaic	0.03	0.07	0.07	0.07	0.22	0.22	0.22	0.39	0.39	0.39
Wind	9.62	16.97	16.97	16.97	17.85	17.85	17.93	17.91	17.98	18.42
Total	**97.59**	**105.69**	**105.69**	**105.88**	**107.49**	**107.72**	**108.91**	**108.36**	**110.00**	**116.55**
End-Use Sector[5]										
Conventional Hydropower	0.63	0.63	0.63	0.63	0.63	0.63	0.63	0.63	0.63	0.63
Geothermal	0.00	0.00	0.00	0.00	0.00	0.00	0.00	0.00	0.00	0.00
Municipal Waste[6]	0.33	0.33	0.33	0.33	0.33	0.33	0.33	0.33	0.33	0.33
Wood and Other Biomass	4.49	4.76	4.79	4.96	5.78	5.90	6.63	6.96	7.19	8.68
Solar Photovoltaic	0.18	0.23	0.63	0.63	0.23	0.80	0.80	0.24	2.52	3.60
Total	**5.63**	**5.96**	**6.39**	**6.56**	**6.98**	**7.66**	**8.40**	**8.16**	**10.68**	**13.25**
Generation (billion kilowatthours)										
Electric Power Sector[1]										
Coal	1993	2122	2121	2120	2449	2447	2440	3229	3220	3200
Petroleum	116	83	84	83	90	90	90	94	94	93
Natural Gas	675	796	795	791	920	918	908	751	732	721
Total Fossil	**2784**	**3002**	**3000**	**2995**	**3459**	**3455**	**3438**	**4074**	**4045**	**4015**
Conventional Hydropower	261.89	297.50	297.50	297.50	303.85	303.85	304.30	304.15	304.51	305.64
Geothermal	15.12	17.34	17.34	17.34	18.99	19.79	20.08	20.36	22.66	23.16
Municipal Waste[3]	20.56	21.56	21.56	23.06	24.35	24.42	24.80	24.48	24.95	25.00
Wood and Other Biomass[4]	9.92	42.72	43.29	45.78	48.72	47.47	59.62	51.31	58.21	96.35
Dedicated Plants	5.38	11.07	11.11	11.21	10.80	11.61	19.00	14.63	23.80	68.14
Cofiring	4.53	31.65	32.18	34.57	37.92	35.86	40.62	36.67	34.41	28.21
Solar Thermal	0.54	1.16	1.16	1.16	1.28	1.28	1.28	1.43	1.43	1.43
Solar Photovoltaic	0.01	0.18	0.18	0.18	0.54	0.54	0.54	0.98	0.98	0.98
Wind	14.60	48.25	48.26	48.25	51.37	51.35	51.68	51.56	51.85	53.41
Total Renewable	**322.64**	**428.70**	**429.28**	**433.27**	**449.10**	**448.71**	**462.32**	**454.27**	**464.59**	**505.96**
End-Use Sector[5]										
Total Fossil	105	130	130	130	196	196	196	325	323	322
Conventional Hydropower[7]	3.18	3.18	3.18	3.18	3.18	3.18	3.18	3.18	3.18	3.18
Geothermal	0.00	0.00	0.00	0.00	0.00	0.00	0.00	0.00	0.00	0.00
Municipal Waste[6]	2.75	2.75	2.75	2.75	2.75	2.75	2.75	2.75	2.75	2.75
Wood and Other Biomass	27.91	29.53	29.69	30.70	35.50	36.17	40.45	42.36	43.70	52.43
Solar Photovoltaic	0.33	0.45	1.21	1.21	0.45	1.53	1.53	0.45	4.78	6.78
Total Renewable	**34.18**	**35.91**	**36.84**	**37.85**	**41.88**	**43.63**	**47.92**	**48.75**	**54.41**	**65.15**
Carbon Dioxide Emissions by the Electric Power Sector (million metric tons)[1]										
Coal	1944	2080	2078	2079	2352	2354	2345	2935	2927	2909
Petroleum	100	69	69	69	74	74	74	77	77	77
Natural Gas	319	346	346	344	391	390	387	327	321	317
Other[8]	12	12	12	13	14	14	14	14	14	14
Total	**2375**	**2507**	**2505**	**2505**	**2831**	**2832**	**2820**	**3354**	**3338**	**3317**

[1]Includes electricity-only and combined heat and power plants whose primary business is to sell electricity, or electricity and heat, to the public.
[2]Includes hydrothermal resources only (hot water and steam).
[3]Includes municipal solid waste, landfill gas, and municipal sewage sludge. Incremental growth is assumed to be for landfill gas facilities. All municipal solid waste is included, although a portion of the municipal solid waste stream contains petroleum-derived plastics and other non-renewable sources.
[4]Includes projections for energy crops after 2010.
[5]Includes combined heat and power plants and electricity-only plants in the commercial and industrial sectors; and small on-site generating systems in the residential, commercial, and industrial sectors used primarily for own-use generation, but which may also sell some power to the grid. Excludes off-grid photovoltaics and other generators not connected to the distribution or transmission systems.
[6]Includes municipal solid waste, landfill gas, and municipal sewage sludge. All municipal solid waste is included, although a portion of the municipal solid waste stream contains petroleum-derived plastics and other non-renewable sources.
[7]Represents own-use industrial hydroelectric power.
[8]Includes emissions from geothermal power and nonbiogenic emissions from municipal solid waste.
Note: Totals may not equal sum of components due to independent rounding. Data for 2005 are model results and may differ slightly from official EIA data reports.
Source: Energy Information Administration, AEO2007 National Energy Modeling System runs LOREN07.D120806A, AEO2007.D112106A, and HIREN07.D120806A.

Table D8. Key Results for Regional Renewable Portfolio Standard Case

Capacity, Generation, and Emissions	2005	2010		2020		2030	
		Reference	Regional RPS	Reference	Regional RPS	Reference	Regional RPS
Net Summer Capacity (gigawatts)							
Electric Power Sector[1]							
Conventional Hydropower	79.97	79.99	79.99	80.12	80.15	80.18	80.39
Geothermal[2]	2.28	2.46	2.51	2.79	2.84	3.15	3.17
Municipal Waste[3]	3.23	3.43	3.91	3.80	4.16	3.87	4.17
Wood and Other Biomass[4]	2.06	2.22	2.22	2.37	6.89	3.80	11.82
Solar Thermal	0.40	0.54	0.54	0.58	0.58	0.63	0.63
Solar Photovoltaic	0.03	0.07	0.07	0.22	0.22	0.39	0.39
Wind	9.62	16.97	17.25	17.85	18.55	17.98	18.63
Total	**97.59**	**105.69**	**106.50**	**107.72**	**113.39**	**110.00**	**119.20**
End-Use Sector[5]							
Conventional Hydropower	0.63	0.63	0.63	0.63	0.63	0.63	0.63
Geothermal	0.00	0.00	0.00	0.00	0.00	0.00	0.00
Municipal Waste[6]	0.33	0.33	0.33	0.33	0.33	0.33	0.33
Wood and Other Biomass	4.49	4.79	4.79	5.90	5.90	7.19	7.19
Solar Photovoltaic	0.18	0.63	0.63	0.80	0.80	2.52	2.53
Total	**5.63**	**6.39**	**6.39**	**7.66**	**7.67**	**10.68**	**10.68**
Generation (billion kilowatthours)							
Electric Power Sector[1]							
Coal	1993	2121	2115	2447	2424	3220	3167
Petroleum	116	84	83	90	90	94	94
Natural Gas	675	795	792	918	898	732	725
Total Fossil	**2784**	**3000**	**2990**	**3455**	**3412**	**4045**	**3987**
Conventional Hydropower	261.89	297.50	297.50	303.85	304.01	304.51	305.55
Geothermal	15.12	17.34	17.57	19.79	20.27	22.66	22.97
Municipal Waste[3]	20.56	21.56	25.29	24.42	27.28	24.95	27.34
Wood and Other Biomass[4]	9.92	43.29	47.84	47.47	88.16	58.21	114.47
Dedicated Plants	5.38	11.11	10.95	11.61	46.47	23.80	84.24
Cofiring	4.53	32.18	36.89	35.86	41.69	34.41	30.23
Solar Thermal	0.54	1.16	1.16	1.28	1.28	1.43	1.43
Solar Photovoltaic	0.01	0.18	0.18	0.54	0.54	0.98	0.98
Wind	14.60	48.26	49.21	51.35	53.65	51.85	53.94
Total Renewable	**322.64**	**429.28**	**438.75**	**448.71**	**495.19**	**464.59**	**526.69**
End-Use Sector[5]							
Total Fossil	105	130	130	196	195	323	323
Conventional Hydropower[7]	3.18	3.18	3.18	3.18	3.18	3.18	3.18
Geothermal	0.00	0.00	0.00	0.00	0.00	0.00	0.00
Municipal Waste[6]	2.75	2.75	2.75	2.75	2.75	2.75	2.75
Wood and Other Biomass	27.91	29.69	29.69	36.17	36.18	43.70	43.69
Solar Photovoltaic	0.33	1.21	1.21	1.53	1.53	4.78	4.80
Total Renewable	**34.18**	**36.84**	**36.84**	**43.63**	**43.65**	**54.41**	**54.42**
Carbon Dioxide Emissions by the Electric Power Sector (million metric tons)[1]							
Coal	1944	2078	2072	2354	2332	2927	2886
Petroleum	100	69	69	74	75	77	77
Natural Gas	319	346	344	390	383	321	318
Other[8]	12	12	14	14	15	14	15
Total	**2375**	**2505**	**2499**	**2832**	**2804**	**3338**	**3297**

[1]Includes electricity-only and combined heat and power plants whose primary business is to sell electricity, or electricity and heat, to the public.
[2]Includes hydrothermal resources only (hot water and steam).
[3]Includes municipal solid waste, landfill gas, and municipal sewage sludge. Incremental growth is assumed to be for landfill gas facilities. All municipal solid waste is included, although a portion of the municipal solid waste stream contains petroleum-derived plastics and other non-renewable sources.
[4]Includes projections for energy crops after 2010.
[5]Includes combined heat and power plants and electricity-only plants in the commercial and industrial sectors; and small on-site generating systems in the residential, commercial, and industrial sectors used primarily for own-use generation, but which may also sell some power to the grid. Excludes off-grid photovoltaics and other generators not connected to the distribution or transmission systems.
[6]Includes municipal solid waste, landfill gas, and municipal sewage sludge. All municipal solid waste is included, although a portion of the municipal solid waste stream contains petroleum-derived plastics and other non-renewable sources.
[7]Represents own-use industrial hydroelectric power.
[8]Includes emissions from geothermal power and nonbiogenic emissions from municipal solid waste.
RPS = Regional Portfolio Standard
Note: Totals may not equal sum of components due to independent rounding. Data for 2005 are model results and may differ slightly from official EIA data reports.
Source: Energy Information Administration, AEO2007 National Energy Modeling System runs AEO2007.D112106A, and RGRPS07.D121206C.

Table D9. Natural Gas Supply and Disposition, Oil and Gas Technological Progress Cases
(Trillion Cubic Feet per Year, Unless Otherwise Noted)

Supply, Disposition, and Prices	2005	2010			2020			2030		
		Slow Technology	Reference	Rapid Technology	Slow Technology	Reference	Rapid Technology	Slow Technology	Reference	Rapid Technology
Natural Gas Prices										
(2005 dollars per million Btu)										
Henry Hub Spot Price	8.60	6.43	6.28	6.21	5.85	5.71	5.41	6.88	6.52	5.69
Average Lower 48 Wellhead Price[11]	7.29	5.73	5.59	5.53	5.20	5.07	4.80	6.13	5.80	5.05
(2005 dollars per thousand cubic feet)										
Average Lower 48 Wellhead Price[11]	7.51	5.90	5.76	5.69	5.36	5.22	4.94	6.32	5.98	5.21
Dry Gas Production[1]	**18.23**	**19.31**	**19.35**	**19.42**	**20.09**	**20.79**	**21.71**	**18.66**	**20.53**	**23.45**
Lower 48 Onshore	14.36	15.22	15.22	15.35	14.16	14.66	15.34	13.54	15.13	17.93
Associated-Dissolved	1.43	1.37	1.39	1.41	1.22	1.28	1.33	1.13	1.19	1.21
Non-Associated	12.93	13.85	13.83	13.95	12.94	13.38	14.01	12.41	13.94	16.72
Conventional	4.94	5.33	5.27	5.26	4.32	4.30	4.21	3.77	3.75	3.59
Unconventional	7.99	8.53	8.56	8.69	8.62	9.09	9.80	8.64	10.19	13.13
Lower 48 Offshore	3.41	3.84	3.88	3.82	3.89	4.09	4.34	2.96	3.25	3.36
Associated-Dissolved	0.71	0.91	0.92	0.93	1.01	1.05	1.10	0.77	0.85	0.92
Non-Associated	2.69	2.93	2.96	2.89	2.88	3.04	3.24	2.19	2.40	2.44
Alaska	0.45	0.25	0.25	0.25	2.04	2.05	2.04	2.16	2.16	2.16
Supplemental Natural Gas[2]	0.07	0.07	0.07	0.07	0.07	0.07	0.07	0.07	0.07	0.07
Net Imports	**3.57**	**4.43**	**4.55**	**4.71**	**5.42**	**5.35**	**5.28**	**6.41**	**5.45**	**4.33**
Pipeline[3]	3.01	2.55	2.74	2.93	1.42	1.65	1.88	0.95	0.92	1.05
Liquefied Natural Gas	0.57	1.87	1.81	1.78	4.00	3.69	3.40	5.46	4.53	3.28
Total Supply	**21.87**	**23.81**	**23.97**	**24.20**	**25.59**	**26.21**	**27.06**	**25.14**	**26.06**	**27.85**
Consumption by Sector										
Residential	4.84	5.02	5.03	5.04	5.25	5.27	5.30	5.27	5.31	5.37
Commercial	3.05	3.20	3.22	3.23	3.73	3.75	3.78	4.18	4.24	4.32
Industrial[4]	6.64	7.58	7.63	7.65	7.95	8.02	8.13	8.44	8.65	8.88
Electric Power[5]	5.78	6.29	6.38	6.56	6.73	7.19	7.82	5.42	5.92	7.15
Transportation[6]	0.03	0.06	0.06	0.06	0.09	0.09	0.09	0.12	0.12	0.12
Pipeline Fuel	0.56	0.64	0.64	0.64	0.75	0.76	0.79	0.73	0.77	0.83
Lease and Plant Fuel[7]	1.07	1.07	1.07	1.07	1.14	1.17	1.20	1.04	1.12	1.23
Total	**21.98**	**23.86**	**24.02**	**24.25**	**25.64**	**26.26**	**27.12**	**25.20**	**26.12**	**27.92**
Discrepancy[8]	**-0.11**	**-0.05**	**-0.05**	**-0.05**	**-0.06**	**-0.05**	**-0.06**	**-0.06**	**-0.06**	**-0.07**
Lower 48 End of Year Reserves	**189.91**	**202.59**	**205.23**	**207.86**	**197.27**	**208.32**	**226.43**	**180.39**	**210.60**	**270.22**

[1]Marketed production (wet) minus extraction losses.
[2]Synthetic natural gas, propane air, coke oven gas, refinery gas, biomass gas, air injected for Btu stabilization, and manufactured gas commingled and distributed with natural gas.
[3]Includes any natural gas regasified in the Bahamas and transported via pipeline to Florida.
[4]Includes energy for combined heat and power plants, except those whose primary business is to sell electricity, or electricity and heat, to the public.
[5]Includes consumption of energy by electricity-only and combined heat and power plants whose primary business is to sell electricity, or electricity and heat, to the public. Includes small power producers and exempt wholesale generators.
[6]Compressed natural gas used as a vehicle fuel. Price includes estimated motor vehicle fuel taxes and estimated dispensing costs or charges.
[7]Represents natural gas used in field gathering and processing plant machinery.
[8]Balancing item. Natural gas lost as a result of converting flow data measured at varying temperatures and pressures to a standard temperature and pressure and the merger of different data reporting systems which vary in scope, format, definition, and respondent type. In addition, 2005 values include net storage injections.
Note: Totals may not equal sum of components due to independent rounding. Data for 2005 are model results and may differ slightly from official EIA data reports.
Sources: 2005 supply values: Energy Information Administration (EIA), *Natural Gas Monthly*, DOE/EIA-0130(2006/04) (Washington, DC, April 2006). 2005 consumption based on: EIA, *Annual Energy Review 2005*, DOE/EIA-0384(2005) (Washington, DC, July 2006). **Projections:** EIA, AEO2007 National Energy Modeling System runs OGLTEC07.D112706A, AEO2007.D112106A, and OGHTEC07.D112706A.

Table D10. Liquid Fuels Supply and Disposition, Oil and Gas Technological Progress Cases
(Million Barrels per Day, Unless Otherwise Noted)

Supply, Disposition, and Prices	2005	2010			2020			2030		
		Slow Technology	Reference	Rapid Technology	Slow Technology	Reference	Rapid Technology	Slow Technology	Reference	Rapid Technology
Prices (2005 dollars per barrel)										
Imported Low Sulfur Light Crude Oil[1]	**56.76**	**57.45**	**57.47**	**57.47**	**52.04**	**52.04**	**52.51**	**59.12**	**59.12**	**59.12**
Imported Crude Oil[1]	**49.19**	**51.20**	**51.20**	**51.20**	**46.47**	**46.47**	**46.47**	**51.63**	**51.63**	**51.63**
Crude Oil Supply										
Domestic Crude Oil Production[2]	5.18	5.54	5.67	5.78	5.41	5.89	6.31	4.77	5.39	5.70
Alaska	0.86	0.69	0.69	0.69	0.74	0.74	0.65	0.27	0.27	0.24
Lower 48 Onshore	2.89	2.83	2.93	3.04	2.58	2.94	3.29	2.47	2.92	3.12
Lower 48 Offshore	1.42	2.02	2.05	2.05	2.10	2.21	2.36	2.03	2.20	2.33
Net Crude Oil Imports	10.09	10.09	9.99	9.88	11.68	11.29	10.97	13.61	13.09	12.82
Other Crude Oil Supply	-0.06	0.00	0.00	0.00	0.00	0.00	0.00	0.00	0.00	0.00
Total Crude Oil Supply	**15.22**	**15.63**	**15.66**	**15.66**	**17.09**	**17.19**	**17.28**	**18.38**	**18.47**	**18.51**
Other Petroleum Supply										
Natural Gas Plant Liquids	1.72	1.80	1.80	1.80	1.71	1.76	1.84	1.55	1.72	1.86
Net Petroleum Product Imports[3]	2.48	1.83	1.80	1.79	2.34	2.27	2.16	3.39	3.28	3.23
Refinery Processing Gain[4]	0.99	1.20	1.21	1.21	1.41	1.41	1.38	1.55	1.49	1.47
Other Supply[5]	0.39	1.01	1.02	1.02	1.32	1.31	1.28	1.91	1.88	1.85
Total Primary Supply[6]	**20.79**	**21.47**	**21.49**	**21.49**	**23.87**	**23.94**	**23.95**	**26.77**	**26.84**	**26.91**
Refined Petroleum Products Supplied										
Residential and Commercial	1.26	1.25	1.25	1.25	1.28	1.29	1.29	1.27	1.28	1.28
Industrial[7]	5.07	5.00	5.01	5.01	5.14	5.16	5.16	5.51	5.53	5.54
Transportation	13.87	14.92	14.93	14.93	17.11	17.15	17.17	19.61	19.69	19.75
Electric Power[8]	0.51	0.40	0.40	0.40	0.43	0.43	0.42	0.46	0.45	0.44
Total	**20.75**	**21.57**	**21.59**	**21.59**	**23.96**	**24.03**	**24.04**	**26.86**	**26.95**	**27.02**
Discrepancy[9]	**0.04**	**-0.10**	**-0.10**	**-0.10**	**-0.09**	**-0.09**	**-0.09**	**-0.09**	**-0.11**	**-0.10**
Lower 48 End of Year Reserves (billion barrels)[2]	**16.98**	**18.95**	**19.53**	**20.08**	**18.24**	**19.98**	**21.74**	**15.85**	**17.94**	**18.18**

[1]Weighted average price delivered to U.S. refiners.
[2]Includes lease condensate.
[3]Includes net imports of finished petroleum products, unfinished oils, other hydrocarbons, alcohols, ethers, and blending components.
[4]Represents volumetric gain in refinery distillation and cracking processes.
[5]Includes alcohols, ethers, petroleum product stock withdrawals, domestic sources of blending components, other hydrocarbons, renewable fuels such as biodiesel, natural gas converted to liquid fuel, and coal converted to liquid fuel.
[6]Total crude supply plus natural gas plant liquids, other inputs, refinery processing gain, and net product imports.
[7]Includes consumption for combined heat and power, which produces electricity and other useful thermal energy.
[8]Includes consumption of energy by electricity-only and combined heat and power plants whose primary business is to sell electricity, or electricity and heat, to the public. Includes small power producers and exempt wholesale generators.
[9]Balancing item. Includes unaccounted for supply, losses and gains.
Note: Totals may not equal sum of components due to independent rounding. Data for 2005 are model results and may differ slightly from official EIA data reports.
Sources: 2005 product supplied data based on: Energy Information Administration (EIA), *Annual Energy Review 2005*, DOE/EIA-0384(2005) (Washington, DC, July 2006). 2005 imported low sulfur light crude oil price: EIA, Form EIA-856, "Monthly Foreign Crude Oil Acquisition Report." Other 2005 data: EIA, *Petroleum Supply Annual 2005*, DOE/EIA-0340(2005)/1 (Washington, DC, October 2006). **Projections:** EIA, AEO2007 National Energy Modeling System runs OGLTEC07.D112706A, AEO2007.D112106A, and OGHTEC07.D112706A.

Table D11. Natural Gas Supply and Disposition, Liquefied Natural Gas Supply Cases
(Trillion Cubic Feet per Year, Unless Otherwise Noted)

Supply, Disposition, and Prices	2005	2010			2020			2030		
		Low LNG	Reference	High LNG	Low LNG	Reference	High LNG	Low LNG	Reference	High LNG
Dry Gas Production[1]	**18.23**	**19.72**	**19.35**	**19.07**	**21.46**	**20.79**	**19.59**	**21.61**	**20.53**	**18.75**
Lower 48 Onshore	14.36	15.53	15.22	15.00	15.21	14.66	13.59	16.06	15.13	13.58
Associated-Dissolved	1.43	1.39	1.39	1.39	1.28	1.28	1.28	1.19	1.19	1.19
Non-Associated	12.93	14.14	13.83	13.61	13.93	13.38	12.31	14.87	13.94	12.39
Conventional	4.94	5.40	5.27	5.16	4.44	4.30	3.99	3.97	3.75	3.42
Unconventional	7.99	8.74	8.56	8.44	9.49	9.09	8.32	10.91	10.19	8.97
Lower 48 Offshore	3.41	3.95	3.88	3.83	4.21	4.09	3.95	3.39	3.25	3.01
Associated-Dissolved	0.71	0.92	0.92	0.92	1.05	1.05	1.05	0.87	0.85	0.81
Non-Associated	2.69	3.03	2.96	2.91	3.16	3.04	2.91	2.52	2.40	2.20
Alaska	0.45	0.25	0.25	0.25	2.05	2.05	2.05	2.16	2.16	2.16
Supplemental Natural Gas[2]	0.07	0.07	0.07	0.07	0.07	0.07	0.07	0.07	0.07	0.07
Net Imports	**3.57**	**3.95**	**4.55**	**5.12**	**3.69**	**5.35**	**8.28**	**3.04**	**5.45**	**10.63**
Pipeline[3]	3.01	2.86	2.74	2.64	2.10	1.65	1.22	1.40	0.92	0.82
Liquefied Natural Gas	0.57	1.09	1.81	2.48	1.59	3.69	7.06	1.63	4.53	9.80
Total Supply	**21.87**	**23.75**	**23.97**	**24.27**	**25.22**	**26.21**	**27.94**	**24.72**	**26.06**	**29.45**
Consumption by Sector										
Residential	4.84	5.02	5.03	5.05	5.25	5.27	5.33	5.26	5.31	5.38
Commercial	3.05	3.20	3.22	3.24	3.73	3.75	3.82	4.17	4.24	4.34
Industrial[4]	6.64	7.57	7.63	7.68	7.96	8.02	8.22	8.41	8.65	8.99
Electric Power[5]	5.78	6.23	6.38	6.60	6.29	7.19	8.62	4.90	5.92	8.82
Transportation[6]	0.03	0.06	0.06	0.06	0.09	0.09	0.09	0.12	0.12	0.12
Pipeline Fuel	0.56	0.64	0.64	0.64	0.75	0.76	0.79	0.75	0.77	0.81
Lease and Plant Fuel[7]	1.07	1.08	1.07	1.06	1.20	1.17	1.12	1.16	1.12	1.04
Total	**21.98**	**23.80**	**24.02**	**24.32**	**25.28**	**26.26**	**28.00**	**24.78**	**26.12**	**29.51**
Discrepancy[8]	**-0.11**	**-0.05**	**-0.05**	**-0.05**	**-0.06**	**-0.05**	**-0.06**	**-0.06**	**-0.06**	**-0.06**
Lower 48 End of Year Reserves	**189.91**	**206.13**	**205.23**	**204.31**	**213.39**	**208.32**	**200.56**	**217.27**	**210.60**	**196.06**
Natural Gas Prices										
(2005 dollars per million Btu)										
Henry Hub Spot Price	8.60	6.57	6.28	6.05	5.96	5.71	5.05	7.12	6.52	5.75
Average Lower 48 Wellhead Price[11]	7.29	5.85	5.59	5.37	5.30	5.07	4.47	6.36	5.80	5.10
(2005 dollars per thousand cubic feet)										
Average Lower 48 Wellhead Price[11]	7.51	6.03	5.76	5.54	5.46	5.22	4.60	6.55	5.98	5.26
Delivered Prices										
(2005 dollars per thousand cubic feet)										
Residential	12.80	11.56	11.31	11.10	11.08	10.86	10.22	12.34	11.77	10.99
Commercial	11.54	9.88	9.62	9.40	9.16	8.93	8.30	10.16	9.58	8.82
Industrial[4]	8.41	6.89	6.62	6.40	6.31	6.08	5.44	7.34	6.76	6.01
Electric Power[5]	8.42	6.64	6.40	6.21	6.05	5.93	5.44	6.98	6.51	6.01
Transportation[10]	15.20	14.57	14.38	14.23	13.51	13.36	12.95	14.24	13.86	13.38
Average[11]	**9.94**	**8.34**	**8.07**	**7.85**	**7.80**	**7.54**	**6.86**	**8.97**	**8.33**	**7.46**

[1]Marketed production (wet) minus extraction losses.
[2]Synthetic natural gas, propane air, coke oven gas, refinery gas, biomass gas, air injected for Btu stabilization, and manufactured gas commingled and distributed with natural gas.
[3]Includes any natural gas regasified in the Bahamas and transported via pipeline to Florida.
[4]Includes energy for combined heat and power plants, except those whose primary business is to sell electricity, or electricity and heat, to the public.
[5]Includes consumption of energy by electricity-only and combined heat and power plants whose primary business is to sell electricity, or electricity and heat, to the public. Includes small power producers and exempt wholesale generators.
[6]Compressed natural gas used as vehicle fuel.
[7]Represents natural gas used in field gathering and processing plant machinery.
[8]Balancing item. Natural gas lost as a result of converting flow data measured at varying temperatures and pressures to a standard temperature and pressure and the merger of different data reporting systems which vary in scope, format, definition, and respondent type. In addition, 2005 values include net storage injections.
[9]Represents lower 48 onshore and offshore supplies.
[10]Compressed natural gas used as a vehicle fuel. Price includes estimated motor vehicle fuel taxes and estimated dispensing costs or charges.
[11]Weighted average prices. Weights used are the sectoral consumption values excluding lease, plant, and pipeline fuel.
Note: Totals may not equal sum of components due to independent rounding. Data for 2005 are model results and may differ slightly from official EIA data reports.
Sources: 2005 supply values: Energy Information Administration (EIA), *Natural Gas Monthly*, DOE/EIA-0130(2006/04) (Washington, DC, April 2006). 2005 consumption based on: EIA, *Annual Energy Review 2005*, DOE/EIA-0384(2005) (Washington, DC, July 2006). **Projections:** EIA, AEO2007 National Energy Modeling System runs LOLNG07.D112406A, AEO2007.D112106A, and HILNG07.D112406B.

Table D12. Petroleum Supply and Disposition, ANWR Drilling Case
(Million Barrels per Day, Unless Otherwise Noted)

Supply, Disposition, and Prices	2005	2010		2020		2030	
		Reference	ANWR	Reference	ANWR	Reference	ANWR
Crude Oil							
Domestic Crude Production[1]	5.18	5.67	5.67	5.89	6.32	5.39	6.03
Alaska	0.86	0.69	0.69	0.74	1.15	0.27	0.92
Lower 48 States	4.31	4.98	4.98	5.15	5.16	5.12	5.11
Net Imports	10.09	9.99	9.99	11.29	10.87	13.09	12.49
Other Crude Supply[2]	-0.06	0.00	0.00	0.00	0.00	0.00	0.00
Total Crude Supply	**15.22**	**15.66**	**15.66**	**17.19**	**17.19**	**18.47**	**18.52**
Other Petroleum Supply							
Natural Gas Plant Liquids	1.72	1.80	1.80	1.76	1.79	1.72	1.75
Net Product Imports[3]	2.48	1.80	1.79	2.27	2.26	3.28	3.23
Refinery Processing Gain[4]	0.99	1.21	1.21	1.41	1.43	1.49	1.51
Other Inputs	0.39	1.02	1.02	1.31	1.31	1.88	1.88
Liquids from Coal	0.00	0.00	0.00	0.10	0.10	0.44	0.45
Other[5]	0.39	1.02	1.02	1.21	1.21	1.43	1.43
Total Primary Supply[6]	**20.79**	**21.49**	**21.48**	**23.94**	**23.97**	**26.84**	**26.89**
Refined Petroleum Products Supplied							
by Fuel							
Liquefied Petroleum Gases	2.03	2.22	2.22	2.26	2.26	2.42	2.43
E85[7]	0.00	0.00	0.00	0.01	0.01	0.02	0.02
Motor Gasoline[8]	9.16	9.53	9.53	10.93	10.94	12.53	12.55
Jet Fuel[9]	1.68	1.95	1.95	2.19	2.20	2.27	2.27
Distillate Fuel Oil[10]	4.12	4.53	4.53	5.11	5.12	5.95	5.95
Residual Fuel Oil	0.92	0.79	0.79	0.82	0.82	0.83	0.83
Other[11]	2.84	2.57	2.57	2.70	2.69	2.93	2.92
by Sector							
Residential and Commercial	1.26	1.25	1.25	1.29	1.29	1.28	1.28
Industrial[12]	5.07	5.01	5.01	5.16	5.16	5.53	5.54
Transportation	13.87	14.93	14.93	17.15	17.15	19.69	19.71
Electric Power[13]	0.51	0.40	0.40	0.43	0.43	0.45	0.45
Total	**20.75**	**21.59**	**21.59**	**24.03**	**24.03**	**26.95**	**26.97**
Discrepancy[14]	**0.04**	**-0.10**	**-0.10**	**-0.09**	**-0.06**	**-0.11**	**-0.08**
Imported Low Sulfur Light Crude Oil Price (2005 dollars per barrel)[15]	56.76	57.47	57.47	52.04	52.04	59.12	58.34
Imported Crude Oil Price (2005 dollars per barrel)[15]	49.19	51.20	51.20	46.47	46.47	51.63	51.63
Import Share of Product Supplied (percent)	60.5	54.9	54.9	56.6	54.8	61.0	58.5
Net Expenditures for Imported Crude Oil and Petroleum Products (billion 2005 dollars)	236.65	222.76	222.71	229.80	221.61	300.51	288.83

[1]Includes lease condensate.
[2]Strategic petroleum reserve stock additions plus unaccounted for crude oil and crude stock withdrawals minus crude product supplied.
[3]Includes other hydrocarbons and alcohols.
[4]Represents volumetric gain in refinery distillation and cracking processes.
[5]Includes petroleum product stock withdrawals; domestic sources of blending components, other hydrocarbons, alcohols, and ethers.
[6]Total crude supply plus natural gas plant liquids, other inputs, refinery processing gain, and net product imports.
[7]E85 refers to a blend of 85 percent ethanol (renewable) and 15 percent motor gasoline (nonrenewable). To address cold starting issues, the percentage of ethanol actually varies seasonally. The annual average ethanol content of 74 percent is used for this forecast.
[8]Includes ethanol and ethers blended into gasoline.
[9]Includes only kerosene type.
[10]Includes distillate and kerosene.
[11]Includes aviation gasoline, liquefied petroleum gas, petrochemical feedstocks, lubricants, waxes, asphalt, road oil, still gas, special naphthas, petroleum coke, crude oil product supplied, and miscellaneous petroleum products.
[12]Includes consumption for combined heat and power (CHP), which produces electricity and other useful thermal energy.
[13]Includes consumption of energy by electricity-only and combined heat and power plants whose primary business is to sell electricity, or electricity and heat, to the public. Includes small power producers and exempt wholesale generators.
[14]Balancing item. Includes unaccounted for supply, losses, and gains.
[15]Weighted average price delivered to U.S. refiners.
Note: Totals may not equal sum of components due to independent rounding. Data for 2005 are model results and may differ slightly from official EIA data reports.
Sources: 2005 imported crude oil price and petroleum product supplied based on: Energy Information Administration (EIA), *Annual Energy Review 2005*, DOE/EIA-0384(2005) (Washington, DC, July 2006). 2005 imported low sulfur light crude oil price: EIA, Form EIA-856, "Monthly Foreign Crude Oil Acquisition Report." Other 2005 data: EIA, *Petroleum Supply Annual 2005*, DOE/EIA-0340(2005)/1 (Washington, DC, October 2006). **Projections:** EIA, AEO2007 National Energy Modeling System runs AEO2007.D112106A and ANWR2007.D112706C.

Table D13. Petroleum Supply and Disposition, Expanded Outer Continental Shelf Access Case

Supply, Disposition, and Prices	2005	2010		2020		2030	
		Reference	OCS Access	Reference	OCS Access	Reference	OCS Access
Petroleum Supply							
Crude Oil							
Domestic Crude Production[1]	5.18	5.67	5.67	5.89	5.98	5.39	5.55
Alaska	0.86	0.69	0.69	0.74	0.74	0.27	0.27
Lower 48 States	4.31	4.98	4.98	5.15	5.23	5.12	5.28
Onshore	2.89	2.93	2.93	2.94	2.94	2.92	2.91
Offshore	1.42	2.05	2.05	2.21	2.30	2.20	2.36
Net Imports	10.09	9.99	9.99	11.29	11.28	13.09	12.97
Other Crude Supply[2]	-0.06	0.00	0.00	0.00	0.00	0.00	0.00
Total Crude Supply	**15.22**	**15.66**	**15.66**	**17.19**	**17.26**	**18.47**	**18.52**
Other Petroleum Supply							
Natural Gas Plant Liquids	1.72	1.80	1.80	1.76	1.78	1.72	1.76
Net Product Imports[3]	2.48	1.80	1.80	2.27	2.24	3.28	3.27
Refinery Processing Gain[4]	0.99	1.21	1.21	1.41	1.35	1.49	1.42
Other Inputs	0.39	1.02	1.02	1.31	1.30	1.88	1.88
Liquids from Coal	0.00	0.00	0.00	0.10	0.10	0.44	0.44
Other[5]	0.39	1.02	1.02	1.21	1.21	1.43	1.44
Total Petroleum Supply[6]	**20.79**	**21.49**	**21.48**	**23.94**	**23.93**	**26.84**	**26.85**
Natural Gas Prices							
(2005 dollars per million Btu)							
Henry Hub Spot Price	8.60	6.28	6.27	5.71	5.66	6.52	6.37
Average Lower 48 Wellhead Price[11]	7.29	5.59	5.58	5.07	5.02	5.80	5.67
(2005 dollars per thousand cubic feet)							
Average Lower 48 Wellhead Price[11]	7.51	5.76	5.75	5.22	5.18	5.98	5.84
Natural Gas Supply (trillion cubic feet)							
Dry Gas Production[7]	**18.23**	**19.35**	**19.35**	**20.79**	**21.02**	**20.53**	**21.14**
Lower 48 Onshore	14.36	15.22	15.22	14.66	14.63	15.13	15.15
Associated-Dissolved	1.43	1.39	1.39	1.28	1.28	1.19	1.19
Non-Associated	12.93	13.83	13.84	13.38	13.35	13.94	13.96
Conventional	4.94	5.27	5.27	4.30	4.29	3.75	3.70
Unconventional	7.99	8.56	8.57	9.09	9.06	10.19	10.26
Lower 48 Offshore	3.41	3.88	3.88	4.09	4.35	3.25	3.84
Associated-Dissolved	0.71	0.92	0.92	1.05	1.08	0.85	1.04
Non-Associated	2.69	2.96	2.96	3.04	3.27	2.40	2.80
Alaska	0.45	0.25	0.25	2.05	2.05	2.16	2.16
Supplemental Natural Gas[8]	0.07	0.07	0.07	0.07	0.07	0.07	0.07
Net Imports	**3.57**	**4.55**	**4.55**	**5.35**	**5.20**	**5.45**	**5.20**
Pipeline	3.01	2.74	2.74	1.65	1.65	0.92	0.92
Liquefied Natural Gas[9]	0.57	1.81	1.81	3.69	3.55	4.53	4.28
Total Natural Gas Supply[10]	**21.87**	**23.97**	**23.97**	**26.21**	**26.29**	**26.06**	**26.41**

[1]Includes lease condensate.
[2]Strategic petroleum reserve stock additions plus unaccounted for crude oil and crude stock withdrawals minus crude product supplied.
[3]Includes other hydrocarbons and alcohols.
[4]Represents volumetric gain in refinery distillation and cracking processes.
[5]Includes petroleum product stock withdrawals; domestic sources of blending components, other hydrocarbons, alcohols, and ethers.
[6]Total crude supply plus natural gas plant liquids, other inputs, refinery processing gain, and net product imports.
[7]Marketed production (wet) minus extraction losses.
[8]Synthetic natural gas, propane air, coke oven gas, refinery gas, biomass gas, air injected for Btu stabilization, and manufactured gas commingled and distributed with natural gas.
[9]Includes any natural gas regasified in the Bahamas and transported via pipeline to Florida.
[10]Dry gas production plus supplemental natural gas and net imports.
Note: Totals may not equal sum of components due to independent rounding. Data for 2005 are model results and may differ slightly from official EIA data reports.
Sources: 2005 imported crude oil price and petroleum product supplied based on: Energy Information Administration (EIA), *Annual Energy Review 2005,* DOE/EIA-0384(2005) (Washington, DC, July 2006). 2005 imported low sulfur light crude oil price: EIA, Form EIA-856, "Monthly Foreign Crude Oil Acquisition Report." Other 2005 data: EIA, *Petroleum Supply Annual 2005,* DOE/EIA-0340(2005)/1 (Washington, DC, October 2006). **Projections:** EIA, AEO2007 National Energy Modeling System runs AEO2007.D112106A and OCSACC.D112706A.

Table D14. Ethanol Supply, Disposition, and Prices, Additional Ethanol Cases

Supply, Disposition, and Prices	2005	2015				2030			
		Reference	Low Cost Ethanol	High Price	Low Cost Ethanol and High Price	Reference	Low Cost Ethanol	High Price	Low Cost Ethanol and High Price
Ethanol Supply (billion gallons)									
Domestic Production									
Corn Based	3.9	11.1	11.1	11.1	12.0	13.6	11.5	14.2	22.3
Cellulose Based	0.0	0.2	0.2	0.2	0.3	0.2	3.9	0.6	10.1
Total Domestic Production	**3.9**	**11.4**	**11.4**	**11.4**	**12.3**	**13.8**	**15.4**	**14.8**	**32.5**
Ethanol Imports	0.1	0.4	0.5	0.5	0.5	0.8	0.8	0.8	1.5
Total Ethanol Supply	**4.0**	**11.7**	**11.9**	**11.9**	**12.8**	**14.6**	**16.2**	**15.6**	**33.9**
Ethanol Consumption (billion gallons)									
Used in E85[1]	0.0	0.0	0.2	0.1	1.0	0.2	1.4	2.4	20.6
Used in Gasoline Blending	4.0	11.7	11.7	11.8	11.9	14.4	14.8	13.2	13.3
Total Ethanol Consumption	**4.0**	**11.7**	**11.9**	**11.9**	**12.8**	**14.6**	**16.2**	**15.6**	**33.9**
Total Motor Gasoline Consumption (billion gallons)	140.4	156.1	156.9	146.0	146.3	192.1	195.9	160.5	154.3
Light-Duty Vehicle Energy Consumption (quadrillion Btu)	16.36	17.99	18.03	16.83	16.90	22.66	22.49	19.50	19.84
Ethanol Percent of Motor Gasoline Pool (percent)[2]	2.9	7.5	7.5	8.1	8.7	7.6	8.2	9.5	18.6
Transportation Sector Indicators									
Light-Duty Vehicle Miles Travelled (billion miles)	2655	3125	3124	2978	2976	4226	4219	3924	3899
New Car Efficiency (miles per gallon)[3]	30.0	32.4	32.5	33.2	33.1	33.7	33.6	36.2	35.8
New Light Truck Efficiency (miles per gallon)[3]	21.8	24.7	24.3	25.3	24.8	26.5	25.8	28.2	27.2
Light-Duty Stock Efficiency (miles per gallon)[4]	19.6	20.6	20.5	20.9	20.8	22.2	22.2	23.9	23.5
E85-Capable Vehicle Sales (thousands)	613	1819	13401	1576	12887	2030	17045	3778	16816
Total Vehicle Sales (thousands)	16235	17268	17270	16838	16828	20187	20176	20076	20063
E85-Capable Vehicle Stock (millions)	4	17	48	16	46	31	229	37	225
Total Vehicle Stock (millions)	220	263	263	259	259	316	316	312	312
Energy Prices (2005 dollars)									
(dollars per gallon)									
Imported Low Sulfur Light Crude Oil Price[5]	1.35	1.19	1.19	1.89	1.89	1.41	1.41	2.38	2.38
Imported Crude Oil Price[5]	1.17	1.06	1.06	1.76	1.76	1.23	1.23	2.21	2.21
Transportation Sector Motor Gasoline[6]	2.32	1.95	1.95	2.69	2.70	2.15	2.17	3.20	3.23
E85[1]	2.17	1.87	1.88	2.08	2.12	2.00	1.91	2.38	2.49
Ethanol Wholesale Price	1.80	1.66	1.68	1.70	1.75	1.70	1.61	1.84	2.28
(dollars per million Btu)									
Imported Low Sulfur Light Crude Oil Price[5]	9.50	8.34	8.35	13.31	13.31	9.89	9.89	16.75	16.76
Imported Crude Oil Price[5]	8.23	7.46	7.46	12.36	12.36	8.64	8.64	15.55	15.55
Transportation Sector Motor Gasoline[6]	18.64	16.06	16.10	22.19	22.31	17.76	17.85	26.42	26.68
E85[1]	23.10	20.09	20.18	22.32	22.74	21.50	20.53	25.54	26.74
Ethanol Wholesale Price	21.65	19.92	20.17	20.44	21.03	20.43	19.37	22.02	27.39

[1]E85 refers to a blend of 85 percent ethanol (renewable) and 15 percent motor gasoline (nonrenewable). To address cold starting issues, the percentage of ethanol actually varies seasonally. The annual average ethanol content of 74 percent is used for this forecast.
[2]Calculated as the amount of ethanol consumed divided by the total amount of ethanol and motor gasoline consumed.
[3]Environmental Protection Agency rated miles per gallon.
[4]Combined car and light truck "on-the-road" estimate.
[5]Weighted average price delivered to U.S. refiners.
[6]Sales weighted average price for all grades. Includes Federal, State, and local taxes.
Note: Totals may not equal sum of components due to independent rounding. Data for 2005 are model results and may differ slightly from official EIA data reports.
Sources: 2005 consumption and imported crude oil price: Energy Information Administration (EIA), *Annual Energy Review 2005*, DOE/EIA-0384(2005) (Washington, DC, July 2006). 2005 imported low sulfur light crude oil price: EIA, Form EIA-856, "Monthly Foreign Crude Oil Acquisition Report." 2005 motor gasoline price based on: EIA, *Petroleum Marketing Annual 2005*, DOE/EIA-0487(2005) (Washington, DC, August 2006). 2005 ethanol prices derived from weekly spot prices in the Oxy Fuel News. 2005 wholesale ethanol prices derived from Bloomburg U.S. average rack price. **Projections:** EIA, AEO2007 National Energy Modeling System runs AEO2007.D112106A, CT_80PCT_7L_RF.D120406A, HP2007.D112106A, and CT_80PCT_7L.D120406A.

Table D15. Key Results for Coal Cost Cases

(Million Short Tons per Year, Unless Otherwise Noted)

Supply, Disposition, and Prices	2005	2015			2030			Growth Rate, 2005-2030		
		Low Cost	Reference	High Cost	Low Cost	Reference	High Cost	Low Cost	Reference	High Cost
Production[1]	**1131**	**1278**	**1266**	**1209**	**1824**	**1691**	**1329**	**1.9%**	**1.6%**	**0.6%**
Appalachia	397	381	371	349	420	373	337	0.2%	-0.3%	-0.7%
Interior	149	188	199	204	275	247	255	2.5%	2.0%	2.2%
West	585	709	697	656	1129	1072	736	2.7%	2.5%	0.9%
Waste Coal Supplied[2]	**13**	**14**	**13**	**13**	**9**	**13**	**17**	**-1.8%**	**-0.0%**	**0.9%**
Net Imports	**-21**	**3**	**5**	**18**	**47**	**68**	**83**	**N/A**	**N/A**	**N/A**
Total Supply[3]	**1124**	**1294**	**1284**	**1239**	**1880**	**1772**	**1428**	**2.1%**	**1.8%**	**1.0%**
Consumption by Sector										
Residential and Commercial	5	5	5	5	5	5	5	-0.3%	-0.3%	-0.3%
Coke Plants	23	21	21	21	21	21	20	-0.4%	-0.5%	-0.6%
Other Industrial[4]	61	63	62	62	64	64	63	0.2%	0.2%	0.1%
Coal-to-Liquids Heat and Power	0	9	8	3	65	57	16	N/A	N/A	N/A
Coal-to-Liquids Liquids Production	0	9	8	3	63	55	15	N/A	N/A	N/A
Electric Power[5]	1039	1186	1178	1146	1662	1570	1310	1.9%	1.7%	0.9%
Total Coal Use	**1128**	**1293**	**1282**	**1239**	**1880**	**1772**	**1428**	**2.1%**	**1.8%**	**0.9%**
Average Minemouth Price[6]										
(2005 dollars per short ton)	23.34	18.68	22.41	27.12	14.30	22.60	41.01	-1.9%	-0.1%	2.3%
(2005 dollars per million Btu)	1.15	0.92	1.11	1.35	0.72	1.15	2.05	-1.9%	-0.0%	2.3%
Delivered Prices[7]										
(2005 dollars per short ton)										
Coke Plants	83.79	66.66	74.51	84.52	57.27	75.55	111.24	-1.5%	-0.4%	1.1%
Other Industrial[4]	47.63	43.03	47.45	52.26	38.01	48.54	65.93	-0.9%	0.1%	1.3%
Coal to Liquids	N/A	11.23	13.79	17.09	15.10	21.89	31.59	N/A	N/A	N/A
Electric Power[5]										
(2005 dollars per short ton)	30.83	28.37	31.84	36.37	24.63	33.52	49.75	-0.9%	0.3%	1.9%
(2005 dollars per million Btu)	1.53	1.42	1.60	1.83	1.23	1.69	2.49	-0.9%	0.4%	2.0%
Average	**32.82**	**29.47**	**33.10**	**37.91**	**24.80**	**33.82**	**50.93**	**-1.1%**	**0.1%**	**1.8%**
Exports[8]	67.10	58.08	64.51	73.77	51.47	63.81	95.87	-1.1%	-0.2%	1.4%
Cumulative Electricity Generating Capacity Additions (gigawatts)[9]										
Coal	0.0	22.6	19.5	15.7	189.2	156.3	77.0	N/A	N/A	N/A
Conventional: Pulverized Coal	0.0	17.9	15.8	12.9	111.5	78.4	44.4	N/A	N/A	N/A
Advanced: IGCC	0.0	4.7	3.6	2.7	77.7	77.9	32.6	N/A	N/A	N/A
Petroleum	0.0	0.2	0.2	0.2	0.3	0.3	0.3	N/A	N/A	N/A
Natural Gas	0.0	32.3	33.1	33.7	92.7	105.2	133.1	N/A	N/A	N/A
Nuclear	0.0	1.8	0.5	1.3	6.0	12.5	42.3	N/A	N/A	N/A
Renewables[10]	0.0	10.7	10.8	10.9	15.5	17.5	20.8	N/A	N/A	N/A
Other	0.0	0.1	0.1	0.1	0.1	0.2	0.2	N/A	N/A	N/A
Total	**0.0**	**67.7**	**64.2**	**61.9**	**303.8**	**291.9**	**273.7**	**N/A**	**N/A**	**N/A**
Liquids from Coal (million barrels per day)	0.00	0.07	0.06	0.02	0.51	0.44	0.12	N/A	N/A	N/A
Labor Productivity										
Coal Mining										
(short tons per miner per hour)	6.36	8.72	6.80	5.25	14.77	7.84	3.38	3.4%	0.8%	-2.5%
Rail: Eastern Railroads (billion freight ton-miles per employee per year)	7.57	10.90	8.91	7.25	18.17	10.61	6.11	3.6%	1.4%	-0.9%
Rail: Western Railroads (billion freight ton-miles per employee per year)	12.19	17.72	14.49	11.79	29.86	17.43	10.05	3.6%	1.4%	-0.8%

Table D15. Key Results for Coal Cost Cases (Continued)
(Million Short Tons per Year, Unless Otherwise Noted)

Supply, Disposition, and Prices	2005	2015			2030			Growth Rate, 2005-2030		
		Low Cost	Reference	High Cost	Low Cost	Reference	High Cost	Low Cost	Reference	High Cost
Cost Indices										
(constant dollar index, 2005=1.000)										
Transportation Rate Multipliers										
Eastern Railroads	1.000	1.043	1.064	1.086	0.968	1.042	1.121	-0.1%	0.2%	0.5%
Western Railroads	1.000	1.029	1.046	1.063	0.971	1.028	1.088	-0.1%	0.1%	0.3%
Equipment Costs										
Mining										
Underground	1.000	0.943	1.032	1.129	0.811	1.032	1.311	-0.8%	0.1%	1.1%
Surface	1.000	0.927	1.016	1.110	0.798	1.016	1.289	-0.9%	0.1%	1.0%
Railroads	1.000	0.905	0.991	1.085	0.743	0.946	1.202	-1.2%	-0.2%	0.7%
Average Coal Miner Wage										
(2005 dollars per hour)	**22.06**	**20.16**	**22.06**	**24.13**	**17.34**	**22.06**	**28.02**	**-1.0%**	**-0.0%**	**1.0%**

[1]Includes anthracite, bituminous coal, and lignite.
[2]Includes waste coal consumed by the electric power and industrial sectors. Waste coal supplied is counted as a supply-side item to balance the same amount of waste coal included in the consumption data.
[3]Production plus net imports plus net storage withdrawals.
[4]Includes consumption for combined heat and power plants, except those plants whose primary business is to sell electricity, or electricity and heat, to the public. Excludes all coal use in the coal to liquids process.
[5]Includes all electricity-only and combined heat and power plants whose primary business is to sell electricity, or electricity and heat, to the public.
[6]Includes reported prices for both open market and captive mines.
[7]Prices weighted by consumption tonnage; weighted average excludes residential and commercial prices, and export free-alongside-ship (f.a.s.) prices.
[8]F.a.s. price at U.S. port of exit.
[9]Cumulative additions after December 31, 2005. Includes all additions of electricity only and combined heat and power plants projected for the electric power, industrial, and commercial sectors.
[10]Includes conventional hydroelectric, geothermal, wood, wood waste, municipal solid waste, landfill gas, other biomass, solar, and wind power. Facilities co-firing biomass and coal are classified as coal.
N/A = Not applicable.
Btu = British thermal unit.
IGCC = Integrated gas combined cycle.
Note: Totals may not equal sum of components due to independent rounding. Data for 2005 are model results and may differ slightly from official EIA data reports.
Sources: 2005 data based on: Energy Information Administration (EIA), *Annual Coal Report 2005*, DOE/EIA-0584(2005) (Washington, DC, October 2006); EIA, *Quarterly Coal Report, October-December 2005*, DOE/EIA-0121(2005/4Q) (Washington, DC, March 2006); Securities and Exchange Commission Form 10K filings (BNSF, Norfolk Southern, and Union Pacific), web site www.sec.gov; CSX Corporation, web site www.csx.com; U.S. Department of Labor, Bureau of Labor Statistics, Average Hourly Earnings of Production Workers: Coal Mining, Series ID : ceu1021210006; and EIA, AEO2007 National Energy Modeling System run AEO2007.D112106A. **Projections:** EIA, AEO2007 National Energy Modeling System runs LCCST07.D112906A, AEO2007.D112106A, and HCCST07.D112906A.

NEMS Overview and Brief Description of Cases

The National Energy Modeling System

The projections in the *Annual Energy Outlook 2007* (*AEO2007*) are generated from the National Energy Modeling System (NEMS) [*1*], developed and maintained by the Office of Integrated Analysis and Forecasting (OIAF) of the Energy Information Administration (EIA). In addition to its use in the development of the *AEO* projections, NEMS is also used in analytical studies for the U.S. Congress, the White House, and other offices within the Department of Energy. The *AEO* projections are also used by analysts and planners in other government agencies and outside organizations.

The projections in NEMS are developed with the use of a market-based approach to energy analysis. For each fuel and consuming sector, NEMS balances energy supply and demand, accounting for economic competition among the various energy fuels and sources. The time horizon of NEMS is the long-term period through 2030, approximately 25 years into the future. In order to represent regional differences in energy markets, the component modules of NEMS function at the regional level: the nine Census divisions for the end-use demand modules; production regions specific to oil, natural gas, and coal supply and distribution; the North American Electric Reliability Council regions and subregions for electricity; and the Petroleum Administration for Defense Districts (PADDs) for refineries.

NEMS is organized and implemented as a modular system. The modules represent each of the fuel supply markets, conversion sectors, and end-use consumption sectors of the energy system. NEMS also includes macroeconomic and international modules. The primary flows of information among the modules are the delivered prices of energy to end users and the quantities consumed by product, region, and sector. The delivered fuel prices encompass all the activities necessary to produce, import, and transport fuels to end users. The information flows also include other data on such areas as economic activity, domestic production, and international petroleum supply.

The Integrating Module controls the execution of each of the component modules. To facilitate modularity, the components do not pass information to each other directly but communicate through a central data structure. This modular design provides the capability to execute modules individually, thus allowing decentralized development of the system and independent analysis and testing of individual modules, and permits the use of the methodology and level of detail most appropriate for each energy sector. NEMS calls each supply, conversion, and end-use demand module in sequence until the delivered prices of energy and the quantities demanded have converged within tolerance, thus achieving an economic equilibrium of supply and demand in the consuming sectors. A solution is reached annually through the long-term horizon. Other variables are also evaluated for convergence, such as petroleum product imports, crude oil imports, and several macroeconomic indicators.

Each NEMS component represents the impacts and costs of legislation and environmental regulations that affect that sector. NEMS accounts for all combustion-related carbon dioxide (CO_2) emissions, as well as emissions of sulfur dioxide, nitrogen oxides, and mercury from the electricity generation sector. NEMS generally represents current legislation and environmental regulations as of October 31, 2006 (such as the Energy Policy Acts of 2005 [EPACT-2005], the Working Families Tax Relief Act of 2004, and the American Jobs Creation Act of 2004) and the costs of compliance with regulations such as the Clean Air Interstate Rule (CAIR) and Clean Air Mercury Rule (CAMR), both of which were finalized and published in 2005, the new corporate average fuel economy (CAFE) standards finalized in March 2006, and the new stationary diesel regulations issued by the U.S. Environmental Protection Agency (EPA) in July 2006. The potential impacts of pending or proposed Federal and State legislation, regulations, or standards—or of sections of legislation that have been enacted but that require funds or implementing regulations that have not been provided or specified—are not reflected in NEMS.

In general, the historical data used for the *AEO2007* projections were based on EIA's *Annual Energy Review 2005*, published in July 2006 [*2*]; however, data were taken from multiple sources. In some cases, only partial or preliminary data were available for 2005. CO_2 emissions were calculated by using CO_2 coefficients from the EIA report, *Emissions of Greenhouse Gases in the United States 2005*, published in November 2006 [*3*].

Historical numbers are presented for comparison only and may be estimates. Source documents should be consulted for the official data values. Footnotes to the *AEO2007* appendix tables indicate the definitions and sources of historical data.

The *AEO2007* projections for years 2006 and 2007 incorporate short-term projections from EIA's September 2006 *Short-Term Energy Outlook* (*STEO*). For short-term energy projections, readers are referred to monthly updates of the *STEO* [*4*].

Component Modules

The component modules of NEMS represent the individual supply, demand, and conversion sectors of domestic energy markets and also include international and macroeconomic modules. In general, the modules interact through values representing the prices or expenditures of energy delivered to the consuming sectors and the quantities of end-use energy consumption.

Macroeconomic Activity Module

The Macroeconomic Activity Module provides a set of macroeconomic drivers to the energy modules, and there is a macroeconomic feedback mechanism within NEMS. Key macroeconomic variables used in the energy modules include gross domestic product (GDP), disposable income, industrial output, new housing starts, new light-duty vehicle sales, interest rates, prices, and employment. The module uses the following models from Global Insight, Inc. (GII): Macroeconomic Model of the U.S. Economy, National Industry Model, and National Employment Model. In addition, EIA has constructed a Regional Economic and Industry Model to project regional economic drivers and a Commercial Floorspace Model to project 13 floorspace types in 9 Census divisions. The accounting framework for industrial output uses the North American Industry Classification System (NAICS).

International Module

The International Module represents the response of world oil markets (supply and demand) to assumed world oil prices. The results/outputs of the module are a set of crude oil and product supply curves that are available to U.S. markets for each case/scenario analyzed. The petroleum import supply curves are made available to U.S. markets through the Petroleum Market Module (PMM) of NEMS in the form of 5 categories of imported crude oil and 17 international petroleum products, including supply curves for oxygenates and unfinished oils. The supply-curve calculations are based on historical market data and a world oil supply/demand balance, which is developed from the *International Energy Outlook 2006*, current investment trends in exploration and development, and long-term resource economics for 221 countries/territories. The oil production estimates include both conventional and unconventional supply recovery technologies.

Residential and Commercial Demand Modules

The Residential Demand Module projects energy consumption in the residential sector by housing type and end use, based on delivered energy prices, the menu of equipment available, the availability of renewable sources of energy, and housing starts. The Commercial Demand Module projects energy consumption in the commercial sector by building type and nonbuilding uses of energy and by category of end use, based on delivered prices of energy, availability of renewable sources of energy, and macroeconomic variables representing interest rates and floorspace construction.

Both modules estimate the equipment stock for the major end-use services, incorporating assessments of advanced technologies, including representations of renewable energy technologies and the effects of both building shell and appliance standards. The Commercial Demand Module incorporates combined heat and power (CHP) technology. The modules also include projections of distributed generation. Both modules incorporate changes to "normal" heating and cooling degree-days by Census division, based on State-level population projections. The Residential Demand Module projects that the average square footage of both new construction and existing structures is increasing, based on trends in the size of new construction and the remodeling of existing homes.

Industrial Demand Module

The Industrial Demand Module projects the consumption of energy for heat and power and for feedstocks and raw materials in each of 21 industries, subject to the delivered prices of energy and macroeconomic variables representing employment and the value of shipments for each industry. As noted in the description of the Macroeconomic Activity Module, the value of shipments is based on NAICS. The industries are classified into three groups—energy-intensive manufacturing, non-energy-intensive manufacturing, and nonmanufacturing. Of the 8 energy-intensive industries, 7 are modeled in the Industrial Demand Module, with components for boiler/steam/ cogeneration, buildings, and process/assembly use of energy. Bulk chemicals are further disaggregated to organic, inorganic, resins, and agricultural chemicals. A representation of cogeneration and a recycling component are also included. The use of energy for petroleum refining is modeled in the PMM, and the projected consumption is included in the industrial totals.

Transportation Demand Module

The Transportation Demand Module projects consumption of fuels in the transportation sector, including petroleum products, electricity, methanol, ethanol, compressed natural gas, and hydrogen, by transportation mode, vehicle vintage, and size class, subject to delivered prices of energy fuels and macroeconomic variables representing disposable personal income, GDP, population, interest rates, and industrial shipments. Fleet vehicles are represented separately to allow analysis of the Clean Air Act Amendments of 1990 (CAAA90) and other legislation and legislative proposals. The module also includes a component to assess the penetration of alternative-fuel vehicles.

The air transportation component explicitly represents the industry practice of parking aircraft to reduce operating costs and the movement of aircraft from passenger to cargo markets as aircraft age [*5*]. For air freight shipments, the model represents fuel use in narrow-body and wide-body aircraft. An infrastructure constraint limits overall growth in passenger and freight air travel to levels commensurate with industry-projected infrastructure expansion and capacity growth.

Electricity Market Module

The Electricity Market Module represents generation, transmission, and pricing of electricity, subject to delivered prices for coal, petroleum products, natural gas, and biofuels; costs of generation by all generation plants, including capital costs; macroeconomic variables for costs of capital and domestic investment; enforced environmental emissions laws and regulations; and electricity load shapes and demand. There are three primary submodules—capacity planning, fuel dispatching, and finance and pricing. Nonutility generation, distributed generation, and transmission and trade are modeled in the planning and dispatching submodules.

All specifically identified CAAA90 compliance options that have been promulgated by the EPA are explicitly represented in the capacity expansion and dispatch decisions; those that have not been promulgated (e.g., fine particulate proposals) are not incorporated. All financial incentives for power generation expansion and dispatch specifically identified in EPACT2005 have been implemented. Several States, primarily in the Northeast, have recently enacted air emission regulations that affect the electricity generation sector. Where firm State compliance plans have been announced, regulations are represented in *AEO2007*.

Renewable Fuels Module

The Renewable Fuels Module (RFM) includes submodules representing renewable resource supply and technology input information for central-station, grid-connected electricity generation technologies, including conventional hydroelectricity, biomass (wood, energy crops, and biomass co-firing), geothermal, landfill gas, solar thermal electricity, solar photovoltaics (PV), and wind energy. The RFM contains renewable resource supply estimates representing the regional opportunities for renewable energy development. Investment tax credits for renewable fuels are incorporated, as currently legislated in the Energy Policy Act of 1992 (EPACT1992) and EPACT2005. EPACT1992 provides a 10-percent tax credit for business investment in solar energy (thermal non-power uses as well as power uses) and geothermal power; those credits have no expiration date. EPACT2005 increases the tax credit to 30 percent for solar energy systems installed before January 1, 2008 (which has since been extended to January 1, 2009, but is not reflected in the *AEO2007* projections).

Production tax credits for wind, geothermal, landfill gas, and some types of hydroelectric and biomass-fueled plants are also represented. They provide a tax credit of up to 1.9 cents per kilowatthour for electricity produced in the first 10 years of plant operation. At the time *AEO2007* was completed, new plants coming on line before January 1, 2008, were eligible to receive the credit. Subsequently—after *AEO2007* modeling runs were completed—the deadline was extended to January 1, 2009. Significant changes made for *AEO2007* in the accounting of new renewable energy capacity resulting from State renewable portfolio standard (RPS) programs, mandates, and goals are described in *Assumptions to the Annual Energy Outlook 2007* [*6*].

Oil and Gas Supply Module

The Oil and Gas Supply Module represents domestic crude oil and natural gas supply within an integrated framework that captures the interrelationships among the various sources of supply: onshore, offshore, and Alaska by both conventional and unconventional techniques, including natural gas recovery from coalbeds and low-permeability formations of sandstone and shale. This framework analyzes cash flow and profitability to compute investment and drilling for each of the supply sources, based on the prices for crude oil and natural gas, the domestic recoverable resource base, and the state of technology. Oil and gas production functions are computed at a level of 12 supply regions, including 3 offshore and 3 Alaskan regions. The module also represents foreign sources of natural gas, including pipeline imports and exports to Canada and Mexico, and liquefied natural gas (LNG) imports and exports.

Crude oil production quantities are input to the PMM in NEMS for conversion and blending into refined petroleum products. Supply curves for natural gas are input to the Natural Gas Transmission and Distribution Module for use in determining natural gas prices and quantities. International LNG supply sources and options for construction of new regasification terminals in Canada, Mexico, and the United States as well as expansions of existing U.S. regasification terminals are represented, based on the projected regional costs associated with international natural gas supply, liquefaction, transportation, and regasification and world natural gas market conditions.

Natural Gas Transmission and Distribution Module

The Natural Gas Transmission and Distribution Module represents the transmission, distribution, and pricing of natural gas, subject to end-use demand for natural gas and the availability of domestic natural gas and natural gas traded on the international market. The module tracks the flows of natural gas and determines the associated capacity expansion requirements in an aggregate pipeline network, connecting the domestic and foreign supply regions with 12 demand regions. The flow of natural gas is determined for both a peak and off-peak period in the year. Key components of pipeline and distributor tariffs are included in separate pricing algorithms.

Petroleum Market Module

The PMM projects prices of petroleum products, crude oil and product import activity, and domestic refinery operations (including fuel consumption), subject to the demand for petroleum products, the availability and price of imported petroleum, and the domestic production of crude oil, natural gas liquids, and alcohol and biodiesel fuels. The module represents refining activities in the five PADDs. It explicitly models the requirements of CAAA90 and the costs of automotive fuels, such as conventional and reformulated gasoline, and includes biofuels production for blending in gasoline and diesel.

AEO2007 represents the nationwide phase-in of gasoline with an annual average sulfur content of 30 ppm between 2005 and 2007, regulations that limit the sulfur content of highway diesel fuel to 15 ppm starting in mid-2006 and of all non-road and locomotive/marine diesel to 15 ppm by mid-2012, and the renewable fuels standard of 7.5 billion gallons by 2012. Demand growth and regulatory changes necessitate capacity expansion for refinery processing units. For those investments, a financing ratio of 60 percent equity and 40 percent debt is assumed, with a hurdle rate and an after-tax return on investment of about 9 percent [*7*]. End-use prices are based on the marginal costs of production, plus markups representing product marketing and distribution costs and State and Federal taxes [*8*]. Refinery capacity expansion at existing sites is permitted in all five refining regions modeled. *AEO2007* accounts for the phasing out of methyl tertiary butyl ether (MTBE) as a result of decisions made by the petroleum industry to discontinue MTBE blending with gasoline.

Fuel ethanol and biodiesel are included in the PMM, because they are commonly blended into petroleum products. The module allows ethanol blending into gasoline at 10 percent by volume or less, as well as E85, a blend of up to 85 percent ethanol by volume. Ethanol is produced primarily in the Midwest from corn or other starchy crops, and it is expected to be produced from cellulosic material in other regions in the future. Biodiesel is produced from soybean oil or yellow grease (primarily, recycled cooking oil). Both soybean oil biodiesel and yellow grease biodiesel are assumed to be blended into highway diesel.

Alternative fuels such as coal-to-liquids (CTL) and gas-to-liquids (GTL) are modeled in the PMM, based on their economics relative to competing feedstocks and products. CTL facilities are likely to be built at locations close to coal supply and water sources, where liquid products and electricity could also be distributed to nearby demand regions. GTL facilities may be built on the North Slope of Alaska but would compete with the Alaska Natural Gas Transportation System for available natural gas resources.

Ethanol production is modeled from two feedstocks: corn and cellulosic materials such as switchgrass and poplar. Corn-based ethanol plants are numerous (more than 80 in operation, producing more than 4 billion gallons annually) and are based on a well-known technology that converts sugars into ethanol. Ethanol from cellulosic sources is a new technology with no full-sized plants constructed. The two sources are modeled to compete on an economic basis to meet the EPACT2005 mandate.

Coal Market Module

The Coal Market Module (CMM) simulates mining, transportation, and pricing of coal, subject to end-use demand for coal differentiated by heat and sulfur content. U.S. coal production is represented in the CMM by 40 separate supply curves—differentiated by region, mine type, coal rank, and sulfur content. The coal supply curves include a response to capacity utilization of mines, mining capacity, labor productivity, and factor input costs (mining equipment, mining labor, and fuel requirements). Projections of U.S. coal distribution are determined by minimizing the cost of coal supplied, given coal demands by demand region and sector, accounting for minemouth prices, transportation costs, existing coal supply contracts, and sulfur and mercury allowance costs. Over the projection horizon, coal transportation costs in the CMM are projected to vary in response to changes in railroad productivity and the cost of rail transportation equipment and diesel fuel.

The CMM produces projections of U.S. steam and metallurgical coal exports and imports, in the context of world coal trade. The CMM determines the pattern of world coal trade flows that minimizes the production and transportation costs of meeting a specified set of regional world coal import demands, subject to constraints on export capacities and trade flows. The international coal market component of the module computes trade in 3 types of coal for 17 export and 20 import regions. U.S. coal production and distribution are computed for 14 supply and 14 demand regions.

Annual Energy Outlook 2007 Cases

Table E1 provides a summary of the cases used to derive the *AEO2007* projections. For each case, the table gives the name used in this report, a brief description of the major assumptions underlying the projections, a designation of the mode in which the case was run in NEMS (either fully integrated, partially integrated, or standalone), and a reference to the pages in the body of the report and in this appendix where the case is discussed. The following sections describe the cases listed in Table E1. The reference case assumptions for each sector are described at web site www.eia.doe.gov/oiaf/aeo/assumption. Regional results and other details of the projections are available at web site www.eia.doe.gov/oiaf/aeo/supplement.

Macroeconomic Growth Cases

In addition to the *AEO2007* reference case, the low economic growth and high economic growth cases were developed to reflect the uncertainty in projections of economic growth. The alternative cases are intended to show the effects of alternative growth assumptions on energy market projections. The cases are described as follows:

- The *low economic growth case* assumes lower growth rates for population (0.5 percent per year), non-farm employment (0.6 percent per year), and productivity (1.9 percent per year), resulting in higher prices and interest rates and lower growth in industrial output. In the low economic growth case, economic output as measured by real GDP increases by 2.3 percent per year from 2005 through 2030, and growth in real GDP per capita averages 1.8 percent per year.
- The *high economic growth case* assumes higher growth rates for population (1.2 percent per year), nonfarm employment (1.3 percent per year), and productivity (2.8 percent per year). With higher productivity gains and employment growth, inflation and interest rates are lower than in the reference case, and consequently economic output grows at a higher rate (3.4 percent per year) than in the reference case (2.9 percent). GDP per capita grows by 2.2 percent per year, compared with 2.1 percent in the reference case.

Price Cases

The world oil price in *AEO2007* is defined as the average price of low-sulfur, light crude oil imported into the United States. The low-sulfur, light crude oil price is similar to prices for the light sweet crude oil contract traded on the New York Mercantile Exchange. *AEO2007* also includes a projection of the U.S. annual average imported refiners' acquisition cost of crude oil, which is more representative of the average cost of all crude oils used by refiners.

Table E1. Summary of the *AEO2007* cases

Case name	Description	Integration mode	Reference in text	Reference in Appendix E
Reference	Baseline economic growth (2.9 percent per year from 2005 through 2030), world oil price, and technology assumptions. Complete projection tables in Appendix A.	Fully integrated	—	—
Low Economic Growth	Gross domestic product grows at an average annual rate of 2.3 percent from 2005 through 2030. Other assumptions are the same as in the reference case. Partial projection tables in Appendix B.	Fully integrated	p. 68	p. 211
High Economic Growth	Gross domestic product grows at an average annual rate of 3.4 percent from 2005 through 2030. Other assumptions are the same as in the reference case. Partial projection tables in Appendix B.	Fully integrated	p. 68	p. 211
Low Price	More optimistic assumptions for worldwide crude oil and natural gas resources than in the reference case. World light, sweet crude oil prices are $36 per barrel in 2030, compared with $59 per barrel in the reference case (2005 dollars). Other assumptions are the same as in the reference case. Partial projection tables in Appendix C.	Fully integrated	p. 34	p. 215
High Price	More pessimistic assumptions for worldwide crude oil and natural gas resources than in the reference case. World light, sweet crude oil prices are about $100 per barrel in 2030. Other assumptions are the same as in the reference case. Partial projection tables in Appendix C.	Fully integrated	p. 34	p. 215
Residential: 2006 Technology	Future equipment purchases based on equipment available in 2006. Existing building shell efficiencies fixed at 2006 levels. Partial projection tables in Appendix D.	With commercial	p. 74	p. 215
Residential: High Technology	Earlier availability, lower costs, and higher efficiencies assumed for more advanced equipment. Building shell efficiencies for new construction meet ENERGY STAR requirements after 2010. Partial projection tables in Appendix D.	With commercial	p. 74	p. 215
Residential: Best Available Technology	Future equipment purchases and new building shells based on most efficient technologies available by fuel. Building shell efficiencies for new construction meet the criteria for most efficient components after 2006. Partial projection tables in Appendix D.	With commercial	p. 75	p. 215
Commercial: 2006 Technology	Future equipment purchases based on equipment available in 2006. Building shell efficiencies fixed at 2006 levels. Partial projection tables in Appendix D.	With residential	p. 75	p. 215
Commercial: High Technology	Earlier availability, lower costs, and higher efficiencies assumed for more advanced equipment. Building shell efficiencies for new and existing buildings increase by 8.75 and 6.25 percent, respectively, from 2003 values by 2030. Partial projection tables in Appendix D.	With residential	p. 75	p. 215
Commercial: Best Available Technology	Future equipment purchases based on most efficient technologies available by fuel. Building shell efficiencies for new and existing buildings increase by 10.5 and 7.5 percent, respectively, from 2003 values by 2030. Partial projection tables in Appendix D.	With residential	p. 76	p. 215
Industrial: 2006 Technology	Efficiency of plant and equipment fixed at 2006 levels. Partial projection tables in Appendix D.	Standalone	p. 79	p. 216

Table E1. Summary of the *AEO2007* cases (continued)

Case name	Description	Integration mode	Reference in text	Reference in Appendix E
Industrial: High Technology	Earlier availability, lower costs, and higher efficiencies assumed for more advanced equipment. Partial projection tables in Appendix D.	Standalone	p. 79	p. 216
Transportation: 2006 Technology	Efficiencies for new equipment in all modes of travel fixed at 2006 levels. Partial projection tables in Appendix D.	Standalone	p. 81	p. 216
Transportation: High Technology	Reduced costs and improved efficiencies assumed for advanced technologies. Partial projection tables in Appendix D.	Standalone	p. 81	p. 216
Electricity: Low Nuclear Cost	New nuclear capacity assumed to have 10 percent lower capital and operating costs in 2030 than in the reference case. Partial projection tables in Appendix D.	Fully integrated	p. 85	p. 217
Electricity: High Nuclear Cost	Costs for new nuclear technology assumed not to improve from 2006 levels in the reference case. Partial projection tables in Appendix D.	Fully integrated	p. 85	p. 217
Electricity: High Fossil Technology	Costs and efficiencies for advanced fossil-fired generating technologies improve by 10 percent in 2030 from reference case values. Partial projection tables in Appendix D.	Fully integrated	p. 87	p. 217
Electricity: Low Fossil Technology	New advanced fossil generating technologies assumed not to improve over time from 2006. Partial projection tables in Appendix D.	Fully integrated	p. 87	p. 217
Renewable Fuels: Low Renewables	New renewable generating technologies assumed not to improve over time from 2006. Partial projection tables in Appendix D.	Fully integrated	p. 86	p. 217
Renewable Fuels: High Renewables	Levelized cost of energy for nonhydropower renewable generating technologies declines by 10 percent in 2030 from reference case values. Partial projection tables in Appendix D.	Fully integrated	p. 86	p. 217
Renewable Fuels: Regional RPS	Represents various State renewable portfolio standard (RPS) programs, with targets aggregated on a regional basis. Assumes full compliance with targets, as limited by statutory authorizations for State funding, where applicable. Partial projection tables in Appendix D.	Fully integrated	p. 87	p. 218
Oil and Gas: Rapid Technology	Cost, finding rate, and success rate parameters adjusted for 50-percent more rapid improvement than in the reference case. Partial projection tables in Appendix D.	Fully integrated	p. 91	p. 218
Oil and Gas: Slow Technology	Cost, finding rate, and success rate parameters adjusted for 50-percent slower improvement than in the reference case. Partial projection tables in Appendix D.	Fully Integrated	p. 91	p. 218
Oil and Gas: High LNG	LNG imports exogenously set to 30 percent more than the results from the low price case, with remaining assumptions from the reference case. Partial projection tables in Appendix D.	Fully integrated	—	p. 218
Oil and Gas: Low LNG	LNG imports exogenously set to 30 percent less than the results from the high price case, with remaining assumptions from the reference case. Partial projection tables in Appendix D.	Fully integrated	—	p. 218
Oil and Gas: OCS Access	Drilling moratorium assumed to expire in 2012 for oil and natural gas exploration and development in the Atlantic, Pacific, and Eastern Gulf of Mexico Outer Continental Shelf. Partial projection tables in Appendix D.	Fully integrated	p. 51	p. 218

Table E1. Summary of the *AEO2007* cases (continued)

Case name	Description	Integration mode	Reference in text	Reference in Appendix E
Oil and Gas: ANWR	Federal oil and gas leasing permitted in the Arctic National Wildlife Refuge starting in 2007. Partial projection tables in Appendix D.	Fully integrated	—	p. 218
Petroleum Market: Lower Cost Ethanol, Reference Energy Price	Capital costs of cellulosic ethanol plants decline by 26 percent and operating costs decline by 20 percent by 2018 from reference case values in 2012. Biomass supply assumed to have greater availability than in the reference case, at reference case prices. Assumed policies enacted that make market penetration of flex-fuel vehicles exceed 80 percent by 2016, and increase fuel dispensing availability for E85 as it becomes more competitive. Uses reference cases energy prices. Partial projection tables in Appendix D.	Fully Integrated	p. 98	p. 219
Petroleum Market: Lower Cost Ethanol, High Energy Price	Capital costs of cellulosic ethanol plants decline by 26 percent and operating costs decline by 20 percent by 2018 from reference case values in 2012. Biomass supply assumed to have greater availability than in the reference case, at reference case prices. Assumed policies enacted that make market penetration of flex-fuel vehicles exceed 80 percent by 2016, and increase fuel dispensing availability for E85 as it becomes more competitive. Uses high price case energy prices. Partial projection tables in Appendix D.	Fully integrated	p. 98	p. 219
Coal Market: Low Coal Cost	Productivity for coal mining and coal transportation assumed to increase more rapidly than in the reference case. Coal mining wages, mine equipment, and coal transportation equipment costs assumed to be lower than in the reference case. Partial projection tables in Appendix D.	Fully integrated	p. 99	p. 219
Coal Market: High Coal Cost	Productivity for coal mining and coal transportation assumed to increase more slowly than in the reference case. Coal mining wages, mine equipment, and coal transportation equipment costs assumed to be higher than in the reference case. Partial projection tables in Appendix D.	Fully integrated	p. 99	p. 219
Integrated 2006 Technology	Combination of the residential, commercial, industrial, and transportation 2006 technology cases, electricity low fossil technology case, low renewables case, and high nuclear cost case. Partial projection tables in Appendix D.	Fully integrated	—	p. 219
Integrated High Technology	Combination of the residential, commercial, industrial, and transportation high technology cases, electricity high fossil technology case, high renewables case, and low nuclear cost case. Partial projection tables in Appendix D.	Fully integrated	—	p. 219

The historical record shows substantial variability in world oil prices, and there is arguably even more uncertainty about future prices in the long term. *AEO2007* considers three price cases (reference, low price, and high price) to allow an assessment of alternative views on the course of future oil and natural gas prices. In the reference case, world oil prices moderate from 2006 levels through 2015 before beginning to rise to $59 per barrel in 2030 (2005 dollars). The low and high price cases define a wide range of potential price paths (from $36 to $100 per barrel in 2030). The two cases reflect different assumptions about the availability of world oil and natural gas resources; they do not assume changes in behavior by the Organization of the Petroleum Exporting Countries (OPEC). Because the low and high price cases are not directly integrated with a world economic model, the impact of world oil prices on international economies is not accounted for directly.

- The *reference case* represents EIA's current judgment regarding the expected behavior of OPEC producers in the long term, adjusting production to keep world oil prices in a range of $50 to $60 per barrel, in keeping with OPEC's stated goal of keeping potential competitors from eroding its market share. Because OPEC (and particularly the Persian Gulf nations) is expected to be the dominant supplier of oil in the international market over the long term, its production choices will significantly affect world oil prices.
- The *low price case* assumes that world crude oil and natural gas resources, including OPEC's, are 15 percent higher than assumed in the reference case.
- The *high price case* assumes that world crude oil and natural gas resources, including OPEC's, are 15 percent lower than assumed in the reference case.

Buildings Sector Cases

In addition to the *AEO2007* reference case, three standalone technology-focused cases using the Residential and Commercial Demand Modules of NEMS were developed to examine the effects of changes to equipment and building shell efficiencies.

For the residential sector, the three technology-focused cases are as follows:

- The *2006 technology case* assumes that all future equipment purchases are based only on the range of equipment available in 2006. Existing building shell efficiencies are assumed to be fixed at 2006 levels (no further improvements). For new construction, building shell technology options are constrained to those available in 2006.
- The *high technology case* assumes earlier availability, lower costs, and higher efficiencies for more advanced equipment [*9*]. For new construction, building shell efficiencies are assumed to meet ENERGY STAR requirements after 2010.
- The *best available technology case* assumes that all future equipment purchases are made from a menu of technologies that includes only the most efficient models available in a particular year for each fuel, regardless of cost. For new construction, building shell efficiencies are assumed to meet the criteria for most efficient components after 2006.

For the commercial sector, the three technology-focused cases are as follows:

- The *2006 technology case* assumes that all future equipment purchases are based only on the range of equipment available in 2006. Building shell efficiencies are assumed to be fixed at 2006 levels.
- The *high technology case* assumes earlier availability, lower costs, and/or higher efficiencies for more advanced equipment than in the reference case [*10*]. Building shell efficiencies for new and existing buildings in 2030 are assumed to be 8.75 percent and 6.25 percent higher, respectively, than their 2003 levels—a 25-percent improvement relative to the reference case.
- The *best available technology case* assumes that all future equipment purchases are made from a menu of technologies that includes only the most efficient models available in a particular year for each fuel, regardless of cost. Building shell efficiencies for new and existing buildings in 2030 are assumed to be 10.5 percent and 7.5 percent higher, respectively, than their 2003 values—a 50-percent improvement relative to the reference case.

The Residential and Commercial Demand Modules of NEMS were also used to complete the high renewables and low renewables cases, which are discussed in more detail as part of the Renewables Fuels Cases section below. In combination with assumptions for electricity generation from renewable fuels in the electric power sector and industrial sector, these sensitivities analyze the impact of changes in generating technologies that use renewable fuels and in the availability of renewable energy sources. For the Residential and Commercial Demand Modules:

- The *high renewables case* assumes greater improvements in residential and commercial PV systems than in the reference case. The high renewables assumptions result in capital cost estimates for 2030 that are approximately 10 percent lower than reference case costs for distributed PV technologies.
- The *low renewables case* assumes that costs and performance levels for residential and commercial PV systems remain constant at 2006 levels through 2030.

Industrial Sector Cases

In addition to the *AEO2007* reference case, two standalone cases using the Industrial Demand Module of NEMS were developed to examine the effects of less rapid and more rapid technology change and adoption. Because these are standalone cases, the energy intensity changes discussed in this section exclude the refining industry. Energy use in the refining industry is solved as part of the PMM in NEMS. The Industrial Demand Module was also used as part of an integrated high renewables case. For the industrial sector:

- The *2006 technology case* holds the energy efficiency of plant and equipment constant at the 2006 level over the projection period. In this case, delivered energy intensity falls by 1.0 percent annually between 2005 and 2030, as compared with 1.5 percent in the reference case. Changes in aggregate energy intensity may result both from changing equipment and production efficiency and from changing composition of industrial output. Because the level and composition of industrial output are the same in the reference, 2006 technology, and high technology cases, any change in energy intensity in the two technology cases is attributable to efficiency changes. The 2006 technology case was run with only the Industrial Demand Module, rather than in fully integrated NEMS runs. Consequently, no potential feedback effects from energy market interactions were captured.
- The *high technology case* assumes earlier availability, lower costs, and higher efficiency for more advanced equipment [*11*] and a more rapid rate of improvement in the recovery of biomass byproducts from industrial processes (0.7 percent per year, as compared with 0.4 percent per year in the reference case). The same assumption is incorporated in the integrated high renewables case, which focuses on electricity generation. While the choice of 0.7-percent recovery is an assumption of the high technology case, it is based on the expectation that there would be higher recovery rates and substantially increased use of CHP in that case. Delivered energy intensity falls by 1.7 percent annually in the high technology case.

Transportation Sector Cases

In addition to the *AEO2007* reference case, two standalone cases using the Transportation Demand Module of NEMS were developed to examine the effects of less rapid technology change and adoption and more rapid technology change and adoption. For the transportation sector:

- The *2006 technology case* assumes that new vehicle fuel efficiencies remain constant at 2006 levels through the projection horizon, unless emissions and/or efficiency regulations require the implementation of technology that affects vehicle efficiency. For example, the new light truck CAFE standards require an increase in fuel economy through 2011, and increases in heavy truck emissions standards are required through 2010 [*12*]. As a result, the technology available for light truck efficiency improvement is frozen at 2011 levels, and the technology available to heavy trucks is frozen at 2010 levels.
- In the *high technology case*, the characteristics of light-duty conventional and alternative-fuel vehicles reflect more optimistic assumptions about incremental improvements in fuel economy and costs [*13*]. In the air travel sector, the high technology case reflects lower costs for improved thermodynamics, advanced aerodynamics, and weight-reducing materials. In the freight truck sector, the high technology case assumes more incremental improvement in fuel efficiency for engine and emissions control technologies [*14*]. More optimistic assumptions for fuel efficiency improvements are also made for the rail and shipping sectors.

Both cases were run with only the Transportation Demand Module rather than as fully integrated NEMS runs. Consequently, no potential macroeconomic feedback on travel demand was captured, nor were changes in fuel prices incorporated.

Electricity Sector Cases

In addition to the reference case, four integrated cases with alternative electric power assumptions were developed to analyze uncertainties about the future costs and performance of new generating technologies. Two of the cases examine alternative assumptions for nuclear power technologies, and two examine alternative assumptions for fossil fuel technologies. Reference case values for technology characteristics are determined in consultation with industry and government specialists; however, there is always uncertainty surrounding newer, untested designs. The electricity cases analyze what could happen if costs of advanced designs were either higher or lower than assumed in the reference case. The cases are fully integrated to allow feedback between the potential shifts in fuel consumption and fuel prices.

Nuclear Technology Cases

- The cost assumptions for the *low nuclear cost case* reflect a 10-percent reduction in the capital and operating costs for advanced nuclear technology in 2030, relative to the reference case. The reference case projects a 17-percent reduction in the capital costs of nuclear power plants from 2006 to 2030. The low nuclear cost case assumes a 25-percent reduction between 2006 and 2030.
- The *high nuclear cost case* assumes that capital costs for the advanced nuclear technology do not decline during the projection period but remain fixed at the 2006 levels assumed in the reference case.

Fossil Technology Cases

- In the *high fossil technology case*, capital costs, heat rates, and operating costs for advanced coal and natural gas generating technologies are assumed to be 10 percent lower than reference case levels in 2030. Because learning is assumed to occur in the reference case, costs and performance in the high case are reduced from initial levels by more than 10 percent. Heat rates in the high fossil technology case fall to between 15 and 22 percent below initial levels, and capital costs are reduced by 20 to 24 percent between 2006 and 2030, depending on the technology.
- In the *low fossil technology case*, capital costs and heat rates for coal gasification combined-cycle units and advanced combustion turbine and combined-cycle units do not decline during the projection period but remain fixed at the 2006 values assumed in the reference case.

Details about annual capital costs, operating and maintenance costs, plant efficiencies, and other factors used in the high and low fossil technology cases are described in the detailed assumptions, which are available at web site www.eia.doe.gov/oiaf/aeo/assumption.

Renewable Fuels Cases

In addition to the *AEO2007* reference case, two integrated cases with alternative assumptions about renewable fuels were developed to examine the effects of less aggressive and more aggressive improvement in renewable technologies. Also included is an integrated case estimating the potential impacts of various State RPS or similar programs. The cases are as follows:

- In the *low renewables case*, capital costs, operating and maintenance costs, and performance levels for wind, solar, biomass, and geothermal resources are assumed to remain constant at 2007 levels through 2030.
- In the *high renewables case*, the levelized costs of energy for nonhydroelectric generating technologies using renewable resources are assumed to decline to 10 percent below the reference case costs for the same resources in 2030. For most renewable resources, lower costs are accomplished by reducing the capital costs of new plant construction. To reflect recent trends in wind energy cost reductions, however, it is assumed that wind plants ultimately achieve the 10-percent cost reduction through a combination of performance improvement (increased capacity factor) and capital cost reductions. Biomass supplies are also assumed to be 10 percent greater for each supply step. Other generating technologies and projection assumptions remain unchanged from those in the reference case. In the high renewables case, the rate of improvement in recovery of biomass byproducts from industrial processes is also increased.
- Many States have implemented RPS or similar renewable generation goals or mandates. Because of the significant variability among State programs and uncertainty regarding actual implementation provisions, the impacts of the programs are not

included in the *AEO2007* reference case. The *regional RPS case* examines the potential impact of the various State and regional RPS or RPS-like programs in place as of September 1, 2006. Program targets are aggregated as necessary, based on the electricity regions used in NEMS. The analysis assumes that limits on credit trading prices or other discretionary limits on requirements to comply using renewable generation are not limiting; however, statutory constraints on State financing of required renewable capacity are considered. Because of the regional representation of the RPS programs, it is assumed that otherwise eligible generation from anywhere within the primary electricity region serving an affected State will be allowed to satisfy the RPS obligation, but that generation from outside that region will not. In recognition of the tight market coupling between the ECAR and MAAC electricity regions (both of which are substantially served by the PJM Interconnect transmission market), wind energy resources in ECAR are assumed to serve RPS requirements in MAAC States (there are no RPS requirements in ECAR States not otherwise serving MAAC , but most MAAC States have RPS programs). Otherwise, all technology and market assumptions are the same as those in the *AEO-2007* reference case.

Oil and Gas Supply Cases

Two alternative technology cases were created to assess the sensitivity of the projections to changes in the assumed rates of progress in oil and natural gas supply technologies. In addition, high and low LNG cases were developed to examine the impacts of variations in LNG imports on the domestic natural gas market. Two final cases examine the potential impacts of the lifting of current moratoria on oil and natural gas exploration and production (E&P) in specified areas of Alaska and the offshore.

- In the *rapid technology case*, the parameters representing the effects of technological progress on finding rates, drilling, lease equipment and operating costs, and success rates for conventional oil and natural gas drilling in the reference case are increased by 50 percent. A number of key E&P technologies for unconventional natural gas are also increased by 50 percent in the rapid technology case. Key Canadian supply parameters were also modified to simulate the assumed impacts of more rapid oil and natural gas technology penetration on the Canadian supply potential. All other parameters in the model are kept at the reference case values, including technology parameters for other modules, parameters affecting foreign oil supply, and assumptions about imports and exports of LNG and natural gas trade between the United States and Mexico. Specific detail by region and fuel category is presented in *Assumptions to the Annual Energy Outlook 2007*, which will be available at web site www.eia.doe.gov/oiaf/aeo/assumption in early 2007.

- In the *slow technology case*, the parameters representing the effects of technological progress on finding rates, drilling, lease equipment and operating costs, and success rates for conventional oil and natural gas drilling in the *AEO2007* reference case are reduced by 50 percent. A number of key E&P technologies for unconventional natural gas are also reduced by 50 percent in the slow technology case. Key Canadian supply parameters are also modified to simulate the assumed impacts of slow oil and natural gas technology penetration on Canadian supply potential. All other parameters in the model are kept at the reference case values.

- The *high LNG case* exogenously specifies LNG imports at levels 30 percent higher than projected in the low price case. The intent is to project the potential impact on domestic markets if LNG imports turn out to be higher than projected in the reference case.

- The *low LNG case* exogenously specifies LNG imports at levels 30 percent lower than projected in the high price case. The intent is to project the potential impact on domestic markets if LNG imports turn out to be lower than projected in the reference case.

- The *OCS access case* assumes that current moratoria on oil and natural gas exploration and development drilling in the Atlantic, Pacific, and Eastern Gulf of Mexico Federal Outer Continental Shelf will expire in 2012. The *AEO2007* reference case assumes that the moratoria will continue throughout the projection period.

- The *ANWR case* assumes that the U.S. Congress will approve leasing on Federal lands in the 1002 area of the Arctic National Wildlife Refuge for oil and natural gas E&P. In the reference case, drilling is not allowed in the 1002 area.

Petroleum Market Cases

In addition to the *AEO2007* reference case, two additional integrated cases were developed to evaluate the impacts of more optimistic assumptions about biomass supplies and progress in the development of cellulosic ethanol technologies on the production and use of cellulosic ethanol. Two ethanol cases were analyzed.

- The *lower cost ethanol, reference energy price case* uses the energy prices from the *AEO2007* reference case.
- The *lower cost ethanol, high energy price case* uses the energy prices from the *AEO2007* high price case.

In each case, it is assumed that technological progress reduces the reference case capital cost of cellulosic ethanol technology in 2018 by about 27 percent and the operating costs in 2018 by about 20 percent from their reference case values in 2012. As in the high renewables case, the supply curve for cellulosic ethanol is shifted in each projection year relative to the reference case, making larger quantities of cellulose available at any given price earlier than in the reference case. It is also assumed that Federal policies will increase the market penetration of flex-fuel vehicles beyond 80 percent of all new light-duty vehicles sold by 2016, that E85 fuel dispensing availability will increase as E85 becomes cost competitive, and that consumers will base their fuel purchase decisions on the relative economics and availability of E85 and gasoline.

Coal Market Cases

Two alternative coal cost cases examine the impacts on U.S. coal supply, demand, distribution, and prices that result from alternative assumptions about mining productivity, labor costs, and mine equipment costs on the production side, and railroad productivity and rail equipment costs on the transportation side. The alternative productivity and cost assumptions are applied in every year from 2007 through 2030. For the coal cost cases, adjustments to the reference case assumptions for coal mining and railroad productivity are based on variations in growth rates observed in the data for those industries since 1980. The low and high coal cost cases represent fully integrated NEMS runs, with feedback from the macroeconomic activity, international, supply, conversion, and end-use demand modules.

- In the *low coal cost case*, average annual productivity growth rates for coal mining and railroad productivity are 2.9 percent and 2.3 percent higher, respectively, than in the *AEO2007* reference case. On the mining side, adjustments to reference case productivity are applied at the supply curve level, while adjustments to railroad productivity are made at the regional level. Coal mining wages and mine equipment costs, which remain constant in real dollars in the reference case, are assumed to decline by approximately 1.0 percent per year in real terms in the low coal cost case. Railroad equipment costs, which are projected to decrease by 0.2 percent per year in constant dollars in the reference case, are assumed to decrease at a faster rate of 1.2 percent per year.
- In the *high coal cost case*, average annual productivity growth rates for coal mining and railroad productivity are 2.9 percent and 2.3 percent lower, respectively, than in the *AEO2007* reference case. Coal mining wages and mine equipment costs are assumed to increase by approximately 1.0 percent per year in real terms. Railroad equipment costs are assumed to increase by 0.7 percent per year.

Additional details about the productivity, wage, and equipment cost assumptions for the reference and alternative coal cost cases are provided in Appendix D.

Integrated Technology Cases

In addition to the sector-specific cases described above, two technology cases combine the assumptions from other technology cases to analyze the impacts of more rapid and slower technology improvement rates.

- The *integrated 2006 technology case* combines the assumptions from the residential, commercial, industrial, and transportation 2006 technology cases, the electricity low fossil technology case, the low renewables case, and the high nuclear cost case.
- The *integrated high technology case* combines the assumptions from the residential, commercial, industrial, and transportation high technology cases, the electricity high fossil technology case, the high renewables case, and the low nuclear cost case.

Endnotes

1. Energy Information Administration, The *National Energy Modeling System: An Overview 2003*, DOE/EIA-0581(2003) (Washington, DC, March 2003).
2. Energy Information Administration, *Annual Energy Review 2005*, DOE/EIA-0384(2005) (Washington, DC, July 2006).
3. Energy Information Administration, *Emissions of Greenhouse Gases in the United States 2005*, DOE/EIA-0573(2005) (Washington, DC, November 2006).
4. Energy Information Administration, *Short-Term Energy Outlook*, web site www.eia.doe.gov/emeu/steo/pub/contents.html. Portions of the preliminary information were also used to initialize the NEMS Petroleum Market Module projection.
5. Jet Information Services, Inc., *World Jet Inventory Year-End 2003* (Woodinville, WA, March 2004); and personal communication from Bill Francoins (Jet Information Services) and Thomas C. Hoang (Boeing).
6. Energy Information Administration, *Assumptions to the Annual Energy Outlook 2007*, DOE/EIA-0554 (2007) (Washington, DC, to be published).
7. The hurdle rate for a CTL plant is assumed to be 12.3 percent because of the higher economic risk associated with the technology.
8. For gasoline blended with ethanol, the tax credit of 51 cents (nominal) per gallon of ethanol is assumed to be extended through 2030, based on the fact that the ethanol tax credit has been continuously in force for the past 25 years and was recently extended from 2007 to 2010 by the American Jobs Creation Act of 2004.
9. High technology assumptions are based on Energy Information Administration, *EIA—Technology Forecast Updates—Residential and Commercial Building Technologies—Advanced Adoption Case* (Navigant Consulting, Inc., September 2004) and *EIA—Technology Forecast—Residential and Commercial Building Technologies—Advanced Case Residential and Commercial Llighting, Ccommercial Rrefrigeration, and Commercial Ventilation Technologies* (Navigant Consulting, Inc., January 2006).
10. High technology assumptions are based on Energy Information Administration, *EIA—Technology Forecast Updates—Residential and Commercial Building Technologies—Advanced Adoption Case* (Navigant Consulting, Inc., September 2004) and *EIA—Technology Forecast—Residential and Commercial Building Technologies—Advanced Case Residential and Commercial Lighting, Commercial Refrigeration, and Commercial Ventilation Technologies* (Navigant Consulting, Inc., January 2006).
11. These assumptions are based in part on Energy Information Administration, *Industrial Technology and Data Analysis Supporting the NEMS Industrial Model* (FOCIS Associates, October 2005).
12. National Highway Traffic Safety Administration, *Average Fuel Economy Standards for Light Trucks Model Years 2008-2011*, 49 CFR Parts 523, 533, and 537, Docket No. 2005-22223, RIN 2127-AJ61 (Washington, DC, August 2005).
13. Energy Information Administration, *Documentation of Technologies Included in the NEMS Fuel Economy Model for Passenger Cars and Light Trucks* (Energy and Environmental Analysis, September 2003).
14. A. Vyas, C. Saricks, and F. Stodolsky, *Projected Effect of Future Energy Efficiency and Emissions Improving Technologies on Fuel Consumption of Heavy Trucks* (Argonne, IL: Argonne National Laboratory, 2001).

Regional Maps

F1. United States Census Divisions

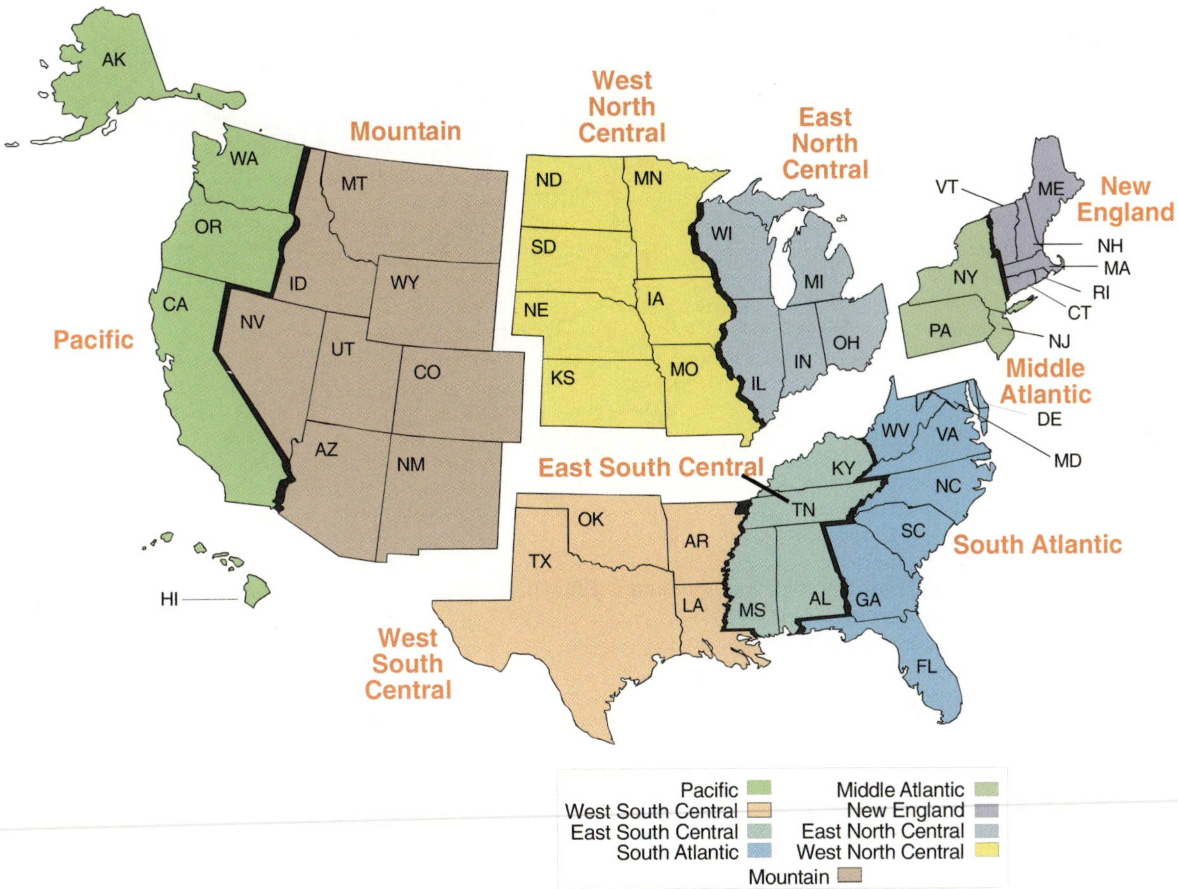

Source: Energy Information Administration. Office of Integrated Analysis and Forecasting.

F2. Electricity Market Module Regions

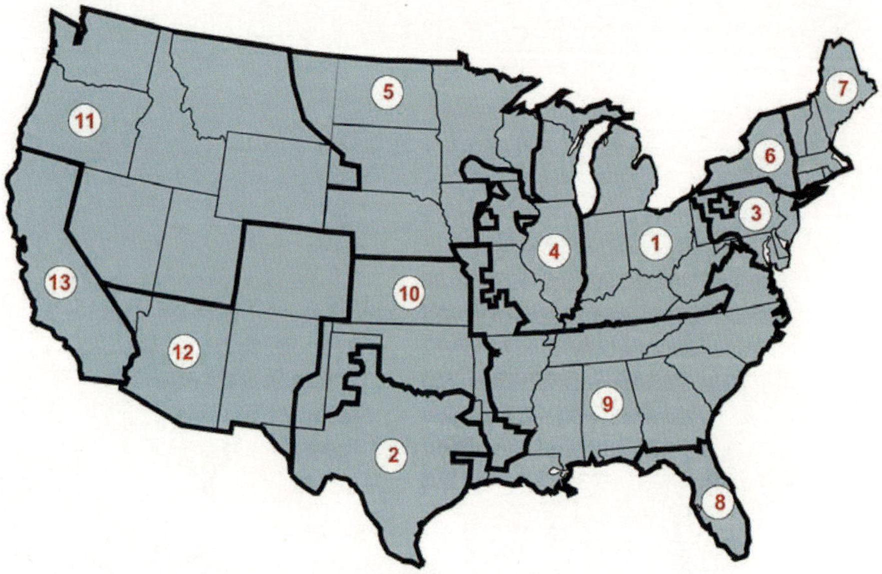

1 East Central Area Reliability Coordination Agreement (ECAR)
2 Electric Reliability Council of Texas (ERCOT)
3 Mid-Atlantic Area Council (MAAC)
4 Mid-America Interconnected Network (MAIN)
5 Mid-Continent Area Power Pool (MAPP)
6. New York (NY) Southern Nevada (RA)
7. New England (NE)
8 Florida Reliability Coordinating Council (FL)
9 Southeastern Electric Reliability Council (SERC)
10 Southwest Power Pool (SPP)
11 Northwest Power Pool (NWP)
12. Rocky Mountain Power Area, Arizona, New Mexico,
13 California (CA)

Source: Energy Information Administration. Office of Integrated Analysis and Forecasting.

F3. Oil and Gas Supply Model Regions

Source: Energy Information Administration. Office of Integrated Analysis and Forecasting.

F4. Natural Gas Transmission and Distribution Model Regions

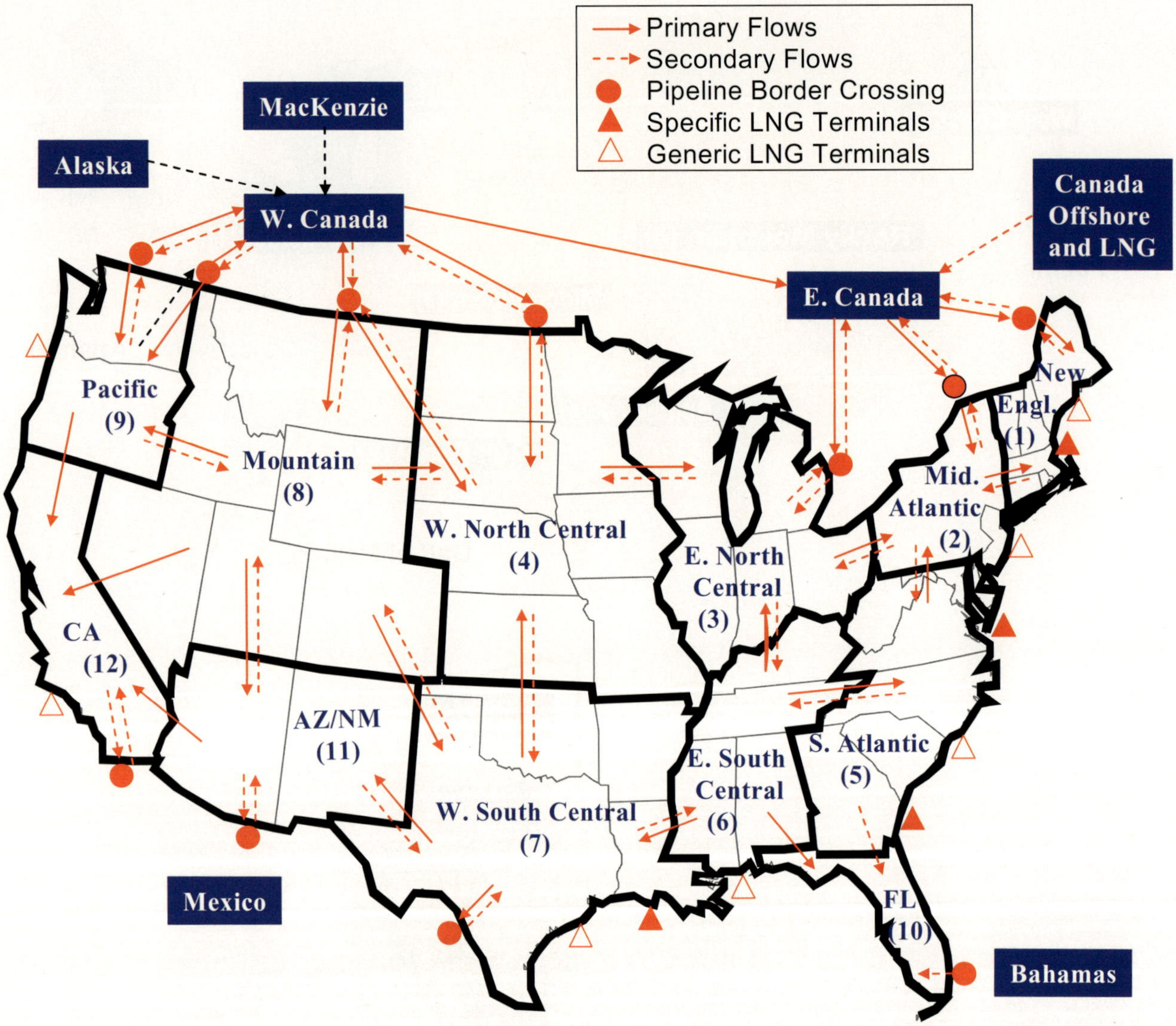

Source: Energy Information Administration. Office of Integrated Analysis and Forecasting.

F5. Petroleum Administration for Defense Districts

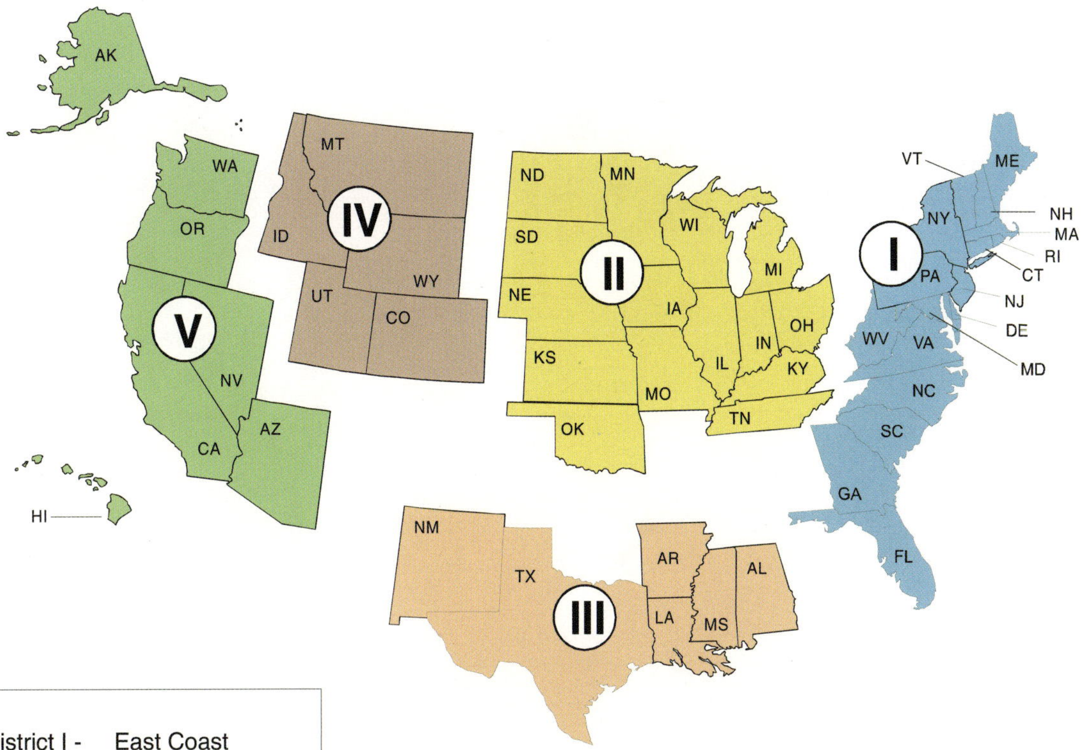

PAD District I - East Coast
PAD District II - Midwest
PAD District III - Gulf Coast
PAD District IV - Rocky Mountain
PAD District V - West Coast

Source: Energy Information Administration. Office of Integrated Analysis and Forecasting.

F6. Coal Supply Regions

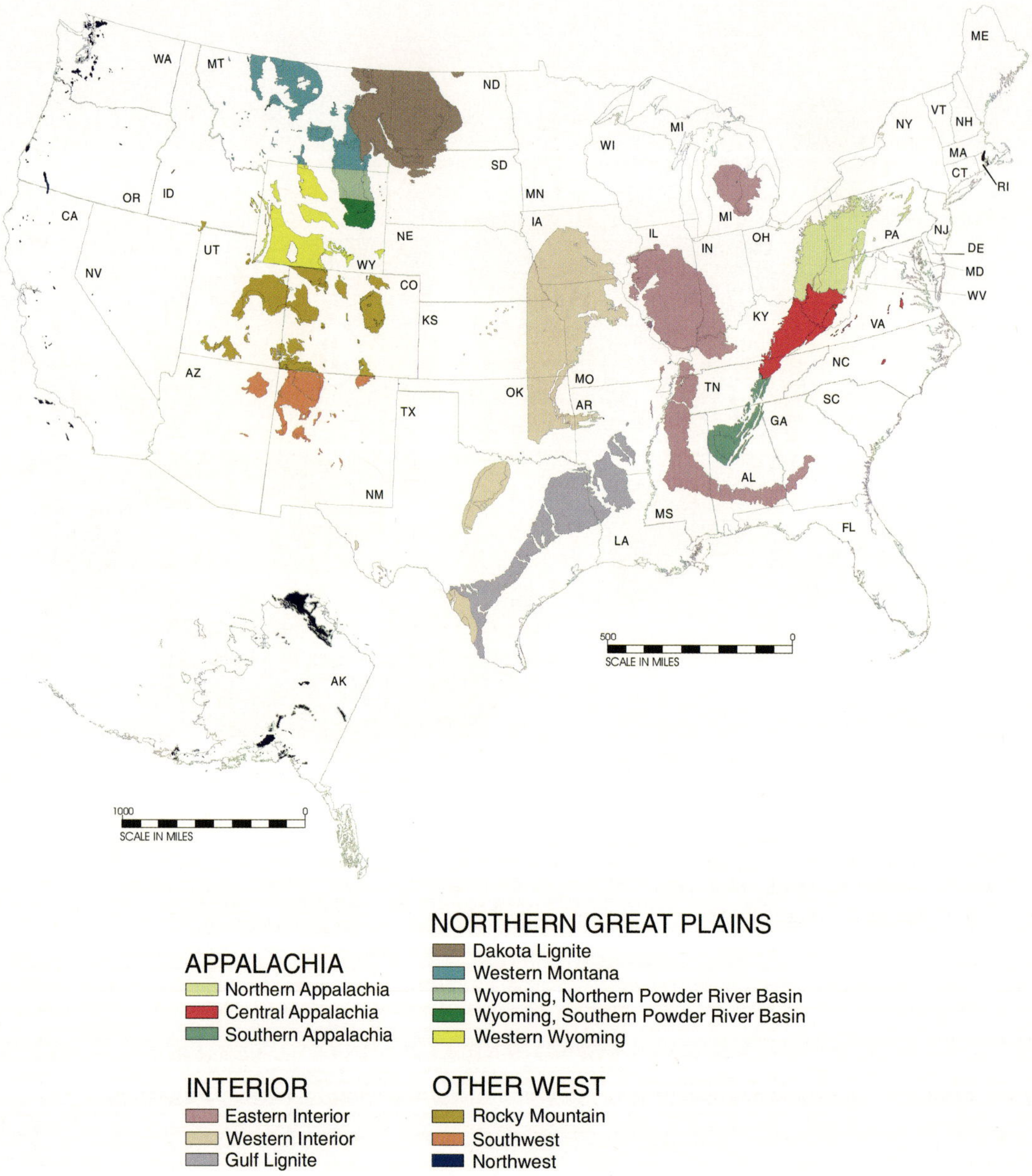

Source: Energy Information Administration. Office of Integrated Analysis and Forecasting.

F7. Coal Demand Regions

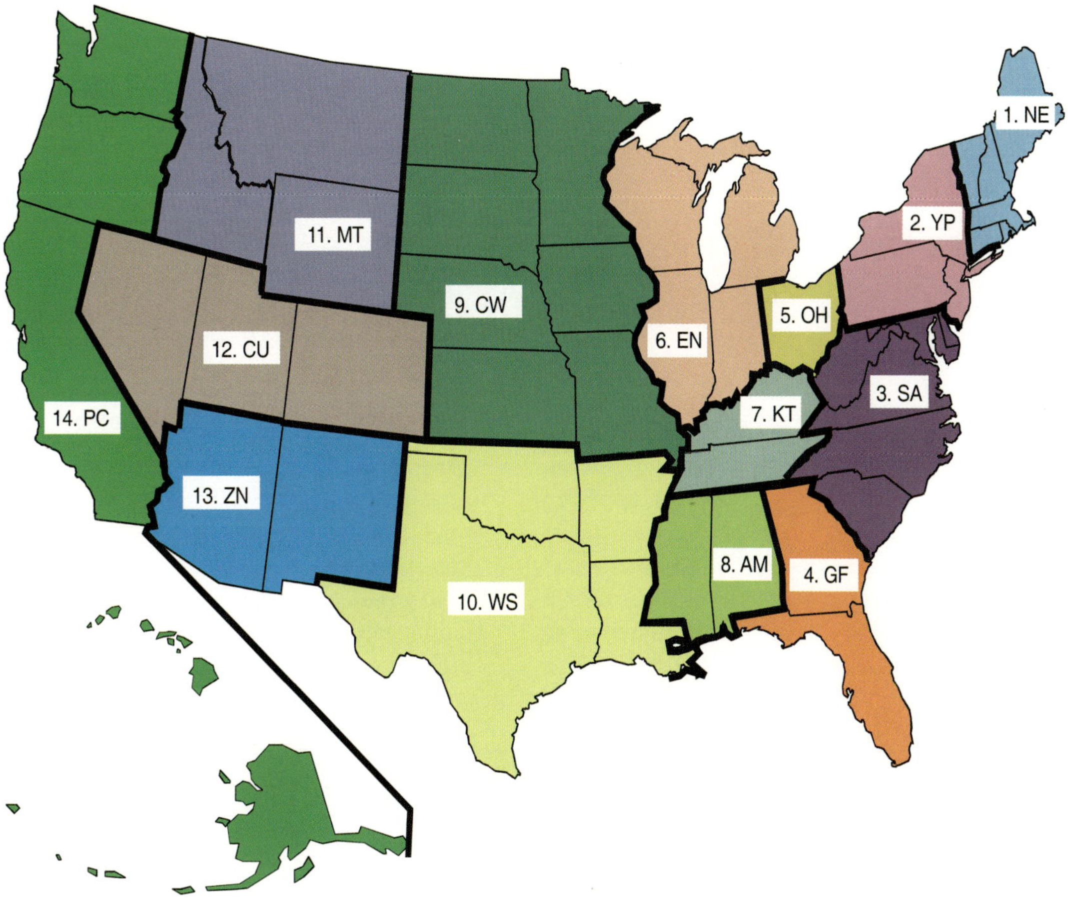

Region Code	Region Content
1. NE	CT,MA,ME,NH,RI,VT
2. YP	NY,PA,NJ
3. SA	WV,MD,DC,DE,VA,NC,SC
4. GF	GA,FL
5. OH	OH
6. EN	IN,IL,MI,WI
7. KT	KY,TN

Region Code	Region Content
8. AM	AL,MS
9. CW	MN,IA,ND,SD,NE,MO,KS
10. WS	TX,LA,OK,AR
11. MT	MT,WY,ID
12. CU	CO,UT,NV
13. ZN	AZ,NM
14. PC	AK,HI,WA,OR,CA

Source: Energy Information Administration. Office of Integrated Analysis and Forecasting.

Conversion Factors

Table G1. Heat Rates

Fuel	Units	Approximate Heat Content
Coal[1]		
Production	million Btu per short ton	20.363
Consumption	million Btu per short ton	20.231
Coke Plants	million Btu per short ton	26.291
Industrial	million Btu per short ton	22.178
Residential and Commercial	million Btu per short ton	22.264
Electric Power Sector	million Btu per short ton	19.970
Imports	million Btu per short ton	25.009
Exports	million Btu per short ton	25.431
Coal Coke	million Btu per short ton	24.800
Crude Oil		
Production	million Btu per barrel	5.800
Imports[1]	million Btu per barrel	5.977
Petroleum Products		
Consumption[1]	million Btu per barrel	5.373
Motor Gasoline[1]	million Btu per barrel	5.218
Jet Fuel	million Btu per barrel	5.670
Distillate Fuel Oil[1]	million Btu per barrel	5.799
Residual Fuel Oil	million Btu per barrel	6.287
Liquefied Petroleum Gas[1]	million Btu per barrel	3.620
Kerosene	million Btu per barrel	5.670
Petrochemical Feedstocks[1]	million Btu per barrel	5.523
Unfinished Oils	million Btu per barrel	5.825
Imports[1]	million Btu per barrel	5.496
Exports[1]	million Btu per barrel	5.741
Natural Gas Plant Liquids		
Production[1]	million Btu per barrel	3.724
Natural Gas[1]		
Production, Dry	Btu per cubic foot	1,030
Consumption	Btu per cubic foot	1,030
End-Use Sectors	Btu per cubic foot	1,030
Electric Power Sector	Btu per cubic foot	1,029
Imports	Btu per cubic foot	1,024
Exports	Btu per cubic foot	1,009
Electricity Consumption	Btu per kilowatthour	3,412

Btu = British thermal unit.

[1]Conversion factors vary from year to year. Values correspond to those published by EIA for 2005 and may differ slightly from model results.

Sources: Energy Information Administration (EIA), *Annual Energy Review 2005*, DOE/EIA-0384(2005) (Washington, DC, July 2006), and EIA, AEO2007 National Energy Modeling System run AEO2007.D112106A.

The Energy Information Administration
2007 EIA Energy Outlook, Modeling, and Data Conference

Renaissance Hotel, Washington, DC *March 28, 2007*

8:30 a.m. - 8:45	**Opening Remarks - *Guy F. Caruso*,** *Administrator*, Energy Information Administration
8:50 a.m. - 9:20	**Overview of the *Annual Energy Outlook 2007* - *John Conti*,** *Director*, Office of Integrated Analysis and Forecasting, Energy Information Administration
9:25 a.m. - 10:25	**Keynote Address: Alternative Transportation Fuels and Technologies - *Keith Collins*,** USDA (invited)
10:40 a.m. - 12:10	**Concurrent Sessions A** **1. Creating a Path to Home Grown Liquid Fuels: Technologies and Initiatives** **2. Varying Views on the Future of the Natural Gas Market** **3. Short-Term Transportation Fuels Outlook**
1:40 p.m. - 3:10	**Concurrent Sessions B** **1. Renewable Incentive Policy Options** **2. Unconventional Oil Supply: How Much, When, and To What Effect?** **3. Electricity 2008—The Changing Face of EIA's Electric Power Surveys**
3:25 p.m. - 4:55	**Concurrent Sessions C** **1. Greenhouse Gas Reduction Strategies** **2. Opportunities and Issues for Ethanol and Biodiesel in the U.S. Transportation Market** **3. International Mid-to-Long-Term Outlooks**

Hotel

The conference will be held at the *Renaissance Hotel*, (202) 898-9000. The *Renaissance Hotel* is located at 999 Ninth Street, NW, Washington, DC 20001, near the Gallery Place Metro station. **EIA has not reserved rooms and is unable to provide discounted room rates.**

Information

For information, contact Peggy Wells, Energy Information Administration, at (202) 586-8845, peggy.wells@eia.doe.gov.

Conference Handouts

Handouts provided in advance by the conference speakers will be posted online as available, at www.eia.doe.gov/oiaf/aeo/conf/handouts.html. **They will not be provided at the conference.**

Conference Registration

Conference registration is free, but space is limited.
Please register by March 21, 2007.

Register online at www.eia.doe.gov/oiaf/aeo/conf/

Or mail or fax this form to:
Peggy Wells
Energy Information Administration, EI-84
1000 Independence Avenue, SW
Washington, DC 20585
Phone: (202) 586-8845
Fax: (202) 586-3045

Or register by e-mail to peggy.wells@eia.doe.gov.

Please provide the information requested below:

Name: ______________________
Title: ______________________
Organization: ______________________
Address: ______________________
____
Phone: ______________________
Fax: ______________________
E-mail: ______________________

Please indicate which sessions you will be attending:

- ❐ Opening Remarks/Overview/Keynote Address

Concurrent Sessions A

- ❐ Creating a Path to Home Grown Liquid Fuels
- ❐ Varying Views on the Natural Gas Market
- ❐ Short-Term Transportation Fuels Outlook

Concurrent Sessions B

- ❐ Renewable Incentive Policy Options
- ❐ Unconventional Oil Supply
- ❐ Electricity 2008—The Changing Face of EIA's Electric Power Surveys

Concurrent Sessions C

- ❐ Greenhouse Gas Reduction Strategies
- ❐ Opportunities and Issues for Ethanol and Biodiesel in the U.S. Transportation Market
- ❐ International Mid-to-Long-Term Outlooks